# Environmental Air Analysis

# Environmental Air Analysis

Arvind Kumar

RANDOM PUBLICATIONS
NEW DELHI (INDIA)

**Environmental Air Analysis**

---

ISBN 978-93-5111-965-4


Published in 2016 in India by

**RANDOM PUBLICATIONS**

4376-A/4B, Gali Murari Lal, Ansari Road
New Delhi-110 002
Phone : +9111-43580356, 011-23289044, 011-43142548
e-mail: sales@randompublications.com,
info@randompublications.com, randomexports@gmail.com
Reprinted 2021

*Type Setting by* : Friends Media, Delhi-110089
*Digitally Printed at* : Replika Press Pvt. Ltd.

# Preface

Air pollution occurs when the air contains gases, dust, fumes or odour in harmful amount. That is, amounts which could be harmful to the health or comfort of humans and animals or which could cause damage to plants and materials. The substances that cause air pollution are called pollutants. Pollutants that are pumped into our atmosphere and directly pollute the air are called primary pollutants. Primary pollutant examples include carbon monoxide from car exhausts and sulfur dioxide from the combustion of coal.

Air is obviously the first and foremost susceptible component of our environment prone to pollution. The occurrence of air pollution was not perceived as a major problem in most countries until the late 1950 and 1960. It was then usually seen as a local problem in urban areas and industrial areas.

Only after 1960s have air pollution evolved as a problem of regional and global dimension. Consequently there is ever increasing need of information on sources, effects and control of air pollution.

The degradation and deterioration of the environment and the deleterious consequences it is having on mankind in general has become a subject of alarm and concern to environmentalists and scientists all over the world. The manner in which the planet is being depleted of its resources and the subsequent effect on climatic conditions leading to much misery and loss of lives, both human and animal, and lost of plant diversity, is engaging the attention of all those concerned with the preservation of the environment.

The Ambient Air Quality Standards establish the concentration at which the pollutant is known to cause adverse health effects to sensitive groups within the population, such as children and the elderly. Both the California and federal governments have adopted health-based standards for the criteria pollutants, which include ozone, particulate matter and carbon monoxide. In general, the air quality standards are expressed as a measure of the amount of pollutant per unit volume of air.

This book will be useful to both teachers and students, administrators and planners interested in air pollution and its management. This book will be useful for post-graduate students of environmental sciences, life sciences, environmental engineering, researchers and teachers.

*– Author*

# Contents

# 1

# Introduction

Air testing for specific chemical compounds is an investigative tool used to characterize the nature and extent of contaminants in air and to determine whether contaminant sources affect indoor air quality. The purpose of this document is to outline the recommended procedure for testing indoor air for volatile chemicals. This document provides guidance for preparing sampling locations and collecting samples for laboratory analysis to ensure the integrity of the test results and allow for meaningful interpretation of the data. The steps discussed include; pre-sampling inspection and preparation of buildings, product inventories, and the collection and analysis of samples.

## GUIDANCE

### PRE-SAMPLING INSPECTION

A pre-sampling inspection should be performed prior to each sampling event to identify conditions that may affect or interfere with the proposed testing. The inspection should evaluate the type of structure, floor layout, physical conditions, and airflows of the building(s) being studied. The inspection information should be identified on the attached Indoor Air Quality Questionnaire and Building Characteristics form. In addition, potential sources of chemicals of concern should be evaluated within the building by conducting a product inventory.

The primary objective of the product inventory is to identify potential air sampling interference by characterizing the occurrence and use of chemicals and products throughout the building, keeping in mind the goal of the investigation and site specific contaminants of concern. For example, it is not necessary to provide detailed information for each individual container of like items. However it is necessary to indicate that "20 bottles of perfume" or "12 cans of latex paint" were present with containers in good condition. This information is used to help formulate the indoor environment profile. Each room on the floor of the building being tested and on lower floors, if possible, should be inspected and an inventory provided. This is important because even products stored in another area of a building can affect the air of the room being tested. For example, when testing for a petroleum spill, all indoor sources of petroleum hydrocarbons should be scrutinized. These can include household and commercial products containing volatile organic compounds (VOCs), petroleum products including fuel from gasoline-operated equipment, unvented space heaters and heating oil tanks, storage and/or recent use of petroleum-based finishes and paints or products containing petroleum distillates. This information should be detailed on the Product Inventory Form.

The presence and description of odours (*e.g.* solvent, moldy) and portable vapor monitoring equipment readings (*e.g.*, photoionization detectors [PIDs] for VOCs, Jerome Mercury Vapor Analyzer for mercury) should be used to help evaluate potential sources. This includes taking readings near products stored or used in the building. Products in buildings should be inventoried every time air is tested to provide an accurate assessment of the potential contribution of volatile chemicals. If available, chemical ingredients of interest should be recorded for each product. If the ingredients are not listed on the label, record the product's exact and full name, and the manufacturer's name, address and phone number, if available. In some cases, Material Safety Data Sheets may be useful for identifying confounding sources

## PREPARATION OF BUILDING

Potential interference from products or activities releasing volatile chemicals may need to be controlled. Removing the source from the indoor environment prior to testing is the most effective means of reducing the interference. Ensuring that containers are tightly sealed may be acceptable. When testing for VOCs, containers should be tested with a PID to determine whether VOCs are leaking. The inability to eliminate potential interference may be justification for not testing, especially when testing for similar compounds at low levels. The investigator should consider the possibility that chemicals may adsorb onto porous materials and may take time to dissipate.

In some cases, the goal of the testing is to evaluate the impact from products used or stored in the building (*e.g.*, pesticide misapplications, school renovation projects). If the goal of testing is to determine whether products are an indoor volatile chemical contaminant source, then removing these sources does not apply. Once interfering conditions are corrected (if applicable), ventilation may be needed prior to testing to eliminate residual contamination in the indoor air. If ventilation is appropriate, it should be completed 24 hours or more prior to the scheduled sampling time. Where applicable, ventilation can be accomplished by operating the building's heating ventilation and air conditioning (HVAC) system to maximize outside air intake. Opening windows and doors and operating exhaust fans may also help or may be needed if the building has no HVAC system. Air samples are sometimes designed to represent typical exposure in a mechanically ventilated building, and the operation of HVAC systems during sampling should be noted. In general, the building's HVAC system should be operating under normal conditions. Unnecessary building ventilation should be avoided within the 24 hours prior to and during testing. During colder months, heating systems should be operating under normal occupied conditions (*i.e.*, 65°-75° F) for at least 24 hours prior to and during the scheduled sampling time.

Depending on the goal of the indoor air sampling, some situations may warrant deviation from the above protocol regarding building ventilation. In such instances, building conditions and sampling efforts should be understood and noted within the framework and scope of the investigation.

## COLLECTION OF SAMPLES

Air samples should be collected from an adequate number of locations to understand likely sources of volatile chemicals and to assess potential exposure to occupants in various locations. In private residences, air samples should be collected from the basement, first floor living space, and from outdoors. In settings with diurnal occupancy patterns such as schools and office buildings, samples should be collected during normally occupied periods to be representative of typical exposure. However, in special circumstances it may be necessary to collect air samples at other times in order to minimize disruptions to normal building activities. Sample collection intakes should be located to approximate the breathing zone for building occupants (*i.e.*, three feet above the floor level where occupants are normally seated or sleep).

To ensure that an air sample is representative of the conditions being tested and to avoid undue influence from sampling personnel, samples should be collected for at least a one-hour period, and personnel should avoid lingering in the immediate area of the sampling device while samples are being collected. If the goal of the sampling is to represent average concentrations over longer time periods then longer duration sampling periods may be appropriate. The sampling team members should avoid actions (*e.g.*, fueling vehicles, using permanent marking pens) that can cause sample interference in the field.

Sample collection techniques vary depending on the analytical method(s) being used, and sample flow rates must conform to the specifications in the sample collection method. Some methods specify collecting samples in duplicate (*e.g.*, Passive Sampling Devices for tetrachloroethene). Sampling personnel should be completely familiar with the sampling protocol for the particular method being used.

## QUALITY ASSURANCE/QUALITY CONTROL

Extreme care should be taken during all aspects of sample collection to ensure that high-quality data are obtained. Appropriate QA/QC measures must be followed for sample collection and laboratory analysis. Items that should be addressed in sampling protocols include sampling techniques, certified-clean sampling apparatus, appropriate sample holding times, temperatures, and pressures. In addition, laboratory accession procedures must be followed including; field documentation (sample collection information and locations), chain of custody, field blanks, field sample duplicates and laboratory duplicates, as appropriate.

## SAMPLING INFORMATION

Detailed information must be gathered at the time of sampling to document conditions prior to and during sampling to aid in interpretation of the test results. The information should be recorded on the building inventory form along with the date and the investigator's initials. Floor plan sketches should be drawn for each floor and should include the floor layout with sample locations, chemical storage areas, garages, doorways, stairways, location of basement sumps, HVAC systems including air supplies and returns, compass orientation (north) and any other pertinent information. In addition, observations such as odours, PID readings, and airflow patterns should be recorded on the building inventory form. Smoke tubes or other devices are helpful and should be used to confirm pressure relationships and air flow patterns, especially between floor levels and between suspected contaminant sources and other areas. The NYSDOH Wadsworth Laboratories requires that information on odours and PID readings also be recorded on the associated sample accession forms for VOC analyses.

Outdoor plot sketches should include the building site, area streets, outdoor sample location, the location of potential interference (*e.g.*, gas stations, factories, lawn mowers), wind direction and compass orientation (north).

## SAMPLE ANALYSIS

New York State Law requires laboratories analyzing environmental samples from New York State to have current Environmental Laboratory Approval Programme (ELAP) certification for the appropriate analyte/matrix combinations. Samples must be analyzed by methods that can achieve minimum reporting limits to allow for comparison to background levels (halogenated VOCs are typically 1 microgram per cubic meter ($\mu$g/m3) or less). The laboratory should verify that they are capable of detecting the appropriate target compounds and can report them at the appropriate reporting limit (typically 1 $\mu$g/m3 or less). Check with an ELAP representative at 518-485-5570 or by e-mail at elap@health.state.ny.us for questions about a laboratory's current certification status.

Indoor air sampling to evaluate potential impacts from chemical contaminant sources (*i.e.*, old spills, soil vapor, groundwater) should generally include the contaminant(s) of concern and potential breakdown products (*e.g.*,

1,1,1-trichloroethane analysis should also include 1,1-dichloroethane, 1,1-dichloroethene, cis-1,2-dichloroethene, trans-1,2-dichloroethene, chloroethane and vinyl chloride). Petroleum products are often a mixture of many individual compounds. Specific aromatic and aliphatic compounds can be good indicators for individual petroleum products (*e.g.*, gasoline, diesel, fuel oil, and kerosene). The primary aromatic compounds benzene, toluene, ethylbenzene, xylenes (BTEX), and trimethylbenzenes should be included in all analyses. Analytical methods using a mass spectrometer detector allow for the identification and quantitation of aromatic and aliphatic hydrocarbons and for oxygenated compounds such as ethanol and methyl tertiary butyl ether (MTBE). Analyzing for specific indicator compounds as suggested below can aid in differentiating potential petroleum sources.

Indicator compounds for gasoline may include BTEX, trimethylbenzene isomers, the appropriate oxygenate additives (MTBE, ethanol, etc.), and the individual C-4 to C-8 aliphatics (*e.g.*, hexane, cyclohexane, dimethylpentane, and 2,2,4-trimethylpentane [iso-octane]).

Indicator compounds for middle distillate fuels (#2 fuel oil, diesel, and kerosene) may include n-nonane, n-decane, n-undecane, n-dodecane, ethylbenzene, xylenes, trimethylbenzene isomers, tetramethylbenzene isomers, naphthalene, 1-methylnaphthalene, and 2-methylnaphthalene.

Indicator compounds for manufactured gas plant (MGP) wastes may include ethylbenzene, xylenes, trimethylbenzene isomers, tetramethylbenzene isomers, thiophenes, indane, indene and naphthalene. Indicator compounds for natural gas or liquefied petroleum (LP) gas may include propane, propene, butane, iso-butane, iso-pentane and n-pentane. Natural gas and LP gas also contain higher molecular weight aliphatic, olefinic, and some aromatic compounds, but at levels much lower than the listed indicator compounds.

## VAPOR AND AIR ANALYSIS

Eurofins Lancaster Laboratories Environmental offers clients the confidence to make important decisions regarding air data and the expertise and experience to address all air testing needs. Our multiple systems provide analysis using TO-15 Full Scan, TO-15 SIM, TO-14 for Volatiles in Air by GC/MS and TO-13 for Semivolatiles in Air by GC/MS.

*Testing available:*

- TO-14A/TO-15
- TO-15 SIM
- Tedlar/SUMMA/Flow Controllers
- TO-13
- EPA 18/25
- BTEX, MTBE, Methane
- GRO

- Helium
- $O_2$ / $CO_2$

**WHY CHOOSE EUROFINS LANCASTER LABORATORIES ENVIRONMENTAL?**

- As a pioneer in air analysis, Eurofins Lancaster Laboratories Environmental has been providing volatiles in air analysis to our clients for more than 20 years with the implementation of our first TO-14 analysis in 1992.
- We offer a number of methods appropriate for air analysis and have specialized equipment to meet the low limits often specified for indoor air analysis.
- Eurofins Lancaster Laboratories Environmental is certified for air analysis by EPA Method TO-15 in eight states that offer the certification, including New Jersey. Most states do not offer air accreditation.
- We offer the highest level of data quality available in the industry and provide certification to the method detection limit (MDL) for our Summa canisters.
- Through use of an advanced Summa canister cleaning system that decreases the time necessary to clean, certify and ship a Summa canister, we can quickly fill our clients requests.

**WHY IS AIR ANALYSIS IMPORTANT TO YOU?**

The EPA published guidance in the Federal Register (FRL-7414-4 11/29/02) detailing how to screen sites to determine if a vapor intrusion pathway exists and if so whether or not a sensitive receptor may be affected. More than 30 states have some type of vapor intrusion policies in place, and states are re-evaluating sites with No Further Action (NFA) letters to determine if vapor intrusion concerns exist.

Evaluating the risk of vapor intrusion begins with inputting site-specific data into a site conceptual model (SCM) to determine if the migration of vapors from a contamination source may reach a receptor.

*Key factors of the SCM include:*

- Nature and extent of the source
- J and E Vapor Model (USEPA 2.3)
- Potential exposure pathways
- Potential receptors

There are situations where the conditions of a site logically do not support that a receptor will be affected (*e.g.*, concentrations are low, source is reasonable distance from receptor, modeling shows receptor not at risk). In other situations, there is overwhelming evidence to suggest a receptor may be affected, and sampling will be needed (*e.g.*, free product, high concentrations, source too close to receptor).

**TYPES OF SAMPLES**

There are several types of samples that can be collected to evaluate a potential indoor air concern. Samples can be collected in the building or residence and direct indoor vapors measured. They can be collected using a subslab probe or through the collection of soil gas monitoring points installed outside the structure. In addition, ambient air, duplicates and blanks may be collected and analyzed. Each sample type can provide meaningful data in the evaluation of migrating vapors, and each presents their own set of challenges.

Indoor air samples may provide you the most accurate assessment of the vapors indoors, but it is the most intrusive and will monitor for not only any vapors migrating through the subsurface as well as detect all vapors in the home. Sources such as fuel oil tanks, gas cans, moth balls, smoking, dry-cleaned garments, carpet adhesive and typical household supplies can all affect the results of an indoor air sample.

Subslab and soil vapor samples on the other hand collect a true vapor sample representative of subsurface conditions. The challenges with soil vapor samples include the use of heavy equipment, proper well construction, appropriate flow rates and proper sample collection techniques. In addition, when evaluating soil vapor data, the appropriate attenuation factor must be applied based on sample location and soil type.

**SAMPLE ANALYSIS**

We have provided the environmental community with air analysis from Summa canisters since 1992 and provide clients with the data quality and expertise needed to make important decisions based on our data. We have three GC/MS instruments dedicated to our Air Group providing TO-15, TO-15 SIM and TO-14 analysis. Eurofins Lancaster Laboratories Environmental currently has more than 400 Summa canisters and offers calibrated flow controllers set from 20 minutes to 24 hours, filters and visual pressure gauges. Our primary stock of cans are 6 Litre. We also offer 1-Litre cans. The standard analysis holding time for a Summa can is 30 days from collection. Eurofins Lancaster Laboratories Environmental holds several state certifications, including Florida, Louisiana, New York, New Jersey, Oregon and NELAP. And our analytical data can be provided in whatever units meet your needs. Either ppb(v) or $\mu$g/m3 can be provided, or we can report data in both units on your analysis reports.

**ANALYSES WE OFFER:**

TO-15 – Designed to detect volatile compounds at 1 ppbv and higher. Common Limits of Quanitiation (LOQs) are in the 1-2 ppbv range and Method

Detection Limits (MDLs) are in the 0.2 to 0.5 ppbv range. This is the most recent published air method and includes a 5-level calibration and more stringent quality control parameters than its predecessor TO-14.

TO-15 SIM (Selected Ion Mode) – Primarily used in the analysis of samples collected at indoor air locations. It provides the most sensitivity and lowest reporting limits with MDLs in the 0.006-0.025 ppbv range and LOQs equal to 0.05 ppbv. TO-14 – Designed for higher VOC concentrations with a 3-level calibration range from 10 to 100 ppb. Commonly used for soil vapor samples, this is an older version. Most regulatory agencies have moved to TO-15 as the method of choice.

TO-13A (Modified – Low Flow) – Analysis for semivolatile PAH compounds using GC/MS. Samples are collected using sorbent tubes with XAD resin, submitted to the lab, extracted and then analyzed using GC/MS technology. Lab reporting limits are directly dependant on the amount of sample volume collected onto the sampling tube. TICs/Library Search – Tentatively Identified Compounds (non-target compounds) can be searched, identified and semi-quantitatively reported when samples are analyzed using a standard full scan TO method, including TO-15, TO-14 or TO-13

EPA 18/25 – Gas Chromatography (GC/FID/PID) analysis for BTEX, methane and lightweight hydrocarbon fractions. Samples are collected into Tedlar bags or Summa cans from O and M systems and submitted to the lab for analysis. This is a quick inexpensive way to monitor your systems efficiency at higher concentrations. Tracer Gases – A tracer gas (helium, tetrafluoroethane, isopropanol, 1,2-difluoroethane, butane, propane, etc.) is often used in a shroud in the field to monitor potential leaks in your sampling train. We provide analysis of many of these compounds. Fixed Gases – Methane, $CO_2$, $O_2$ analysis can be indicators of natural attenuation.

## SAMPLING EQUIPMENT, PREPARATION AND HANDLING

Summa canisters prepared at Eurofins Lancaster Laboratories Environmental can be individually certified to our method detection limit. Whether by full scan or by SIM, we'll certify it clean before you receive it in the field. Each can is tracked through the use of unique can identification tags.

Our Summa can certification process starts with the cleaning process, which includes the repeated filling and evacuation of humidified nitrogen through three, heated cycles and three, non-heated cycles. Upon completion of our cleaning, we certify that the can maintains its pressure by filling the can with 30 psi and letting it sit for 24 hours. We then measure the pressure in the can to confirm the pressure is within 10 percent of its original pressure. Each can is then analyzed exactly like a field sample on the GC/MS, and all target compounds must be less than our method detection limit (MDL) for our can to be certifed. After our cans are certified, we evacuate them to negative 30 inches of mercury, and they are ready for the field. Flow controllers are also cleaned and certified using a similar process. The flow control valves are set up on the Summa cans to collect approximately 5 Litres of sample in a 6-Litre can. This ensures a consistent vacuum during your entire field collection event. We also install vacuum gauges to monitor the sampling process. Filters are provided on the flow controller to ensure no unwanted dust or particles affect the performance of the flow contoller and disrupt the integrity of the sample.

All cans received from the field are confirmed for their pressure and must read a detectable vacuum. If no vacuum exists or if the vacuum is less than negative 10 inches of mercury, you will be notified immediately to determine if you would like to proceed with analysis.

## TRANSPORTATION AND TOXIC AIR POLLUTANTS

Pollutants are generated by a wide variety of sources and enter the air, water, and soil through different media. Toxic air pollutants-also known as Hazardous Air Pollutants or HAPs-are those that are known to cause or suspected of causing cancer or other serious health ailments.

The Clean Air Act Amendments of 1990 listed 188 HAPs and addressed the need to control toxic emissions from transportation. In 2001, EPA issued its first Mobile Source Air Toxics Rule, which identified 21 mobile source air toxic (MSAT) compounds as being hazardous air pollutants that required regulation. A subset of six of these MSAT compounds were identified as having the greatest influence on health and included benzene, 1,3-butadiene, formaldehyde, acrolein, acetaldehyde, and diesel particulate matter (DPM). More recently, EPA issued a second MSAT Rule in February 2007, which generally supported the findings in the first rule and provided additional recommendations of compounds having the greatest impact on health. The rule also identified several engine emission certification standards that must be implemented. Unlike the criteria pollutants, toxics do not have National Ambient Air Quality Standards (NAAQS) making evaluation of their impacts more subjective.

To address stakeholders concerns and requests for MSAT analysis during project development and alternative analysis, FHWA developed the Interim Guidance on Air Toxic Analysis in NEPA Documents. The guidance provides a tiered approach for analyzing MSAT in NEPA documents. Depending on the specific project circumstances, FHWA has identified three levels of analysis: No analysis for projects with no potential for meaningful MSAT effects; Qualitative analysis for projects with low potential MSAT effects; or Quantitative analysis to differentiate alternatives for projects with higher potential MSAT effects. The FHWA's ongoing work in air toxics includes a research programme to better understand and quantify the contribution of mobile sources to air emissions, the establishment of policies for addressing mobile source emissions in environmental reports, and the assessment of scientific literature on health impacts associated with motor vehicle emissions.

Transportation agencies are interested in knowing the emissions burden associated with mobile sources. This understanding is a primary goal of FHWA's research. Critical components of our work are development and evaluation of methods to analyze and forecast future emissions from facilities and the fleets that are useful for decision makers and stakeholders to discern differences between transportation alternatives during the environmental documentation phase of project delivery.

## AIR POLLUTION ACCIDENTS AND EPISODES

The public concern is also based on news stories on air pollution accidents and episodes reported by the media. It is important for us to look at these pollution episodes.

During a 3 day fog in 1930, 60 people died in Meuse Valley, Belgium, while 592 people died in Manchester, England in 1931 during a 9 day fog. The 1948 plant emissions and atmospheric conditions in Donora, Penn. USA caused a 4 day fog and 7000 people were reported sick and 20 people died. The 4 day fog of 1952 in London, England resulted in 4000 deaths and concentration levels were several times higher than the current air quality standards in the United States. To read the September, 1998, EPA announcement of the final rule to protect Eastern US from Smog,

A four hour release of methyl isocyanate at a chemical plant owned by Union Carbide in 1984 killed 2800 people in Bhopal, India and opened the eyes of government agencies and public around the world. This Bhopal gas tragedy can be read in a nutshell in the following table:

| Accident | Bhopal Gas Tragedy |
|---|---|
| Location | Bhopal, Madhya Pradesh, India |
| Year | 1984 |
| Pollutant | Methyl isocyanate |
| Physical Properties of Methyl Isocyanate | Methyl isocyanate is a colorless liquid that has a sharp odor.<br>The odor threshold for methyl isocyanate is 2.1 ppm.<br>The chemical formula for methyl isocyanate is $C_2H_3NO$, and the molecular weight is 57.05 g/mol.<br>The vapor pressure for methyl isocyanate is 348 mm Hg at 20 C. |
| Pathway | Inhalation |
| # of Deaths | 2000 |
| Cause of Death | Primarily : Pulmonary edema<br>Secondary : Respiratory infections such as bronchitis and bronchial pneumonia. |
| Adverse health effects on | More than 170, 000 survivors |
| Reproductive adverse effects | Leucorrhea, pelvic inflammatory disease, excessive menstrual bleeding, and suppression of lactation and also stillbirths and spontaneous abortions |

As a result new regulations and preventive measures were introduced for air toxins. Public appreciation of radio nuclide emissions increased after the accidents at Three Mile Island, U.S.A. and Chernobyl, in ester while Russia. The accident at Chernobyl in 1986 caused 32 deaths and 135,000 people and their livestock had to be removed from the region for several months. The radiation exposure could increase the cancer death rates in USSR and Europe in coming years. The agricultural activities near the plant have been halted. Click the following link to know the response from EPA regarding Chernobyl accident. In the press, air pollution releases from accidents, transportation sources, plants, waste incineration facilities and natural sources receive coverage on a regular basis. The cost of air pollution could easily add up for a nation as well as for an industrial complex. If enough is not done to prevent air pollution problems, it is possible that future generations may see forests on postcards and calendars.

**WHAT IS AN AIR POLLUTANT?**

In our daily life we come across many airborne chemicals. Are all these chemicals termed as air pollutants? This question leads one to define an air pollutant. A contaminant that affects human life, plant life, animal life and property or a contaminant which interferes with the enjoyment of life and property could be termed as an air pollutant. Different countries have different legal definitions for an air pollutant. However, the above definition gives us an idea. The Ohio EPA provides the definition of "Air pollutant" or "air contaminant" as particulate matter, dust, fumes, gas, mist, smoke, vapor or odourous substances, or any combination there of.

***Chemical Composition of Dry Air***

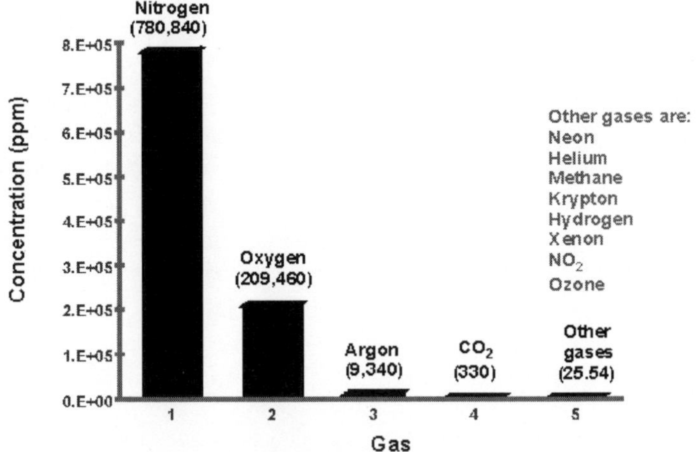


An air pollutant can be defined based on the concentration of chemical present in environment. The composition of clean air is used as a bench mark.

If the concentration of a chemical is above the concentration of chemical present in air, it is then termed as an air pollutant. There are two basic physical forms of air pollutants.

The first is gaseous form. For example, sulfur dioxide, ozone and hydrocarbon vapors exist in the form of a gas. The gases lack definite volume and shape and the molecules are widely separated. The second form of air pollution is particulate matter such as smoke, dust, fly ash and mists.

The pollutants are also classified as primary pollutants and secondary pollutants. The primary pollutants remain in the same chemical form as they are released from a source directly into the atmosphere. For example: sulfur dioxide and hydrocarbons. The secondary pollutants are a result of chemical reaction among two or more pollutants. The production of PAN (Peroxyacetyl Nitrate) during photochemical reactions is an example of secondary pollutant.

## UNITS FOR THE MEASUREMENT OF CONCENTRATION OF AIR POLLUTANTS

*There are two units of measurement. They are as follows:*

- $\mu$g/m3
- ppm (parts per million)

One ppm is 1 part in 1,000,000

Density = (Molecular weight of pollutant in gm/mol) / (Volume in l/mol)

Concentration in ppm = Concentration in mg/m3 / Density in mg/ml

At 00 C and at a pressure of 76 cm of Hg, volume of the air is 22.41 l/mol

To obtain volume at any temperature use gas law which is given by

$$P1V1/T1 = P2V2/T2$$

## COMMON AIR POLLUTANTS

Let us now turn our attention to common air pollutants. EPA has set national air quality standards for six common pollutants (also referred to as "criteria" pollutants): i. sulfur oxides, ii. carbon monoxide, iii. nitrogen dioxide, iv. ozone, v. total suspended particulate matter, and vi. lead. The other class of air pollutants which are of concern are hazardous and toxic air pollutants. The later area is rapidly expanding because of public pressure, concern over adverse health effects and accidental environmental damage. In this chapter seven criteria pollutants and some other pollutants are discussed. To view

- Areas where air pollution levels persistently exceed national air quality standards.
- Locations of air pollution monitoring sites, operated by state and local agencies.
- The 25 largest individual sources of each pollutant in the United States.
- Location of major stationary sources of air pollution.

The following table shows the criteria air pollutant monitors in the US from 1970 to 1990.

| Year | $PM_{10}$ | $O_3$ | $NO_2$ | $SO_2$ | CO |
|---|---|---|---|---|---|
| 1970 | 245 | 1 | 43 | 86 | 82 |
| 1975 | 1120 | 321 | 303 | 827 | 494 |
| 1980 | 1135 | 546 | 375 | 1088 | 511 |
| 1985 | 970 | 527 | 305 | 906 | 458 |
| 1990 | 720 | 627 | 345 | 743 | 493 |

The change in the National Air Quality Concentrations and Emissions (1980-2010) can be found from EPA. The following table shows the percentage change in the National Air Quality Concentrations and Emissions (1988-1997).

| **Pollutant** | **Percentage Decrease in Concentration 1988-1997** | **Percentage Decrease in Emissions 1988-1997** |
|---|---|---|
| Carbon Monoxide | 38 | 25 |
| Lead | 67 | 44 |
| Nitrogen Dioxide ($NO_2$) | 14 | 1($NO_2$) |
| Ozone (Pre-existing NAAQS) (1 hour) | 19 | 20(VOC) |
| Ozone (Revised NAAQS) (8 hour) | 16 | |
| $PM_{10}$ | 26 | 12 |
| Sulfur Dioxide ($SO_2$) | 39 | 12 |

## SULFUR DIOXIDE

Sulfur dioxide is considered a primary precursor of acidic precipitation. The sources of $SO_2$ are natural sources such as volcanoes and manmade sources such as power plants and industrial sources that burn coal or fuel. During the burning of fossil fuels 2 lb of $SO_2$ is produced for each pound of sulfur present in the fuel. It can harm human, and animal lungs, as well as plants and trees. Sulfur dioxide is the main contributor to acid rain. It reacts with the oxygen in the air to become sulfur trioxide, which then reacts with water in the air to form sulfuric acid. Acid rain can slowly kill both animal populations in lakes and rivers and trees and other plants by damaging leaves and root systems. It can deteriorate metal and stone on buildings and statues. The effects of acid rain are not only local, but they can occur hundreds of miles from the sources of sulfur dioxide.

## OZONE

Ozone is a gaseous, secondary pollutant and is formed during photochemical smog in the atmosphere. The interaction of $NO_2$ with VOCs produces ozone in the presence of sunlight. If the air over the city does not move, pollutants become trapped close to the earth's surface forming smog and increasing ozone problems which can lead to breathing problems. High ozone levels at the ground level harm plants, including trees and crop plants, and causes the accelerated deterioration of materials such as rubber and fabrics.

There is another type of ozone problem which came to attention in late sixties. Concerns were expressed on the destruction of ozone layer due to the use of supersonic transports. At present the destruction of ozone layer in the stratosphere due to the use of certain chemical compounds ( chlorofluorocarbons or CFCs, methane etc. ) is an environmental issue.

## TOTAL SUSPENDED PARTICULATE MATTER

TSP is mostly a primary pollutant, but some of it is formed as secondary pollutant. It consists of soot, dust, tiny objects of liquid, and other material. An increase in the incidence of respiratory diseases and gastric cancer has been linked with the increase in particulate level. The natural sources include volcanoes, forest fires, and desert land. Some manmade sources are steel industry, power plants, and flour mills. Agricultural activities also contribute to TSP loading. Particulate gradually settle back to earth and can cause people to cough, get sore throats, or develop other more serious breathing problems. Particulate matter also causes discoloration of buildings and other structures.

## NITROGEN OXIDES

Nitrogen Oxides are formed naturally by bacteria in soil and play an important role in plant growth. However, nitrogen oxides that enter the air through exhaust from vehicles and some power plants can be harmful. They can combine with water to make acid rain, react in the air to produce ozone and other pollutants, or be harmful by themselves as a gas in the air. Nitrogen dioxide is of greatest concern and is brown- red in colour. Nitrogen oxide is relatively less harmful as compared to other oxides of nitrogen.

## CARBON MONOXIDE

It is a colourless, odourless and tasteless gas and affects the central nervous system of humans. The gas is emitted when vehicles burn gasoline and when kerosene and wood stoves are used to heat homes. The gas reduces the ability of hemoglobin to carry oxygen to body tissue. The effects of carbon monoxide include headaches, reduced mental alertness, heart damage; it may even cause death, and it contributes to smog.

## HYDROCARBONS

Hydrocarbons are composed of only hydrogen and carbon. The volatile organic compounds (VOC) are the compounds which take part in atmospheric photochemical process. VOCs are composed of hydrogen and carbon, and may also contain elements such as oxygen, nitrogen, sulfur, chlorine, and fluorine. VOC emissions are produced during combustion and their rate of production is affected by time in combustion chamber, fuel and air mix, temperature, turbulence, pressure and design of chamber. The manmade sources of hydrocarbons include dry cleaning operations, auto paint shop, chemical plants, auto emissions, service stations and waste facilities. The VOCs are used in the manufacture of glue and paints as solvents A summary of the major air pollutants and their effects is given in the following table.

**Table. Major Air Pollutants.**

| POLLUTANT | DESCRIPTION | SOURCE (s) | EFFECTS |
|---|---|---|---|
| Sulfur Dioxide<br>( $SO_2$ ) | gaseous compound made up of sulfur and oxygen | coal-burning power plants and industries<br>industrial boilers and processes<br>coal-burning stoves<br>refineries<br>heaters | eye irritation<br>dead aquatic life<br>lung damage<br>reacts in atmosphere resulting in acidic precipitation<br>deteriorate buildings and statues<br>damage forests |
| Hydrocarbons | Composed of hydrogen and carbon. | dry cleaning operations.<br>auto paint shop<br>chemical plants<br>auto emissions<br>service stations<br>waste facilities | |
| Ozone<br>( $O_3$ ) | gaseous pollutant | vehicle exhaust | lung damage<br>eye irritation<br>respiratory tract problem |
| Particulate Matter | very small particles of soot, dust, or other matter, including tiny droplets of liquids | diesel engines<br>power plants<br>steel industry, flour mills<br>windblown dust<br>wood stoves | damage crops<br>lung damage<br>reduce visibility<br>discolor buildings and statues<br>eye irritation |
| Nitrogen Oxides ( $NO_x$ ) | several gaseous compounds made up of nitrogen and oxygen | vehicles<br>industrial boilers<br>industrial processes<br>power plants<br>commercial and residential heaters<br>coal-burning stoves<br>natural gas pipelines | lung damage<br>forms acid rain, damaging forests, buildings, & statues<br>forms ozone and other pollutants (smog) |
| Lead ( Pb ) | metallic element | vehicles burning leaded gasoline<br>power plants<br>metal refineries | brain, kidney damage<br>contaminated crops and livestock<br>Smog |
| Carbon Monoxide ( CO ) | colorless, odorless gas | vehicles burning gasoline<br>indoor sources include kerosene- or wood-burning stoves<br>dry cleaners | headaches, reduced mental alertness, death<br>heart damage<br>Smog |

The table below lists criteria pollutants and the threshold amounts for designation as large sources.

| Criteria Pollutant Emission Thresholds (Tons/Year) | | |
|---|---|---|
| CO | Carbon Monoxide gas | 1000 |
| NO2 | Nitrogen Dioxide gas | 100 |
| SO2 | Sulfur Dioxide gas | 100 |
| VOC | Volatile Organic Compounds * | 100 |
| PT | Particulate Matter (total) | 100 |
| PM10 | Particulate Matter (<10 μm) | 100 |
| Pb | Lead particles | 5 |

*Note:*

* VOCs are not criteria pollutants, but they are precursors of criteria pollutant ozone (smog).

**LEAD**

Lead is fairly abundant and is derived from ore bearing minerals. The gray metal can be easily molded, formed and worked. It can withstand weathering and chemical erosion. Lead has been used in the manufacture of pipes, paint house hold pottery, gasoline additives and storage batteries. In the U.S. the major source of lead mining is the state of Missouri. Automobiles and leaded gasoline are major sources of atmospheric lead. Lead was more of a problem a few years ago when all vehicles used gasoline with lead additives. When lead gasoline is burned, lead is released into the air. When people or animals breathe lead over a period of time, it accumulates in their bodies and can cause brain or kidney damage. Today most cars use unleaded gasoline, but there is still much leaded gasoline being sold, and lead continues to be a major pollutant, especially in cities.

**CARBON DIOXIDE**

Carbon dioxide emissions have increased significantly during $19^{th}$ century because of the use of coal, oil and natural gas. It finds uses as a refrigerant, in fire extinguishers and in beverage carbonation. Higher concentrations can affect respiratory function and cause excitation followed by depression of the central nervous system. Contact with liquefied $CO_2$ can cause frostbite. Workers briefly exposed to very high concentrations have effects like damage to the retina, sensitivity to light (photophobia), abnormal eye movements, constriction of visual fields, and enlargement of blind spots.

## TOXIC AIR POLLUTANTS

Toxic air pollutants may originate from natural sources such as volcanoes as well as from manmade sources such as stationary and mobile sources. The stationary sources serve as major contributors to air pollution, since they include factories, refineries, or power pollutants, which are constantly emitting pollutants into the atmosphere.

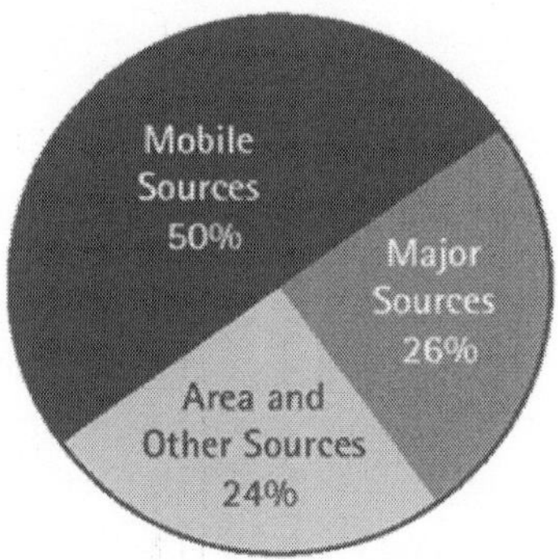


Based on 1996 National Toxics Inventory data, major sources account for about 26 percent of air toxics emissions, smaller area sources and other sources (such as forest fires) for 24 percent, and mobile sources for 50 percent. Accidental releases, which also contribute air toxics to the atmosphere, are not included in these estimates. The Clean Air Act amendments of 1990 provided a list of 189 chemicals to be regulated under the hazardous air pollutant provision of the act. A complete list of chemicals is given in Appendix C. On June 18, 1995, the US EPA has removed Caprolactum from the above list based on a July 1993 petition filed by Allied Signal, BASF, and DSM chemicals North America. The chemical is used in the manufacture of synthetic nylon fibre.

| Rank | Source Category | Emission |
|---|---|---|
| 1 | On-road motor vehicles | 1.52E+06 |
| 2 | Residential wood combustion | 5.25E+05 |
| 3 | Glycol dehydrators | 2.45E+05 |
| 4 | Consumer and commercial product solvent use | 2.22E+05 |
| 5 | Non-road mobile vehicles | 2.09E+05 |
| 6 | Forest fires | 1.91E+05 |
| 7 | Prescribed burning | 1.31E+05 |
| 8 | Industrial wood waste combustion | 9.93E+04 |
| 9 | Dry cleaning | 8.98E+04 |
| 10 | Halogenated solvent cleaning | 5.77E+04 |
| 11 | Utility coal combustion | 3.96E+04 |
| 12 | Gasoline distribution; stage II | 2.27E+04 |
| 13 | Primary aluminum production | 1.80E+04 |
| 14 | Industrial coal combustion | 1.69E+04 |
| 15 | Manufacture of motor vehicles and car bodies | 1.51E+04 |
| 16 | Gasoline distribution, stage 1 | 1.37E+04 |
| 17 | Plastics foam products | 1.36E+04 |
| 18 | Commercial printing, gravure | 1.27E+04 |
| 19 | Pulp mills | 1.21E+04 |
| 20 | Structure fires | 1.18E+04 |

The toxic air pollutants in US are reported to the public via the Toxic Release Inventory (TRI). TRI information is briefly summarized in a word document. For that information. In 1995, the toxic substances released to air, land, water and underground totaled 1.66 billion lb., compared to 1.75 billion lb. in 1994. The total TRI releases in 1995 were 45.6 per cent below baseline year.

## POLLUTANTS FROM VARIOUS ACTIVITIES

The sources may be classified as primary or secondary, mobile or stationary, combustion or non combustion, point or area and natural (biogenic) sources. The following table shows the constituents of atmospheric fine particles (<2.5mm) and their major sources, classified as primary and secondary and further as natural and anthropogenic.

Point sources include stationary facilities that emit sufficient amounts of pollutants to be worth listing. Area sources are all other point sources that individually emit small amount of pollutants. Dry cleaners in a city are an example of area sources. They contribute significantly to pollution as a group. Mobile sources include automobiles, trucks, air planes, ships, boats and lawnmowers. Natural sources are soil, water, vegetables, volcanic eruptions and lightning strikes.

Sources may also be classified the way they generate emissions: transportation, stationary combustion sources, industrial processes, solid waste disposal facilities and miscellaneous. The U.S. Environmental Protection Agency uses the above classification for reporting air emissions to the public.

*The general definitions are as follows:*

- *Transportation sources:* This category includes most emissions produced from transportation sources during the combustion process. The internal combustion engines fueled by gasoline and diesel are the biggest sources in this category. The other sources include trains, ships, lawnmowers, farm tractors, planes, and construction machinery.
- *Stationary combustion sources:* In this category the sources only produce energy and the emission is a result of fuel combustion. The sources include power plants as well as home heating furnaces.
- *Industrial Processes:* The sources which emit pollutants during manufacturing of products are included in this category. Petrochemical plants, petrochemical refining, food and agriculture industries, chemical processing, metallurgical and mineral product factories and wood processing industries are the major sources of air emissions. The smaller sources include dry- cleaning, painting and degreasing operations.
- *Solid Waste Disposal:* This category includes facilities that dispose off unwanted trash. Refuse incineration and open burning are important sources.

- *Miscellaneous:* The sources which do not fit in the above four categories are listed under this title. These sources include forest fires, house fires, agriculture burning, asphalt road paving and coal mining.

The pollutants typically released from the above sources are given in the table.

| SOURCE | TYPICAL AIR POLLUTANTS RELEASED |
|---|---|
| Transportation sources | carbon monoxide ( CO ), lead ( Pb ), nitrogen oxides ( NOx ), ozone ( $O_3$ ) |
| Stationary sources | carbon monoxide ( CO ), lead ( Pb ), nitrogen oxides ( NOx), particulate matter, sulfur dioxide ($SO_2$) |
| Industrial processes | carbon monoxide ( CO ), lead ( Pb ), nitrogen oxides ( NOx), particulate matter, sulfur dioxide ($SO_2$) |
| Solid Waste Disposal | carbon monoxide ( CO ) |
| Miscellaneous ( forest fires, asphalt road paving, coal mining etc. ) | carbon monoxide ( CO ), nitrogen oxides ( NOx), particulate matter, sulfur dioxide ($SO_2$) |

The following table indicates a table serves to illustrate the nationwide primary PM10 emission estimates from mobile and stationary sources, 1985 to 1993.

## AIR POLLUTION INDEX

In order to communicate with the public, regulatory agencies in various countries have developed different air pollution indices. The idea is to translate technical information on concentration levels of various pollutants into a simple and easy to understand language for the public.

U.S. Pollutant Standard Index ( Source U.S. EPA )

| Index Value | Air Quality Level | TSP (24 hr) $10^{-6}$ gm/m$^3$ | $SO_2$ (24 hr) $10^{-6}$ gm/m$^3$ | CO (8 hr) $10^{-3}$ gm/m$^3$ | $O_3$ (24 hr) $10^{-6}$ gm/m$^3$ | $NO_2$ (24 hr) $10^{-6}$ gm/m$^3$ |
|---|---|---|---|---|---|---|
| 500 | Significant Harm | 1000 | 2620 | 57.5 | 1200 | 3750 |
| 400 | Emergency | 875 | 2100 | 46 | 1000 | 3000 |
| 300 | Warning | 625 | 1600 | 34 | 800 | 2260 |
| 200 | Alert | 375 | 800 | 17 | 400[c] | 1130 |
| 200 | NAAQS | 260 | 365 | 10 | 235 | [a] |
| 50 | 50 % of NAAQS | 75[b] | 80[b] | 5 | 80 | [a] |
| 0 | | 0 | 0 | 0 | 0 | [a] |

Associated Health Effects With U.S. Pollutant Standard Index ( Source U.S. EPA )

The U.S. Environmental Protection Agency has developed the pollutant standard index (PSI) for introducing consistency in providing information regarding the air quality throughout the U.S. The system is based on a scale of 0-500. The computed index below 100 indicates that the air quality is within acceptable range. A value over 100 implies potential health problems. The alerts are issued at 200, 300 and 400 levels. Five pollutants (carbon monoxide, sulfur dioxide, total suspended particulate, ozone and nitrogen dioxide) are included in the index. The reported index is based on the highest index value of any of the five pollutants.

The concentration associated with the index value is given in Table, and the health effects are discussed in the following table.

| Index Value | Health Effect Descriptor | General Health Effects | Cautionary Statements |
|---|---|---|---|
| 400 – 500 | Hazardous | Premature death of ill & elderly. Healthy people will experience adverse symptoms that affect their normal activity. | All persons should remain indoors, keeping windows & doors closed. All persons should minimize physical exertion & avoid automobile traffic |
| 300 – 400 | Hazardous | Premature onset of certain diseases in addition to significant aggravation of symptoms & decreased exercise tolerance in healthy persons | Elderly & persons with existing diseases should stay indoors & avoid physical exertion. General population should avoid outdoor activity. |
| 200 – 300 | Very Unhealthy | Significant aggravation of symptoms & decreased exercise tolerance in persons with heart or lung disease with widespread symptoms in the healthy population | Elderly & persons with existing heart or lung disease should stay indoors & reduce physical activity |
| 100 - 200 | Unhealthy | Mild aggravation of symptoms with susceptible persons, with irritation symptoms in the healthy population | Persons with existing heart or respiratory ailments should reduce physical exertion & outdoor activity |
| 50 – 100 | Moderate | | |
| 0 – 50 | Good | | |

a. No index values reported at concentration levels below those specified by " Alert Level " criteria.
b. Annual primary NAAQS
c. 400*10-6 gm/m3 was used instead of the O3 Alert Level of 200*10-6 gm/m3

## EMISSION INVENTORY

Emission Inventory is a study of the pollutant emission estimates from sources in a given area. The development of emission inventory is important for a company as well as for the pollution control agencies. The inventory allows an environmental scientist to locate pollution sources, to define types and amounts of emission from each source, to define physical characteristics of sources, to determine emission frequency and duration of each pollutant exposure, to determine relative contributions to pollution problem in the area due to individual sources or a group of sources, to determine pollution controls needed to protect public health, and to provide a data base for air quality modeling and risk assessment.

- *Development of an Emission Inventory :* The emission inventories have been developed by a plant, local agency, or a nation. The development of a good inventory requires substantial resources and careful planning. Several steps are involved in the development of an emission inventory. The details depend on the area coverage, nature of sources, and purpose. Some of the basic steps are listed below:
- *Planning:* This step defines scope and purpose of inventory. The following points are considered during this step :the pollutants to be enlisted in the inventory are specified along with the methods to collect/estimate data, and the use of this data and the geographical area involved are determined. The legal authority and the responsibility of specific groups within the organization to acquire data is considered along with an assessment of cost, and resources.
- *Data Collection:* This stage follows the plan of action set in planning stage. During this phase emissions are classified, pollutant sources are located and classified, and the quality and quantity of materials handled, processed, or burned in each source is determined.

The data required to develop an inventory for a plant may be collected by mail survey which is the most common and economical technique for developing an initial regional or national emission inventory. Plant inspection is the most accurate method of data collection and is used to examine various processes, to interview with the staff and to carry out source emission testing, if necessary. The manufacturer specifications are also studied. It is more time consuming than mail surveys, and is usually used only at important point sources in the plant. Field surveys are similar to plant surveys, and used mainly to gather data about small area sources. Data is often found in industrial and government files, periodicals, trade journals and scientific publications. These publications often contain process activity level, and control device description. However they do not provide raw emission data, and rely upon estimates of emissions from published data on related sources. This method is usually used as a last resort method.

The kind of information collected during the development of an emission inventory includes general source information ( location, ownership, and nature of business ), process information ( type of equipment, type of reactions ), activity levels (amount of fuel and materials ( input ), amount of production (output) of the plant ), control device information ( type of air pollution control devices )and information required to estimate emissions ( temperature, tank conditions, hours of operation, seasonal variation ). For mobile sources information include year of vehicle, type of vehicle and vehicle.

- *Data Analysis:* After the data have been gathered, several analyses are made to check the accuracy of the collected information and to develop concise information for further use. It includes calculation of emission rate for each pollutant by using :
- Stack monitoring data, when available,
- Use of emission factors from AP-42,
- Mass balance, and
- Engineering calculations.

One or more techniques could be used for emission rate calculations. Continuous monitoring is the most accurate but the most expensive method. Emission rate is the weight of a pollutant emitted per unit time. Emission factor is an estimate of the rate at which a pollutant is released into the atmosphere as a result of some activity. An example of the calculation of emission rate using an emission factor is shown below:

Emission Rate = [ Input ] × [ Emission Factor ] × [Applicable Correction Factors ] × [ Hours of Operation ] × [ Seasonal Variation ]

The values of emission factors can be obtained from AP-42. Average emission factors from many similar facilities are also used in the absence of plant specific data. Area and mobile source emissions are also estimated using emission factors. Biogenic emissions are difficult to calculate. A mathematical model is used to calculate emissions based on the type of vegetation, temperatures, scalar radiation and land use of the area.

- *Reporting Data:* Depending on regulations the information is filed with local, regional and national pollution control agencies. For example in the U.S. data gathered by state agencies are reported to U.S. EPA's computerized National Emissions Data System. NEDS is a computerized data system developed for storage and retrieval of source and emission data, and is used to generate national emissions reports, fuel summary listings, and other data reports.

*The Emission Inventory developed may be used for :*

- Identifying types of pollutants emitted from specific sources,
- Determining the magnitude or amount of emissions from those sources,
- Developing emission distributions in time and space,
- Calculating emission rates under specific plant operating conditions, and
- Finding out the relation of ambient air pollutant concentrations to specific sources.

The US EPA has developed two major emission inventories: The first inventory is for criteria pollutants, and the second inventory is for toxic substances known as (Toxic Release Inventory - TRI ). Both the inventories have been very useful in identifying and solving air pollution problems. New regulations were proposed in the U.S. based on results obtained from these inventories. The fees charged by the state agencies under Title V of the 1990 Clean Air Act Amendments are designed using information from emission inventories. The requests to construct/modify existing and proposed sources are analyzed with the help of emission inventory data.

An example of a regional emission inventory is Great Lakes Regional Air Toxics Emissions Inventory developed by Great Lakes Commission. So far the emission inventory has focused on individual, manufacturing, and commercial sources in eight Great Lakes states. Future plans include the incorporation of 72 chemicals and compounds emitted by mobile sources in the area. Environmental managers can use emission inventory to estimate fees, to eliminate or to improve processes responsible for releasing large amounts of raw material, to implement control technology and to change processes for improving product quality.

## AIR POLLUTION IN ASIA

Asia represents a major source of air pollution as a result of rapid population growth, explosive industrialization, and few environmental regulations.

Due to the use of high sulfur coal to generate energy, the cities in China are heavily polluted by sulfur dioxide and particulate. The average ash content of Chinese coal is 27 per cent. The sulfur content varies upto 5 per cent. The combustion sources include small domestic stoves as well as large industrial plants. China produces over 15 million tons of $SO_2$ and 20 million tons of particulate. Industrial emissions of carbon dioxide and greenhouse gases are emitted in large quantities in China. Nitrogen oxide emissions are likely to increase as the production of cars will increase in China. China employs very little air pollution control technology. Acid rain is an important issue in China.

Guangzhou and Shanghai were mainly polluted by nitrogen dioxide. The critical pollutant in Beijing was the total suspended particle from coal burning, construction projects and dust-raising winds.

Air pollution is a serious problem in major cities in India. Delhi's pollution scenario is India's grimmest, and leads the other metros in vehicular pollution levels. The presence of suspended particulate matter is due to the use of coal in power plants. More than 45 million metric tons of ash is produced annually due to the use of low quality coal. In 23 Indian cities with populations of more than one million, auto exhausts and industrial emissions dangerously cross limits . Recent studies reveal that the number of patients with respiratory diseases and allergies has roughly doubled since the start of the 1990s. In Calcutta winter levels for particulate matter are 12 times above the standards. In Mumbai's (Bombay) "gas chamber", the eastern suburb of Chembur, pollution figures zoom to 10 times above the safe levels. India's metropolitan vehicle population has roughly tripled since 1990. The most damaging pollutants come from petrol driven cars and two wheelers. On July 5, 1997, IPAN (Indian Public Affairs Network) published an article of how the growing catalytic converter use could ease Asian cities' air pollution.

The air quality in Indonesia in deteriorating rapidly with industrial expansion. In the capital city, Jakarta, brownish yellow clouds of lead laden smog are common from 2.5 million vehicles. During several periods in 1997, Malaysia experienced haze conditions due to particulate matter from fires burning in Indonesia. Based upon readings from the Malaysian Air Pollutant Index, the air pollution levels registered in the "unhealthy" and occasionally in the "very unhealthy" ranges. $SO_2$ is a problem area in South Korea and is being controlled by the use of air pollution control equipment.

Vehicle emissions contribute most significantly to Hong Kong's air pollution problems, diesel powered engines being the prime culprit. In 1996 there were some 300,000 vehicles on Hong Kong's road, and one in three were diesel vehicles that covered two third of the mileage recorded by the total vehicle fleet. Today there are around 480,000 vehicles in Hong Kong, with diesel vehicles accounting for about 60 per cent of the overall mileage.

## AIR POLLUTION MANAGEMENT

Air pollution management aims at the elimination, or reduction to acceptable levels, of airborne gaseous pollutants, suspended particulate matter and physical and, to a certain extent, biological agents whose presence in the atmosphere can cause adverse effects on human health (*e.g.*, irritation, increase of incidence or prevalence of respiratory diseases, morbidity, cancer, excess mortality) or welfare (*e.g.*, sensory effects, reduction of visibility), deleterious effects on animal or plant life, damage to materials of economic value to society and damage to the environment (*e.g.*, climatic modifications). The serious hazards associated with radioactive pollutants, as well as the special procedures required for their control and disposal, also deserve careful attention.

The importance of efficient management of outdoor and indoor air pollution cannot be overemphasized. Unless there is adequate control, the multiplication of pollution sources in the modern world may lead to irreparable damage to the environment and mankind.

The objective of this article is to give a general overview of the possible approaches to the management of ambient air pollution from motor vehicle and industrial sources. However, it is to be emphasized from the very beginning that indoor air pollution (in particular, in developing countries) might play an even larger role than outdoor air pollution due to the observation that indoor air pollutant concentrations are often substantially higher than outdoor concentrations. Beyond considerations of emissions from fixed or mobile sources, air pollution management involves consideration of additional factors (such as topography and meteorology, and community and government participation, among many others) all of which must be integrated into a comprehensive programme. For example, meteorological conditions can greatly affect the ground-level concentrations resulting from the same pollutant emission. Air pollution sources may be scattered over a community or a region and their effects may be felt by, or their control may involve, more than one administration. Furthermore, air pollution does not respect any boundaries, and emissions from one region may induce effects in another region by long-distance transport.

Air pollution management, therefore, requires a multidisciplinary approach as well as a joint effort by private and governmental entities.

## SOURCES OF AIR POLLUTION

*The sources of man-made air pollution (or emission sources) are of basically two types:*

- Stationary, which can be subdivided into area sources such as agricultural production, mining and quarrying, industrial, point and area sources such as manufacturing of chemicals, non-metallic mineral products, basic metal industries, power generation and community sources (*e.g.*, heating of homes and buildings, municipal waste and sewage sludge incinerators, fireplaces, cooking facilities, laundry services and cleaning plants)
- Mobile, comprising any form of combustion-engine vehicles (*e.g.*, light-duty gasoline powered cars, light- and heavy-duty diesel powered vehicles, motorcycles, aircraft, including line sources with emissions of gases and particulate matter from vehicle traffic).

In addition, there are also natural sources of pollution (*e.g.*, eroded areas, volcanoes, certain plants which release great amounts of pollen, sources of bacteria, spores and viruses). Natural sources are not discussed in this article.

## TYPES OF AIR POLLUTANTS

Air pollutants are usually classified into suspended particulate matter (dusts, fumes, mists, smokes), gaseous pollutants (gases and vapours) and odours. Some examples of usual pollutants are presented below:

Suspended particulate matter (SPM, PM-10) includes diesel exhaust, coal fly-ash, mineral dusts (*e.g.*, coal, asbestos, limestone, cement), metal dusts and fumes (*e.g.*, zinc, copper, iron, lead) and acid mists (*e.g.*, sulphuric acid), fluorides, paint pigments, pesticide mists, carbon black and oil smoke. Suspended particulate pollutants, besides their effects of provoking respiratory diseases, cancers, corrosion, destruction of plant life and so on, can also constitute a nuisance (*e.g.*, accumulation of dirt), interfere with sunlight (*e.g.*, formation of smog and haze due to light scattering) and act as catalytic surfaces for reaction of adsorbed chemicals. Gaseous pollutants include sulphur compounds (*e.g.*, sulphur dioxide ($SO_2$) and sulphur trioxide ($SO_3$)), carbon monoxide, nitrogen compounds (*e.g.*, nitric oxide (NO), nitrogen dioxide ($NO_2$), ammonia), organic compounds (*e.g.*, hydrocarbons (HC), volatile organic compounds (VOC), polycyclic aromatic hydrocarbons (PAH), aldehydes), halogen compounds and halogen derivatives (*e.g.*, HF and HCl), hydrogen sulphide, carbon disulphide and mercaptans (odours).

Secondary pollutants may be formed by thermal, chemical or photochemical reactions. For example, by thermal action sulphur dioxide can oxidize to sulphur trioxide which, dissolved in water, gives rise to the formation of sulphuric acid mist (catalysed by manganese and iron oxides). Photochemical reactions between nitrogen oxides and reactive hydrocarbons can produce ozone ($O_3$), formaldehyde and peroxyacetyl nitrate (PAN); reactions between HCl and formaldehyde can form bis-chloromethyl ether.

While some odours are known to be caused by specific chemical agents such as hydrogen sulphide ($H_2S$), carbon disulphide ($CS_2$) and mercaptans (R-SH or R1-S-R2) others are difficult to define chemically.

Examples of the main pollutants associated with some industrial air pollution sources are presented in table below (Economopoulos 1993).

**Table. Common Atmospheric Pollutants and their Sources.**

| Index Value | Health Effect Descriptor | General Health Effects | Cautionary Statements |
|---|---|---|---|
| 400 – 500 | Hazardous | Premature death of ill & elderly. Healthy people will experience adverse symptoms that affect their normal activity. | All persons should remain indoors, keeping windows & doors closed. All persons should minimize physical exertion & avoid automobile traffic |
| 300 – 400 | Hazardous | Premature onset of certain diseases in addition to significant aggravation of symptoms & decreased exercise tolerance in healthy persons | Elderly & persons with existing diseases should stay indoors & avoid physical exertion. General population should avoid outdoor activity. |
| 200 – 300 | Very Unhealthy | Significant aggravation of symptoms & decreased exercise tolerance in persons with heart or lung disease with widespread symptoms in the healthy population | Elderly & persons with existing heart or lung disease should stay indoors & reduce physical activity |
| 100 - 200 | Unhealthy | Mild aggravation of symptoms with susceptible persons, with irritation symptoms in the healthy population | Persons with existing heart or respiratory ailments should reduce physical exertion & outdoor activity |
| 50 – 100 | Moderate | | |
| 0 – 50 | Good | | |

## CLEAN AIR IMPLEMENTATION PLANS

Air quality management aims at the preservation of environmental quality by prescribing the tolerated degree of pollution, leaving it to the local authorities and polluters to devise and implement actions to ensure that this degree of pollution will not be exceeded. An example of legislation within this approach is the adoption of ambient air quality standards based, very often, on air quality guidelines (WHO 1987) for different pollutants; these are accepted maximum levels of pollutants (or indicators) in the target area (*e.g.*, at ground level at a specified point in a community) and can be either primary or secondary standards. Primary standards (WHO 1980) are the maximum levels consistent with an adequate safety margin and with the preservation of public health, and must be complied with within a specific time limit; secondary standards are those judged to be necessary for protection against known or anticipated adverse effects other than health hazards (mainly on vegetation) and must be complied "within a reasonable time". Air quality standards are short-, medium- or long-term values valid for 24 hours per day, 7 days per week, and for monthly, seasonal or annual exposure of all living subjects (including sensitive subgroups such as children, the elderly and the sick) as well as non-living objects; this is in contrast to maximum permissible levels for occupational exposure, which are for a partial weekly exposure (*e.g.*, 8 hours per day, 5 days per week) of adult and supposedly healthy workers.

Typical measures in air quality management are control measures at the source, for example, enforcement of the use of catalytic converters in vehicles or of emission standards in incinerators, land-use planning and shut-down of factories or reduction of traffic during unfavourable weather conditions. The best air quality management stresses that the air pollutant emissions should be kept to a minimum; this is basically defined through emission standards for single sources of air pollution and could be achieved for industrial sources, for example, through closed systems and high-efficiency collectors. An emission standard is a limit on the amount or concentration of a pollutant emitted from a source. This type of legislation requires a decision, for each industry, on the best means of controlling its emissions (*i.e.*, fixing emission standards).

*The basic aim of air pollution management is to derive a clean air implementation plan (or air pollution abatement plan) (Schwela and Köth-Jahr 1994) which consists of the following elements:*

- Description of area with respect to topography, meteorology and socioeconomy
- Emissions inventory
- Comparison with emission standards
- Air pollutant concentrations inventory
- Simulated air pollutant concentrations
- Comparison with air quality standards

- Inventory of effects on public health and the environment
- Causal analysis
- Control measures
- Cost of control measures
- Cost of public health and environmental effects
- Cost-benefit analysis (costs of control vs. costs of efforts)
- Transportation and land-use planning
- Enforcement plan; resource commitment
- Projections for the future on population, traffic, industries and fuel consumption
- Strategies for follow-up.

## EMISSIONS INVENTORY; COMPARISON WITH EMISSION STANDARDS

The emissions inventory is a most complete listing of sources in a given area and of their individual emissions, estimated as accurately as possible from all emitting point, line and area (diffuse) sources. When these emissions are compared with emission standards set for a particular source, first hints on possible control measures are given if emission standards are not complied with. The emissions inventory also serves to assess a priority list of important sources according to the amount of pollutants emitted, and indicates the relative influence of different sources—for example, traffic as compared to industrial or residential sources.

The emissions inventory also allows an estimate of air pollutant concentrations for those pollutants for which ambient concentration measurements are difficult or too expensive to perform.

## AIR POLLUTANT CONCENTRATIONS INVENTORY; COMPARISON WITH AIR QUALITY STANDARDS

The air pollutant concentrations inventory summarizes the results of the monitoring of ambient air pollutants in terms of annual means, percentiles and trends of these quantities.

*Compounds measured for such an inventory include the following:*

- Sulphur dioxide
- Nitrogen oxides
- Suspended particulate matter
- Carbon monoxide
- Ozone
- Heavy metals (Pb, Cd, Ni, Cu, Fe, As, Be)
- Polycyclic aromatic hydrocarbons: benzo(a)pyrene, benzo(e)pyrene, benzo(a)anthracene, dibenzo(a,h)anthracene, benzoghi)perylene, coronen

- Volatile organic compounds: n-hexane, benzene, 3-methyl-hexane, n-heptane, toluene, octane, ethyl-benzene xylene (o-,m-,p-), n-nonane, isopropylbenzene, propylbenezene, n-2-/3-/4-ethyltoluene, 1,2,4-/ 1,3,5-trimethylbenzene, trichloromethane, 1,1,1 trichloroethane, tetrachloromethane, tri-/tetrachloroethene.

Comparison of air pollutant concentrations with air quality standards or guidelines, if they exist, indicates problem areas for which a causal analysis has to be performed in order to find out which sources are responsible for the non-compliance. Dispersion modelling has to be used in performing this causal analysis. Devices and procedures used in today's ambient air pollution monitoring are described in "Air quality monitoring".

## SIMULATED AIR POLLUTANT CONCENTRATIONS; COMPARISON WITH AIR QUALITY STANDARDS

Starting from the emissions inventory, with its thousands of compounds which cannot all be monitored in the ambient air for economy reasons, use of dispersion modelling can help to estimate the concentrations of more "exotic" compounds. Using appropriate meteorology parameters in a suitable dispersion model, annual averages and percentiles can be estimated and compared to air quality standards or guidelines, if they exist.

## INVENTORY OF EFFECTS ON PUBLIC HEALTH AND THE ENVIRONMENT; CAUSAL ANALYSIS

Another important source of information is the effects inventory (Ministerium für Umwelt 1993), which consists of results of epidemiological studies in the given area and of effects of air pollution observed in biological and material receptors such as, for example, plants, animals and construction metals and building stones. Observed effects attributed to air pollution have to be causally analysed with respect to the component responsible for a particular effect—for example, increased prevalence of chronic bronchitis in a polluted area. If the compound or compounds have been fixed in a causal analysis (compound-causal analysis), a second analysis has to be performed to find out the responsible sources (source-causal analysis).

## CONTROL MEASURES; COST OF CONTROL MEASURES

Control measures for industrial facilities include adequate, well-designed, well-installed, efficiently operated and maintained air cleaning devices, also called separators or collectors. A separator or collector can be defined as an "apparatus for separating any one or more of the following from a gaseous medium in which they are suspended or mixed: solid particles (filter and dust separators), liquid particles (filter and droplet separator) and gases (gas purifier)".

*The basic types of air pollution control equipment (discussed further in "Air pollution control") are the following:*

- *For particulate matter:* Inertial separators (*e.g.*, cyclones); fabric filters (baghouses); electrostatic precipitators; wet collectors (scrubbers)
- *For gaseous pollutants:* Wet collectors (scrubbers); adsorption units (*e.g.*, adsorption beds); afterburners, which can be direct-fired (thermal incineration) or catalytic (catalytic combustion).

Wet collectors (scrubbers) can be used to collect, at the same time, gaseous pollutants and particulate matter. Also, certain types of combustion devices can burn combustible gases and vapours as well as certain combustible aerosols. Depending on the type of effluent, one or a combination of more than one collector can be used.

The control of odours that are chemically identifiable relies on the control of the chemical agent(s) from which they emanate (*e.g.*, by absorption, by incineration). However, when an odour is not defined chemically or the producing agent is found at extremely low levels, other techniques may be used, such as masking (by a stronger, more agreeable and harmless agent) or counteraction (by an additive which counteracts or partially neutralizes the offensive odour).

It should be kept in mind that adequate operation and maintenance are indispensable to ensure the expected efficiency from a collector. This should be ensured at the planning stage, both from the know-how and financial points of view. Energy requirements must not be overlooked. Whenever selecting an air cleaning device, not only the initial cost but also operational and maintenance costs should be considered. Whenever dealing with high-toxicity pollutants, high efficiency should be ensured, as well as special procedures for maintenance and disposal of waste materials.

The fundamental control measures in industrial facilities are the following: Substitution of materials. Examples: substitution of less toxic solvents for highly toxic ones used in certain industrial processes; use of fuels with lower sulphur content (*e.g.*, washed coal), therefore giving rise to less sulphur compounds and so on.

Modification or change of the industrial process or equipment. Examples: in the steel industry, a change from raw ore to pelleted sintered ore (to reduce the dust released during ore handling); use of closed systems instead of open ones; change of fuel heating systems to steam, hot water or electrical systems; use of catalysers at the exhaust air outlets (combustion processes) and so on.

Modifications in processes, as well as in plant layout, may also facilitate and/or improve the conditions for dispersion and collection of pollutants. For example, a different plant layout may facilitate the installation of a local exhaust system; the performance of a process at a lower rate may allow the use of a certain collector (with volume limitations but otherwise adequate). Process modifications that concentrate different effluent sources are closely related to the volume of effluent handled, and the efficiency of some air-cleaning equipment increases with the concentration of pollutants in the effluent. Both the substitution of materials and the modification of processes may have technical and/or economic limitations, and these should be considered.

Adequate housekeeping and storage. Examples: strict sanitation in food and animal product processing; avoidance of open storage of chemicals (*e.g.*, sulphur piles) or dusty materials (*e.g.*, sand), or, failing this, spraying of the piles of loose particulate with water (if possible) or application of surface coatings (*e.g.*, wetting agents, plastic) to piles of materials likely to give off pollutants. Adequate disposal of wastes. Examples: avoidance of simply piling up chemical wastes (such as scraps from polymerization reactors), as well as of dumping pollutant materials (solid or liquid) in water streams. The latter practice not only causes water pollution but can also create a secondary source of air pollution, as in the case of liquid wastes from sulphite process pulp mills, which release offensive odourous gaseous pollutants.

*Maintenance. Example:* well maintained and well-tuned internal combustion engines produce less carbon monoxide and hydrocarbons.

*Work practices. Example:* taking into account meteorological conditions, particularly winds, when spraying pesticides.

By analogy with adequate practices at the workplace, good practices at the community level can contribute to air pollution control - for example, changes in the use of motor vehicles (more collective transportation, small cars and so on) and control of heating facilities (better insulation of buildings in order to require less heating, better fuels and so on).

Control measures in vehicle emissions are adequate and efficient mandatory inspection and maintenance programmes which are enforced for the existing car fleet, programmes of enforcement of the use of catalytic converters in new cars, aggressive substitution of solar/battery-powered cars for fuel-powered ones, regulation of road traffic, and transportation and land use planning concepts.

Motor vehicle emissions are controlled by controlling emissions per vehicle mile travelled (VMT) and by controlling VMT itself (Walsh 1992). Emissions per VMT can be reduced by controlling vehicle performance - hardware, maintenance - for both new and in-use cars. Fuel composition of leaded gasoline may be controlled by reducing lead or sulphur content, which also has a beneficial effect on decreasing HC emissions from vehicles. Lowering the levels of sulphur in diesel fuel as a means to lower diesel particulate emission has the additional beneficial effect of increasing the potential for catalytic control of diesel particulate and organic HC emissions.

Another important management tool for reducing vehicle evaporative and refuelling emissions is the control of gasoline volatility. Control of fuel volatility can greatly lower vehicle evaporative HC emissions. Use of oxygenated additives in gasoline lowers HC and CO exhaust as long as fuel volatility is not increased.

*Reduction of VMT is an additional means of controlling vehicle emissions by control strategies such as:*

- Use of more efficient transportation modes
- Increasing the average number of passengers per car
- Spreading congested peak traffic loads
- Reducing travel demand.

While such approaches promote fuel conservation, they are not yet accepted by the general population, and governments have not seriously tried to implement them. All these technological and political solutions to the motor vehicle problem except substitution of electrical cars are increasingly offset by growth in the vehicle population. The vehicle problem can be solved only if the growth problem is addressed in an appropriate way.

## COST OF PUBLIC HEALTH AND ENVIRONMENTAL EFFECTS; COST-BENEFIT ANALYSIS

The estimation of the costs of public health and environmental effects is the most difficult part of a clean air implementation plan, as it is very difficult to estimate the value of lifetime reduction of disabling illnesses, hospital admission rates and hours of work lost. However, this estimation and a comparison with the cost of control measures is absolutely necessary in order to balance the costs of control measures versus the costs of no such measure undertaken, in terms of public health and environmental effects.

## TRANSPORTATION AND LAND-USE PLANNING

The pollution problem is intimately connected to land-use and transportation, including issues such as community planning, road design, traffic control and mass transportation; to concerns of demography, topography and economy; and to social concerns (Venzia 1977). In general, the rapidly growing urban aggregations have severe pollution problems due to poor land-use and transportation practices. Transportation planning for air pollution control includes transportation controls, transportation policies, mass transit and highway congestion costs. Transportation controls have an important impact on the general public in terms of equity, repressiveness and social and economic disruption - in particular, direct transportation controls such as motor vehicle constraints, gasoline limitations and motor vehicle emission reductions. Emission reductions due to direct controls can be reliably estimated and verified. Indirect transportation controls such as reduction of vehicle miles travelled by improvement of mass transit systems, traffic flow improvement regulations, regulations on parking lots, road and gasoline taxes, car-use permissions and incentives for voluntary approaches are mostly based on past trial-and-error experience, and include many uncertainties when trying to develop a viable transportation plan.

National action plans incurring indirect transportation controls can affect transportation and land-use planning with regard to highways, parking lots and shopping centres. Long-term planning for the transportation system and the area influenced by it will prevent significant deterioration of air quality and provide for compliance with air quality standards. Mass transit is consistently considered as a potential solution for urban air pollution problems. Selection of a mass transit system to serve an area and different modal splits between highway use and bus or rail service will ultimately alter land-use patterns. There is an optimum split that will minimize air pollution; however, this may not be acceptable when non-environmental factors are considered.

The automobile has been called the greatest generator of economic externalities ever known. Some of these, such as jobs and mobility, are positive, but the negative ones, such as air pollution, accidents resulting in death and injury, property damage, noise, loss of time, and aggravation, lead to the conclusion that transportation is not a decreasing cost industry in urbanized areas. Highway congestion costs are another externality; lost time and congestion costs, however, are difficult to determine. A true evaluation of competing transportation modes, such as mass transportation, cannot be obtained if travel costs for work trips do not include congestion costs.

Land-use planning for air pollution control includes zoning codes and performance standards, land-use controls, housing and land development, and land-use planning policies. Land-use zoning was the initial attempt to accomplish protection of the people, their property and their economic opportunity. However, the ubiquitous nature of air pollutants required more than physical separation of industries and residential areas to protect the individual. For this reason, performance standards based initially on aesthetics or qualitative decisions were introduced into some zoning codes in an attempt to quantify criteria for identifying potential problems.

The limitations of the assimilative capacity of the environment must be identified for long-term land-use planning. Then, land-use controls can be developed that will prorate the capacity equitably among desired local activities. Land-use controls include permit systems for review of new stationary sources, zoning regulation between industrial and residential areas, restriction by easement or purchase of land, receptor location control, emission-density zoning and emission allocation regulations.

Housing policies aimed at making home ownership available to many who could otherwise not afford it (such as tax incentives and mortgage policies) stimulate urban sprawl and indirectly discourage higher-density residential development. These policies have now proven to be environmentally disastrous, as no consideration was given to the simultaneous development of efficient transportation systems to serve the needs of the multitude of new communities being developed. The lesson learnt from this development is that programmes impacting on the environment should be coordinated, and comprehensive planning undertaken at the level where the problem occurs and on a scale large enough to include the entire system.

Land-use planning must be examined at national, provincial or state, regional and local levels to adequately ensure long-term protection of the environment. Governmental programmes usually start with power plant siting, mineral extraction sites, coastal zoning and desert, mountain or other recreational development. As the multiplicity of local governments in a given region cannot adequately deal with regional environmental problems, regional governments or agencies should coordinate land development and density patterns by supervising the spatial arrangement and location of new construction and use, and transportation facilities. Land-use and transportation planning must be interrelated with enforcement of regulations to maintain the desired air quality. Ideally, air pollution control should be planned for by the same regional agency that does land-use planning because of the overlapping externalities associated with both issues.

## ENFORCEMENT PLAN, RESOURCE COMMITMENT

The clean air implementation plan should always contain an enforcement plan which indicates how the control measures can be enforced. This implies also a resource commitment which, according to a polluter pays principle, will state what the polluter has to implement and how the government will help the polluter in fulfilling the commitment.

## PROJECTIONS FOR THE FUTURE

In the sense of a precautionary plan, the clean air implementation plan should also include estimates of the trends in population, traffic, industries and fuel consumption in order to assess responses to future problems. This will avoid future stresses by enforcing measures well in advance of imagined problems.

## STRATEGIES FOR FOLLOW-UP

A strategy for follow-up of air quality management consists of plans and policies on how to implement future clean air implementation plans.

## ROLE OF ENVIRONMENTAL IMPACT ASSESSMENT

Environmental impact assessment (EIA) is the process of providing a detailed statement by the responsible agency on the environmental impact of a proposed action significantly affecting the quality of the human environment

(Lee 1993). EIA is an instrument of prevention aiming at consideration of the human environment at an early stage of the development of a programme or project.

EIA is particularly important for countries which develop projects in the framework of economic reorientation and restructuring. EIA has become legislation in many developed countries and is now increasingly applied in developing countries and economies in transition. EIA is integrative in the sense of comprehensive environmental planning and management considering the interactions between different environmental media. On the other hand, EIA integrates the estimation of environmental consequences into the planning process and thereby becomes an instrument of sustainable development. EIA also combines technical and participative properties as it collects, analyses and applies scientific and technical data with consideration of quality control and quality assurance, and stresses the importance of consultations prior to licensing procedures between environmental agencies and the public which could be affected by particular projects. A clean air implementation plan can be considered as a part of the EIA procedure with reference to the air.

# 2

# Air Pollution

One of the formal definitions of air pollution is as follows— 'The presence in the atmosphere of one or more contaminants in such quality and for such duration as is injurious, or tends to be injurious, to human health or welfare, animal or plant life.' It is the contamination of air by the discharge of harmful substances. Air pollution can cause health problems and it can also damage the environment and property. It has caused thinning of the protective ozone layer of the atmosphere, which is leading to climate change. Modernisation and progress have led to air getting more and more polluted over the years. Industries, vehicles, increase in the population, and urbanization are some of the major factors responsible for air pollution.

The following industries are among those that emit a great deal of pollutants into the air: thermal power plants, cement, steel, refineries, petro chemicals, and mines. Air pollution results from a variety of causes, not all of which are within human control. Dust storms in desert areas and smoke from forest fires and grass fires contribute to chemical and particulate pollution of the air. The source of pollution may be in one country but the impact of pollution may be felt elsewhere. The discovery of pesticides in Antarctica, where they have never been used, suggests the extent to which aerial transport can carry pollutants from one place to another.

## SOURCES OF AIR POLLUTION

The combustion of gasoline and other hydrocarbon fuels in automobiles, trucks, and jet airplanes produces several primary pollutants: nitrogen oxides, gaseous hydrocarbons, and carbon monoxide, as well as large quantities of particulates, chiefly lead. In the presence of sunlight, nitrogen oxides combine with hydrocarbons to form a secondary class of pollutants, the photochemical oxidants, among them ozone and the eye-stinging peroxyacetylnitrate (PAN). Nitrogen oxides also react with oxygen in the air to form nitrogen dioxide, a foul-smelling brown gas.

In urban areas like Los Angeles where transportation is the main cause of air pollution, nitrogen dioxide tints the air, blending with other contaminants and the atmospheric water vapour to produce brown smog. Although the use of catalytic converters has reduced smog-producing compounds in motor vehicle exhaust emissions, recent studies have shown that in so doing the converters produce nitrous oxide, which contributes substantially to global warming. Air may be severely polluted not only by transportation but also by the burning of fossil fuels (oil and coal) in generating stations, factories, office buildings, and homes and by the incineration of garbage.

The massive combustion produces tons of ash, soot, and other particulates responsible for the gray smog of cities like New York and Chicago, along with enormous quantities of sulphur oxides (which also may be result from burning coal and oil). These oxides rust iron, damage building stone, decompose nylon, tarnish silver, and kill plants. Air pollution from cities also affects rural areas for many miles downwind. Every industrial process exhibits its own pattern of air pollution.

Petroleum refineries are responsible for extensive hydrocarbon and particulate pollution. Iron and steel mills, metal smelters, pulp and paper mills, chemical plants, cement and asphalt plants—all discharge vast amounts of various particulates. Uninsulated high-voltage power lines ionize the adjacent air, forming ozone and other hazardous pollutants. Airborne pollutants from other sources include insecticides, herbicides, radioactive fallout, and dust from fertilizers, mining operations, and livestock feedlots.

## Chemical Pollution

In some parts of the world, the bodies of whales and dolphins washing ashore are so highly contaminated that they qualify as toxic waste and have to be specially disposed of.

*There are many different sources of chemical pollution, including*:

- Domestic sewage
- Industrial discharges
- Seepage from waste sites
- Atmospheric fallout
- Domestic run-off
- Accidents and spills at sea
- Operational discharges from oil rigs
- Mining discharges and
- Agricultural run-off.

## Health Effects of Air Pollution

The human health effects of poor air quality are far reaching, but principally affect the body's respiratory system and the cardiovascular system. Individual reactions to air pollutants depend on the type of pollutant a person is exposed to, the degree of exposure, the individual's health status and genetics.

People who exercise outdoors on hot smoggy days increase their exposure to pollutants in the air.The health effects caused by air pollutants may range from subtle biochemical and physiological changes to difficulty breathing, wheezing, coughing and aggravation of existing respiratory and cardiac conditions. These effects can result in increased medication use, increased doctor or emergency room visits, more hospital admissions and even premature death.

- Human Respiratory System
- Human Cardiovascular System
- Heart and Lung Diseases
- Pyramid of Health Effects
- Populations at Risk
- Leading Causes of Hospitalization
- Leading Causes of Death
- Estimating Health Benefits

## AIR POLLUTION AND BIOSPHERE

The biosphere is the biological component of earth systems, which also include the lithosphere, hydrosphere, atmosphere and other "spheres". The biosphere includes all living organisms on earth, together with the dead organic matter produced by them. The biosphere concept is common to many scientific disciplines including astronomy, geophysics, geology, hydrology, biogeography and evolution, and is a core concept in ecology, earth science and physical geography. A key component of earth systems, the biosphere interacts with and exchanges matter and energy with the other spheres, helping to drive the global biogeochemical cycling of carbon, nitrogen, phosphorus, sulphur and other elements. From an ecological point of view, the biosphere is the "global ecosystem", comprising the totality of biodiversity on earth and performing all manner of biological functions, including photosynthesis, respiration, decomposition, nitrogen fixation and denitrification.

The biosphere is dynamic, undergoing strong seasonal cycles in primary productivity and the many biological processes driven by the energy captured by photosynthesis. Seasonal cycles in solar irradiation of the hemispheres is the main driver of this dynamic, especially by its strong effect on terrestrial primary productivity in the temperate and boreal biomes, which essentially cease productivity in the winter time. The biosphere has evolved since the first single-celled organisms originated 3.5 billion years ago under atmospheric conditions resembling those of our neighbouring planets Mars and Venus, which have atmospheres composed primarily of carbon dioxide.

Billions of years of primary production by plants released oxygen from this carbon dioxide and deposited the carbon in sediments, eventually producing the oxygen-rich atmosphere we know today. Free oxygen, both for breathing and in the stratospheric ozone that protects us from harmful UV radiation, has made possible life as we know it while transforming the chemistry of earth systems forever. As a result of long-term interactions between the biosphere and the other earth systems, there is almost no part of the earth's surface that has not been profoundly altered by living organisms. The earth is a living planet, even in terms of its physics and chemistry. A concept related to, but different from, that of the biosphere, is the Gaia hypotheses, which posits that living organisms have and continue to transform earth systems for their own benefit.

## HISTORY OF THE BIOSPHERE CONCEPT

The term "biosphere" originated with the geologist Eduard Suess in 1875, who defined it as "the place on earth's surface where life dwells". It is Vernadsky's work that redefined ecology as the science of the biosphere and placed the biosphere concept in its current central position in earth systems science.

## AIR QUALITY STANDARDS

This nation-wide programme was initiated in 1984. As on March 31, 1995, the network comprised 290 stations covering over 90 towns/cities distributed over 24 States and 4 Union Territories. The National Ambient Air Quality Monitoring network is operated through the respective States Pollution Control Boards, the National Environmental Engineering Research Institute, Nagpur and also through the CPCB. The pollutants monitored are Sulphur dioxide, Nitrogen dioxide and Suspended Particulate Matter besides the meteorological parametres, like wind speed and direction, temperature and humidity.

In addition to the three conventional parametres, NEERI monitors special parametres, like Ammonia, Hydrogen Sulphide, Respirable Suspended Particulate Matter and Polyaromatic Hydrocarbons. Based on Annual Mean Concentration of $SO_2$, $NO_2$ and SPM and the Notified Ambient Air Quality Standards, the Ambient Air Quality Status is described in terms of Low, Moderate, High and Critical for Industrial, Residential and mixed use areas of Cities/Towns in different States/UTs.

## PRIMARY POLLUTANTS

Describing a primary pollutant, the Environmental Protection Agency states it is "emitted into the atmosphere directly from the source of the pollutant and retains the same chemical form." The secondary variety forms as a result of chemical reactions that take place after a substance is released into the air. Understanding the characteristics of pollutants as well as how they get into the environment is crucial to understanding the effects.

## PARTICULATE MATTER

Encompassing things as varied as dust, ash, smoke, sand and mist, PM or particulates occur both naturally and as the result of human actions. Natural sources include volcanic eruptions, wind-blown dust and wildfires. Anthropogenic source examples are automobiles, industrial processes, wood burning stoves and dust kicked up by construction projects. PM size is measured in microns. The most problematic are particulates 10 microns or less. These are small enough to be inhaled deep into the lungs, causing and aggravating asthma and other breathing disorders.

## SULPHUR DIOXIDE

Introduced to the air mainly from the burning of coal, sulphur dioxide is a primary pollutant and the cause of acid rain. Normal water has a pH of about 5.5; $SO_2$ can reduce this to as low as 4.0. The phenomenon was one of the first recognized primary regional pollutants, as the effects go beyond the cities where the industrial polluting processes are taking place. Acid rain acidifies lakes and streams, and damages trees and soils, among other negative consequences. Direct health effects also include eye, nose and throat irritation as well as headaches, nausea and dizziness.

## NITROGEN OXIDES

Burning of fossil fuels is a big contributor to nitrogen oxide pollution. Automobiles and industrial smelting processes are also major culprits. Nitrogen dioxide is common and can at times be seen over cities as a reddish brown gas. The EPA points out that NOx "reacts with ammonia, moisture, and other compounds to form nitric acid and related particles." These particles contribute to all sorts of respiratory illnesses.

Nitrogen is a nutrient, which when introduced in high quantities to water bodies, contributes to increased biochemical oxygen demand and eutrophication, described as the enrichment of water by plant nutrients. Nitrous oxide is a greenhouse gas.

## VOLATILE ORGANIC COMPOUNDS

VOCs are extremely reactive organic compounds. Harmful ground-level ozone is made from a reaction between NOx and VOCs when exposed to sunlight. Other health and environmental problems like VOCs entering aquifers are also a concern. Automobiles, industrial processes, cleaning solvents and paints are some of the leading VOC emitters.

## CARBON MONOXIDE

Being odourless, colourless and tasteless makes the highly toxic gas CO extremely dangerous. Sources include incomplete combustion, malfunctioning furnaces, space heaters and fireplaces. According to the University of Virginia Health System, CO not only prevents your body from using oxygen properly, but damages the central nervous system.

**WARNING**

Carbon monoxide detectors are essential for every building, considering the multitude of potential sources and the inherently dangerous nature of the gas if undetected.

**SULPHUR DIOXIDE**

Sulphur dioxide is the chemical compound with the formula $SO_2$. It is produced by volcanoes and in various industrial processes. Since coal and petroleum often contain sulphur compounds, their combustion generates sulphur dioxide unless the sulphur compounds are removed before burning the fuel. Further oxidation of $SO_2$, usually in the presence of a catalyst such as $NO_2$, forms $H_2SO_4$, and thus acid rain. Sulphur dioxide emissions are also a precursor to particulates in the atmosphere. Both of these impacts are cause for concern over the environmental impact of these fuels.

## NITROGEN OXIDE

In general chemistry, nitrogen oxide is a term for a couple of different combinations of the elements - nitrogen and oxygen. The two common variants of nitrogen oxide are nitric oxide and nitrogen dioxide respectively, where both have a single nitrogen atom attached to either one or two oxygen atoms. Other kinds of nitrogen-oxygen combinations such as nitrous oxide are also sometimes referred to as a nitrogen oxide. The two kinds of common nitrogen oxides are sometimes notated NOx, where x represents a variable for the one or two parts of oxygen in the molecule.

A lot of NOx is produced from combustion in traditional engines, and gets distributed into the atmosphere. Scientists are looking at nitrogen oxide emissions as part of an overall chemical process that changes the air around us, and affects the response of the environment to worsening air quality. One issue in studying atmospheric health is the result of NOx linking up with a class of chemicals called Volatile Organic Compounds or VOCs. Although manufacturers are trying to limit the commercial applications of VOCs, they are still present in many consumer materials. When NOx blends with VOCs, it creates ozone.

This can lead to "smog" in the air, and cause health problems for some residents of a particular area where the ozone is excessive. Local and federal agencies are looking at trying to limit the proliferation of nitrogen oxide elements into the atmosphere. The U.S. Environmental Protection Agency has issued a full set of information resources for the effects of nitrogen oxide variations on its web site, for educating the general public on what these chemical elements are, and what they can do. The international Kyoto Protocol that limits certain types of emissions sometimes called "greenhouse gases" is contemplating the inclusion of nitrogen oxides in their list of targeted emissions.

In addition, various studies continue to look at how engines can regulate the production of nitrogen oxide as a by-product of combustion. The gas nitrous oxide is sometimes thrown in with the above molecular types, although it should technically be called dinitrous oxide. This gas is used as a type of limited anesthetic, and as a propellant for some types of consumer products. Other types of nitrogen/oxygen combinations also include two nitrogen atoms. Items like these, such as dinitrous trioxide, are generally unstable and are not commonly found in nature.

## CARBON MONOXIDE

Carbon monoxide also called carbonous oxide, is a colourless, odourless and tasteless gas which is slightly lighter than air. It is highly toxic to humans and animals in higher quantities, although it is also produced in normal animal metabolism in low quantities, and is thought to have some normal biological functions. It consists of one carbon atom and one oxygen atom, connected by a covalent double bond and a dative covalent bond.

It is the simplest oxocarbon, and is an anhydride of formic acid. In coordination complexes the carbon monoxide ligand is called carbonyl. Carbon monoxide is produced from the partial oxidation of carbon-containing compounds; it forms when there is not enough oxygen to produce carbon dioxide such as when operating a stove or an internal combustion engine in an enclosed space. In presence of oxygen, carbon monoxide burns with a blue flame, producing carbon dioxide.

Coal gas, which was widely used before the 1960s for domestic lighting, cooking and heating despite its toxicity, had carbon monoxide as a primary constituent. Some processes in modern technology, such as iron smelting, still produce carbon monoxide as a byproduct.

Worldwide, the largest source of carbon monoxide is natural in origin, due to photochemical reactions in the troposphere which generate about $5 \times 10^{12}$ kilograms per year.

Other natural sources of CO include volcanoes, forest fires, and other forms of combustion. In biology, carbon monoxide is naturally produced by the action of heme oxygenase 1 and 2 on the heme from hemoglobin breakdown. This process produces a certain amount of carboxyhemoglobin in normal persons, even if they do not breathe any carbon monoxide.

Following the first report that carbon monoxide is a normal neurotransmitter in 1993, as well as one of three gases that naturally modulate inflammatory responses in the body carbon monoxide has received a great deal of clinical attention as a biological regulator. In many tissues, all three gases are known to act as anti-inflammatories, vasodilators and encouragers of neo-vascular growth. Clinical trials of small amounts of carbon monoxide as a drug, are on-going.

## MONITORING AIR POLLUTION

The Ambient Air Quality Standards establish the concentration at which the pollutant is known to cause adverse health effects to sensitive groups within the population, such as children and the elderly. Both the California and federal governments have adopted health-based standards for the *criteria pollutants*, which include ozone, particulate matter and carbon monoxide. In general, the air quality standards are expressed as a measure of the amount of pollutant per unit volume of air. For example, the particulate matter standard is expressed as micrograms of particulate matter per cubic meter of air (ug/$m^3$) and the ozone standard is expressed as parts per million (ppm).

### OZONE MONITORING

Ozone is a colourless gas with a pungent odour. It is the chief component of urban smog. Ozone is not directly emitted as a pollutant, but is formed in the atmosphere when precursor emissions, hydrocarbons and nitrogen oxides, react in the presence of sunlight. Generally, low wind speeds or stagnant air coupled with warm temperatures and cloudless skies provide for the optimum conditions. As a result, summer is generally the peak ozone season. Because of the reaction time involved, peak ozone concentrations often occur far downwind of the precursor emissions. Therefore, ozone is a regional pollutant that often impacts a widespread area. In addition to adverse health effects, ozone causes damage to open vegetation, building surfaces, exposed rubber surfaces, and certain exposed plastics.

Meteorology (weather) and topography (the lay of the land) play major roles in ozone formation. When the weather is warm and the winds are light, a vertical downward motion of air and a natural cooling of the earth's surface act together to form an inversion that traps pollutants.

Sunlight then causes a chemical reaction between the hydrocarbons and nitrogen oxides to form ozone. The Sacramento Valley is shaped like an elongated bowl. Temperature inversion layers can clamp a lid on the bowl, allowing air pollution to rise to unhealthy levels. Weather conditions cause air pollution concentrations to fluctuate widely from day to day and season to season. Topography alone gives the NSVAB great potential for trapping and accumulating air pollutants. The strong inversions typical of NSVAB summers are caused by subsidence, the slow sinking of air causing compressional warming. The surface inversions typical of winter are formed primarily at night as air is cooled when it comes in contact with the earth's cold surface. These are called radiation inversions. Temperature inversions prevent pollutants from rising and being diluted vertically.

Thus, pollutants remain trapped in the layer of air where people breathe. Summer subsidence inversions occur on over 90 per cent of summer days; they persist throughout the day and tend to intensify during the afternoon.

Winter radiation inversions occur on over 70 per cent of winter nights, but are usually destroyed by daytime heating, bringing a rapid improvement in air quality by afternoon. Both types of inversion mechanisms may operate at any time of the year, and in the fall both may occur together to produce the heaviest pollution potential. Recognizing the adverse health impacts of daylong exposure, the United States Environmental Protection Agency promulgated an 8-hour ozone standard in 1997 as a successor to the 1-hour standard, which was established in 1979.

**Table. Ambient Air Quality Standards For Ozone**

| State Ozone Standard | National Ozone Standards |
|---|---|
| 0.09 ppm for 1 hour— | 0.12 ppm for 1 hour—not to be exceeded more than once per year |
| Not to be exceeded | 0.08 ppm for 8 hours—not to be exceeded based on the fourth highest concentration averaged over 3 years. |

Litigation delayed the implementation of the national 8-hour ozone standard proposed in 1997. EPA issued a proposed rule in May 2003. The proposed rule does not identify, or designate, areas that do not meet the new standard. Designations for attainment and non-attainment areas will occur by April 15, 2004, under a separate process.

## OZONE SUMMARY

The placement of the air monitoring stations operating from 2000 through 2002 in the NSVAB. The placement of the ozone monitors appears evenly distributed throughout the NSVAB. Currently there are eleven ozone monitors operating in the NSVAB. Shasta County has three monitors, one located in Redding, one in Anderson and one in Lassen Volcanic Park; Butte County has two monitors, one located in Chico and one located in Paradise; Sutter County has three monitors, located in Yuba City, Pleasant Grove and one on the Sutter Buttes; Tehama County has two monitors, one in downtown Red Bluff and one on the Tuscan Buttes; Glenn County has one monitor in Willows; and Colusa County has one monitor in the town of Colusa. The State standard allows only one exceedance per year on average at any site within the Air District in the preceding three-year period.

This is meant to take into account year-to-year weather fluctuation and any exceptional exceedances. The California Air Resources Board has established three categories of exceptional exceedances: "exceptional events" "extreme concentration events"; and "unusual concentration events". Ozone trends are variable and unique for each district within the NSVAB. During the past three-year period, the Butte County Paradise monitor, and the Tehama County Red Bluff monitor experienced the highest number of ozone violations in the basin. Ozone concentrations in the NSVAB have remained relatively constant over the past three years while population and vehicle miles traveled (VMT) have increased during the same period. Shasta County ozone violations significantly decreased in the past three years. The decreases in ozone concentrations are largely due to favourable meteorological conditions during this time period. As explained transport of Pollutants, ozone violations in the NSVAB have been classified as transport from the Broader Sacramento Area. The California Air Resources Board (ARB) has defined the impacts of transported air pollution from the Broader Sacramento Area to air districts in the Northern Sacramento Valley.

## PM10 MONITORING

### PARTICULATE MATTER

Refers to particles with an aerodynamic diameter of 10 microns or smaller. For comparison, the diameter of a human hair is about 50 to 100 microns. PM10 is a mixture of substances that includes: elements such as carbon, lead, and nickel; compounds such as nitrates, organic compounds, and sulfates; and complex mixtures such as soil and diesel exhaust. These substances occur in the form of solid particles or as liquid droplets. Primary particles are emitted directly into the atmosphere.

Secondary particles result from gases that are transformed into particles through physical and chemical processes in the atmosphere. PM2.5 includes a subgroup of particles that are less than 2.5 microns in aerodynamic diameter. Fine particulate matter poses an increased health risk because it can be deposited deep into the lung and may contain substances that are particularly harmful to human health.

The EPA promulgated two new national PM2.5 standards in 1997. EPA plans to make final designations by December 15, 2004 based on data from 2001–2003, to reflect the most recent three years of data.

### PM10 SUMMARY

Appendix B, Particulate Matter Tables and Graphs depict, by county, three-year PM10 air quality statistics including: Maximum 24-hour Concentration; Maximum Annual Geometric and Arithmetic Mean; Estimated Days Above State 24-hour Standard; and Days Above National 24-hour Standard. PM10 trends are also unique and variable for each district within the NSVAB. In comparison to ozone, PM10 concentrations do not relate well to growth in population or increased vehicle usage. High PM10 concentrations do not always occur in high population areas. Again, weather and topography play an important role in the fluctuation of air pollution concentrations from day to day and season to season.

In the past three years the Yuba City, Glenn and Chico monitoring stations had the highest number of estimated days above the State PM10 standard. Yuba

City and Red Bluff had the highest annual averages. The NSVAB has had only one national 24-hour PM10 standard exceedance since 1987.

This national exceedance occurred in Colusa County in 1999 and was significantly influenced by wildfires in the area. Because many of the sources that contribute to ozone also contribute to PM10, future ozone emission controls may improve PM10 air quality.

### EMISSION INVENTORY

The California Air Pollution Control and Air Quality Management Districts and the California Air Resources Board (ARB) develop the emission inventory and associated emissions projections jointly. The California Emission Forecasting System (CEFS) is the computer tool used to develop the projections; the emission estimates are based on the most currently available growth and control data. For mobile sources, CEFS integrates the emission estimates from the EMFAC model. The emission projections are based on the 1999 inventory with updates as of November 2002. In the following tables are forecast emissions for the Sacramento Valley Air Basin for Reactive Organic Gases (ROG) and Oxides of Nitrogen (NOx) for several source categories. The annual average emissions are reported in tons per day for the years 2010, 2015 and 2020. The projected emissions show a downtrend for both ROG and NOx, which are the precursor emissions for ozone.

## AMBIENT AIR MONITORING

### DEFINITION

Ambient air monitoring measures levels of contamination in outdoor air, or in the air that people breathe. The levels of pollution measured in the ambient air reflect the combined influences of many different nearby sources, and even some distant ones.

### AMBIENT AIR MONITORING NEAR LANDFILLS

The main reason ambient air monitoring is performed at or near landfills is to evaluate worker and community exposure concerns regarding releases of toxic chemicals to the air. However, because federal regulations currently do not require ambient air monitoring to be performed in the vicinity of municipal solid waste landfills, no ambient air monitoring data are available for many landfills. This is especially true for smaller landfills and those that have not generated extensive community health concerns.

In some cases, a landfill may be considered a hazardous waste site under federal and state regulations. At these sites, regulatory agencies or the landfill owner and operator may collect ambient air data. At other landfills, states may operate ambient air monitoring stations near landfills to measure concentrations of some or all of EPA's criteria pollutants (carbon monoxide, lead, nitrogen dioxide, ozone, particulate matter, and sulfur dioxide). If organic compounds are of concern, continuous monitoring for total hydrocarbons is also possible (but to obtain speciated data, sampling and analysis is usually needed). These pollutants, however, originate from many sources in addition to landfills, and their monitoring data often are viewed as an indicator of general air quality, rather than as the influence of any one particular source (*e.g.*, a landfill).

## MEASURING AMBIENT AIR CONCENTRATION

Ambient air concentrations are generally measured according to specifications set forth in an ambient air monitoring plan. Though the content of these plans varies from project to project, the plans typically address at least the following critical elements of ambient air monitoring:

### Chemicals Selected for Monitoring

One of the first decisions environmental professionals make when developing an ambient air monitoring plan is to select the chemicals to be monitored—a decision that is largely influenced by the purpose of conducting monitoring in the first place. For example, at sites where potential exposure to landfill gas is of concern, monitoring typically focuses on NMOCs, rather than on metals or particulate matter. At sites where windblown dust is an issue, monitoring would likely also consider particulate matter.

Results from soil gas, near-surface, and emissions monitoring data, if available, may be a useful guide for selecting chemicals to consider in air monitoring programmes. The programmes should attempt to measure as many of the chemicals detected in the soil gas and emissions as possible, but especially the most toxic chemicals with the highest concentrations and emission rates. lists some of the more prevalent NMOCs in landfill gas. The EPA's compilation of Air Pollutant Emissions Factors (known as AP-42), provides typical concentrations of more than 40 NMOCs and inorganic compounds in MSW landfill gas. If no site-specific data are available, the substances on this list may provide a starting point. Realistically, however, ambient air monitoring for the scores of chemicals that landfills emit is a prohibitively expensive endeavor. From a practical standpoint, selection of chemicals for monitoring is determined by weighing several factors, such as cost, chemical toxicity, and the availability of sampling methods that can reliably measure ambient air concentrations of a given chemical.

### Sampling Methods

After sponsoring many years of research into ambient air monitoring, EPA has approved several different types of sampling and analytical methods for a long list of common air pollutants. For criteria pollutants (carbon monoxide, lead, nitrogen dioxide, ozone, particulate matter, and sulfur dioxide), EPA has published a list of sampling devices that are capable of measuring concentrations both accurately and precisely for comparison to its national ambient air quality standards (NAAQS). When possible, use of EPA-approved methods is encouraged, because the approval is based on extensive testing of the accuracy and precision of ambient air monitoring. In some cases, however, EPA-approved methods might not be available for certain chemicals or monitoring frequencies (*e.g.*, the compendium documents do not address many of the continuous sampling devices that are available), and use of other methods might be necessary. In these cases, extra care should be taken to ensure that the selected methods are capable of generating high- quality data.

The monitoring methods selected for use in a given programme determine the detection limits for each chemical. The detection limit is the lowest concentration at which the method can reliably measure a chemical's ambient air concentration. For ambient air monitoring data to be useful to the environmental health official, all efforts should be made to use methods with detection limits that are lower than or comparable to ambient air concentrations that would be of health concern.

**Ambient Air Monitoring Locations**

One of the most important elements of developing an ambient air monitoring programme is selecting monitoring locations. With strategically chosen locations, monitoring programmes can generate data of great usefulness for the environmental health professional. Poorly chosen locations, in contrast, can cause monitoring programmes to generate data that offer little insight into air quality in neighbourhoods of concern. In general, monitoring locations are selected according to the goal of the sampling programme. If the goal is to address community concerns, monitoring locations should include residential neighbourhoods at downwind locations nearest the landfill and other places where people might be exposed to landfill gases.

Many additional concerns should be considered when selecting monitoring locations. For perspective on the extent to which landfill emissions affect air quality, simultaneous monitoring at locations upwind and downwind of the landfill of concern is advised. It is equally as critical to review the surroundings of monitoring stations to ensure that local sources of air pollution will not bias a monitor's readings. As examples, monitoring alongside busy roadways or atop industrial facilities will likely generate results indicative of emissions from these sources, even if a landfill is nearby. Schools, parks, and churches generally make excellent choices for monitoring locations because they have few sources of emissions on their premises, they often have sources of electricity readily available, and they typically are located in or near residential neighbourhoods.

**Monitoring Schedules**

Ambient air monitoring plans should specify both the frequency and duration of the proposed monitoring, and both factors should be considered when interpreting data. The frequency of monitoring is often determined by the available sampling methods. Continuous methods provide an ongoing account of air quality, but these methods usually measure levels of only one pollutant; periodic monitoring is typically, though not always, conducted by collecting 24-hour averaged samples on either a 6-day or 12-day cycle. These frequencies ensure that ambient air samples will be collected on every day of the week over a long-term programme.

Sometimes 8- or 12-hour sampling is conducted. Though useful for occupational exposures, such as on the landfill, such sampling may miss significant off-site releases affecting nearby residents during non-working hours, such as the predawn hour when landfill gas odours are not diluted and diffused by strong wind. The duration of monitoring is also an important consideration. Because landfill emissions might exhibit significant seasonal variations, monitoring for a year or longer is needed to accurately estimate the long-term average concentrations of air pollutants.

Further, landfill emissions can change from year to year for various reasons, such as increases or decreases in daily disposal rates, changes in waste mix and moisture, landfill closure, and installation of pollution controls. As a result, monitoring results collected when a landfill actively received wastes might not be representative of air quality after the landfill closes. Use of long-term monitoring at fixed locations, when funds to conduct such monitoring are available, is the best approach for evaluating ongoing effects of landfill air emissions on local air quality.

**Data Quality Parameters**

In ambient air monitoring plans, data quality objectives will be specified for the programme. Data quality objectives provide a goal for exactly how accurate, precise, and complete a data set must be. In general, ambient air monitoring programmes should strive to collect and analyse air samples in accordance with their method's data quality specifications. Though these specifications vary from method to method, measurement accuracy and precision of better than 50 per cent is usually feasible for most methods. A sampling completeness (defined as the per cent of attempted sampling events that are successful) of better than 90 per cent is desired.

**WHAT DO AMBIENT AIR MONITORING DATA**

As noted earlier, ambient air monitoring data characterize levels of contaminants in the air that people breathe. Because these data are almost always the best metric for exposure concentrations at landfill sites, it is extremely important that environmental health professionals interpret ambient air monitoring data critically. At a minimum, you should ask yourself thequestions below when reviewing these data to ensure that they are truly representative of exposure concentrations.

## CONSIDER WHEN REVIEWING AMBIENT AIR MONITORING DATA

### Chemicals Selected for Monitoring

- What chemicals were selected for monitoring?
- Do these include the chemicals of concern identified by residents, regulators, and public health officials?
- Do the chemicals selected for monitoring include those expected to be emitted in greatest quantities from the landfill and/or those that are most toxic chemicals in the emissions?
- Are there any data gaps in the chemicals selected for monitoring?

### Sampling Methods

- Were EPA-approved sampling methods selected? If not, why?
- Are the selected methods recommended for measuring the chemicals selected for monitoring?
- Are the selected methods capable of achieving detection limits comparable to or lower than ambient air concentrations that would be of public health concern?

### Meteorologic Data

- Is there an on-site meteorologic station providing data on wind direction, speed, rainfall, and atmospheric pressure?
- If there is no on-site meteorologic station, how far away is the closest reporting station and how relevant is the data from that station to the site and community?
- Are there known prevailing wind patterns at the site or in the community that may affect changes in contaminant flow pattern, such as the canyon winds in Southern California or seashore wind patterns?

### Ambient Air Monitoring Location

- Was monitoring performed at both upwind (sometimes labeled background) and downwind locations?
- Do you have reason to believe that ambient air concentrations of certain pollutants were higher in residential areas other than those selected for monitoring?
- Are the monitoring locations considerably removed from other emissions sources (*e.g.*, industrial facilities or heavily traveled roadways) that might bias the air quality measurements?

**Monitoring Schedules**

- Was monitoring continuous or periodic?
- If periodic, what monitoring frequency was selected? Was this frequency sufficient for characterizing fluctuations in emissions from the landfill and other sources?
- Over what duration was monitoring performed?
- Is this duration sufficient for characterizing seasonal fluctuations in air quality?
- Was monitoring performed at a time when landfill emissions were considered to be relatively high (*e.g.,* when the landfill was active) or relatively low (*e.g.*, after emissions controls were installed)?
- Was monitoring performed at any period when people complained about odours, such as predawn hours or evenings?

**Data Quality Parameters**

- What per cent of attempted sampling events were successful?
- How accurate were the reported sampling results
- How precise were the reported sampling results?
- EPA maintains an extensive database of ambient air monitoring results that have been submitted to the Agency over the last 30 years. This database, called the Aerometric Information Retrieval System (AIRS), might include ambient air monitoring data for landfill sites that you will review.

Data summary reports for two of EPA's nationwide ambient air monitoring programmes can be found in these reports that are useful for determining whether concentrations measured at a given site are unusually high or low when compared to concentrations at other locations, but these comparisons should be made with caution. When reviewing monitoring data and considering the questions above, you should remember that ambient air monitoring data characterize levels of contamination that result from a combination of many nearby emissions sources, and these data do not characterize influences from any one source (*e.g.,* a landfill) alone. In fact, ambient air monitoring conducted in urban environments will almost certainly identify elevated concentrations of many chemicals (*e.g.,* benzene and 1,3-butadiene) that originate primarily from mobile sources and emissions from gasoline stations.

Failure to consider these other sources might cause you to reach biased conclusions regarding air quality near landfill sites. Perhaps the best way to determine whether a particular landfill is the primary source of a pollutant is to examine whether ambient air concentrations decrease markedly from a source. Chemicals with concentrations that vary little with changing wind directions or with increased distance from a landfill likely do not originate primarily from the site of concern, though exceptions may exist.

## AIR POLLUTION CHEMISTRY

Some air pollutants that are released into the atmosphere by man-made activities pose environmental and health risks directly. These primary pollutants include carbon monoxide, particulate matter, nitrogen oxides and lead, emitted from exhausts of road vehicles. Additional impacts, however, result from the conversion of primary pollutants by a complex series of chemical reactions in the atmosphere, to secondary pollutants, many of which are potentially more harmful than their precursors. Since much of the pollutant chemistry is driven by the presence of sunlight, the secondary products are commonly referred to as photochemical pollutants. A well-known secondary photochemical pollutant is ozone ($O_3$). Its formation results from the sunlight-initiated oxidation (reaction with oxygen) of volatile organic compounds (VOCs) such as benzene in the presence of nitrogen oxides (NOx), mostly nitric oxide (NO) and nitrogen dioxide ($NO_2$). Once formed, ozone is scavenged by NO, and in the absence of other competing reactions, a "photostationary state" is formed where concentrations of NO, $NO_2$ and $O_3$ are all inter-related. In rural areas away from major sources of NO, such as urban road transport, ozone scavenging by NO is lower, and consequently ozone concentrations in the atmosphere are higher. The primary pollutants sulphur dioxide and nitrogen oxides also undergo chemical transformation as they are dispersed in the atmosphere, forming sulphuric acid and nitric acid respectively, which may be deposited downwind as acid rain.

### ASTHMA

Asthma is a common disease in the UK, affecting more than 3 million people. It can occur in people of any age, but is most likely to develop in children by the age of 5 and in adults during their 30s. People over 65 are also prone to the disease. Asthma is an allergic condition that is often inherited by children from their parents. This type of asthma is often related to eczema and hay fever, with 50 per cent of adults and 80 per cent of children who have other allergies commonly developing asthma. Symptoms of asthma in a person can increase or decrease in severity through life. Approximately 1 in 3 children who suffer from asthma will have no symptoms by the time they reach adulthood.

Asthma affects the airways and disrupts the transport of air in and out of the lungs. Asthma sufferers have sensitive airways which become inflamed and narrowed under certain conditions. The inflammation is caused by the body's immune system, in order to counteract the irritant. When inflammation occurs it becomes difficult for oxygen to reach the lungs. Consequently asthmatics may experience difficulty with breathing. Numerous factors can be responsible for triggering an asthma attack. Asthmatics are usually allergic to more than one trigger and their asthma symptoms may vary from wheeziness, to shortness of breath, chest tightening or the over production of mucus.

Both indoor and outdoor air pollution, natural and man-made, can trigger asthma attacks. Common indoor pollutant triggers include the dustmite, mould and cigarette smoke. As much as 85 per cent of children that experience asthmatic symptoms are allergic to dustmite. Cigarette smoke is damaging to everyone's airways, but can be particularly bad for people with asthma. Smoke causes the airways to narrow, making it more difficult to breathe.

Outdoors, natural plant, grass and tree pollen can performance as triggers in some asthmatics. Man-made pollution may also be detrimental to asthmatics. Although it has not been proven that a link exists between air pollution and asthma, certain pollutants, including sulphur dioxide, nitrogen dioxide and ozone are known to restrict the airways and make it more difficult for asthma sufferers to breathe. Fine particulate matter is also suspected to be a lung irritant.

Although there is no cure for asthma at present, inhalers can be used to reduce the severity of the symptoms. Preventer inhalers may be used to build up a long-term resistance to asthma triggers, whilst reliever inhalers are used as an instantaneous means of relieving asthmatic symptoms, by relaxing the muscles controlling the airways.

## CARBON MONOXIDE

Carbon monoxide (CO) is a colourless, odourless, poisonous gas produced when fuels containing carbon are burned where there is too little oxygen. It also forms as a result of burning fuels at too high a temperature. It burns in air or oxygen with a blue flame and is slightly lighter than air. In the presence of an adequate supply of $O_2$ most carbon monoxide produced during combustion is immediately oxidised to carbon dioxide ($CO_2$). However, this is not the case in spark ignition engines in motor cars, especially under idling and deceleration conditions. Thus, the major source of atmospheric carbon monoxide is road transport. Smaller contributions come from processes involving the combustion of organic matter, for example in power stations and waste incineration.

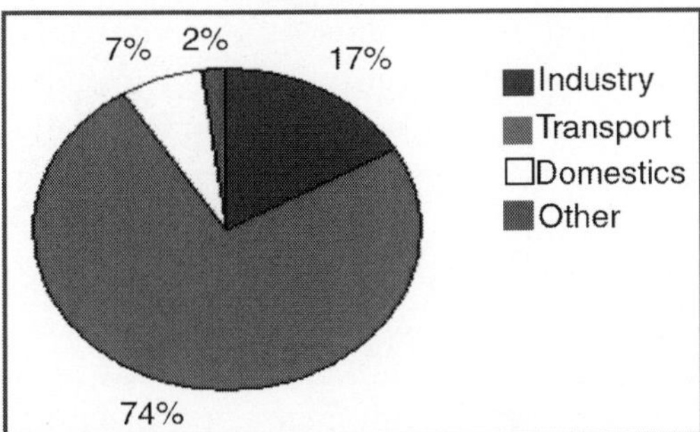


**Fig.** Carbon Monoxide Emissions in The UK 4758 Thousand Tonnes

Natural background levels of carbon monoxide fall in the range of 10-200 parts per billion. Levels in urban areas are highly variable, depending upon weather conditions and traffic density. Concentrations are generally less than

10 parts per million (ppm) but can be as high as 500 ppm. Carbon monoxide is poisonous when inhaled because it combines with haemoglobin, the oxygen-carrying substance in red blood cells. The hemoglobin then cannot take up oxygen from the air. Lack of oxygen causes cells and tissues to die.

**CARS**

The motor vehicle engine emits many types of pollutants including nitrogen oxides (NOx), volatile organic compounds (VOCs), carbon monoxide (CO), carbon dioxide ($CO_2$), particulates, sulphur dioxide ($SO_2$) and lead. Emissions of such pollutants may lead to poor air quality, posing environmental and health risks. Individually, a vehicle engine is not a particularly important source of pollution. Collectively however, they represent a major source of air pollutants inmost developed countries, including the UK.

Emissions are related to use of the engine, mainly the fuel type and the temperature of fuel combustion. If the engine is efficient, then the products of combustion will be mainly carbon dioxide and water. However, at low speeds and idling engines are inefficient and therefore the products of incomplete combustion dominate, for example carbon monoxide and VOCs from petrol engines, and carbon monoxide, VOCs, particulates and smoke from diesel engines. As the temperature of combustion increases, the efficiency of conversion to carbon dioxide and water increases. However, impurities in the fuel such as nitrogen are oxidised to nitrogen dioxide. At high temperatures atmospheric nitrogen ($N_2$) is also oxidised to nitrogen dioxide.

Nitrogen oxides pollution from vehicles can be significantly reduced by fitting a catalytic converter to the exhaust system. This is a relatively low cost method of pollution control which has little effect on vehicle performance and fuel consumption. All new cars sold in Britain from January 1993 onwards have catalytic converters. Diesel fuel contains more energy per litre than petrol and coupled with the fact that diesel engines are more efficient than petrol engines, diesel cars are more efficient to run. Diesel fuel contains no lead and emissions of the regulated pollutants (carbon monoxide, hydrocarbons and nitrogen oxides) are lower than those from petrol cars without a catalyst. However, when compared to petrol cars with a catalyst, diesels have higher emissions of nitrogen oxides and much higher emissions of particulate matter.

Emissions from cars are greatest when an engine is cold. On a cold day a petrol car may take up to 10km to warm up and operate at maximum efficiency; a diesel car may only take 5km. Consequently, diesel cars produce less unburned fuel during a cold start, which will result in lower emissions of carbon monoxide and hydrocarbons. Diesel cars could make a significant impact on air quality in urban areas where most cold starts occur, especially when it is considered that a catalyst on a petrol car would take several minutes to reach its operating temperature. Overall, diesel cars emit less hydrocarbons, carbon monoxide and lead pollution than petrol cars, but produce more noxious gases and significantly more particulates.

## CATALYTIC CONVERTERS

Since January 1993, all new petrol-driven cars sold in the European Union (EU) have been fitted with a catalytic converter (CAT). This is made up of a very thin layer of platinum group metals on a honeycomb structure. The surface area of a typical 3-way CAT covers the equivalent of two football pitches. As exhaust gasses pass through the catalyst a chemical reaction occurs which converts carbon monoxide (CO), volatile organic compounds (VOCs, including hydrocarbons) and nitrogen oxides (NOx) to less harmful compounds (water, nitrogen and carbon dioxide). To work most effectively, a catalytic converter needs to reach an optimum temperature. It may not reach this in a short journey. Devises to pre-warm the catalyst are being developed which improve the overall performance of catalytic converters. The use of catalytic converters leads to a dramatic reduction in the emissions of carbon monoxide, hydrocarbons and nitrogen oxides. However, they also result in an increase in carbon dioxide ($CO_2$) emissions, which do not cause a problem for urban air quality, but may contribute to global warming. The efficiency of a CAT can be as high as 90 per cent. Oxidation catalysts may be fitted to either petrol or diesel cars. The catalyst oxidises the pollutants formed by incomplete combustion to carbon dioxide and water, and is effective for hydrocarbons and carbon monoxide. However, they do not reduce nitrogen oxides emissions.

## CLEAN AIR ACTS

In response to the Great London Smog of December 1952, the Government introduced its first Clean Air Act in 1956. This Act aimed to control domestic sources of smoke pollution by introducing smokeless zones. In these areas, smokeless fuels had to be burnt. The Clean Air Act focused on reducing smoke pollution, but the introduction of cleaner coals and the increased usage of electricity and gas actually helped to reduce sulphur dioxide levels at the same time. In addition, power stations were relocated to more rural areas. As a consequence, air pollution in cities was dramatically reduced.

The Clean Air Act of 1968 introduced the basic principle for the use of tall chimneys for industries burning coal, liquid or gaseous fuels. At the time of this legislation it was recognised that smoke pollution could be controlled, but that sulphur dioxide removal was generally impracticable. Hence, the higher the chimney, the better the dispersal of the air pollution.

## CLEANER FUELS

To replace pollutant fuels (petrol and diesel), cleaner alternative fuels are currently being developed. They include: compressed natural gas (CNG);

liquefied petroleum gas (LPG); city diesel; hydrogen; alcohol fuels; and battery operated vehicles. On a cycle representing congested urban traffic, both LPG and CNG have lower emissions of carbon monoxide (CO) than for petrol-powered vehicles. Indeed, emissions of carbon monoxide from CNG-powered vehicles are of the same order as those emitted by diesel vehicles. However, emissions of hydrocarbons from CNG vehicles are relatively high because of methane, the major component of natural gas. Although methane is a small contributor to the formation of low level ozone it is a major factor in global warming. Emissions of nitrogen oxides and particulates from both LPG- and CNG-powered vehicles are significantly lower than those from diesel vehicles. Moreover, emissions of nitrogen oxides from CNG vehicles are half those from equivalent petrol-engined vehicles. City diesel is a petroleum-based lower emission diesel developed in Sweden but now available in many European Countries including the UK. Exhaust emissions from vehicles fuelled with city diesel compare favourably with exhaust emissions from equivalent vehicles fuelled with conventional diesel. The main benefit of city diesel is that its combustion reduces particulate emissions by 34 - 84 per cent depending on engine type and type of particulate measured. An additional benefit of city diesel is that it is a low sulphur fuel, which is necessary for the optimum running of oxidation catalytic converters. Whilst pollution from road transport may be reduced by using alternative fuels to petrol and diesel, ultimately it is cheaper to improve conventional fuels than to use many of the alternatives, and no investment is needed for new storage tanks and service stations.

## DISPERSION

Many atmospheric factors influence the way air pollution is dispersed, including wind direction and wind speed, type of terrain and heating effects. To better understand how atmosphere processes can affect ground level pollution, atmospheric conditions can be described simply as either stable or unstable, where the stability is determined by wind (which stirs the air) and heating effects (which cause convection currents). Atmospheric stability affects pollution released from ground level and elevated sources differently. In unstable conditions, ground level pollution is readily dispersed thereby reducing ground level concentrations. Elevated emissions, however, such as those released from a chimney, are returned more readily to ground level, leading to higher ground level concentrations. Stable conditions mean less atmospheric mixing and therefore higher concentrations around ground level sources, but better dispersal rates, and therefore lower ground level concentrations, for elevated plumes.

# 3

# Air Pollution's Effects

Air pollution is responsible for major health effects. Every year, the health of countless people is ruined or endangered by air pollution. Many different chemicals in the air affect the human body in negative ways. Just how sick people will get depends on what chemicals they are exposed to, in what concentrations, and for how long. Studies have estimated that the number of people killed annually in the US alone could be over 50,000.Older people are highly vulnerable to diseases induced by air pollution. Those with heart or lung disorders are under additional risk. Children and infants are also at serious risk. Because people are exposed to so many potentially dangerous pollutants, it is often hard to know exactly which pollutants are responsible for causing sickness.

Also, because a mixture of different pollutants can intensify sickness, it is often difficult to isolate those pollutants that are at fault. Many diseases could be caused by air pollution without their becoming apparent for a long time. Diseases such as bronchitis, lung cancer, and heart disease may all eventually appear in people exposed to air pollution. Air pollutants such as ozone, nitrogen oxides, and sulfur dioxide also have harmful effects on natural ecosystems. They can kill plants and trees by destroying their leaves, and can kill animals, especially fish in highly polluted rivers.

## AIR POLLUTION AND ITS EFFECTS ON HEART ATTACK RISK

In an effort to better understand heart disease, researchers like to draw up lists of things linked to an uptick in heart attack rates: earthquakes, heavy meals, outdoor temperatures, participation in war, and, my personal favourite, suffering a favourite team's loss in the Super Bowl. Traffic and air pollution are also on the list, but it turns out they contribute in a much more significant way to heart attack rates than many other things. In fact, a study published last week in the journal Lancet finds that traffic and air pollution account for nearly 12 per cent of heart attacks worldwide.

In comparison, cocaine use accounts for less than 1 per cent. Of course, all of us are exposed to air pollution–which raises an individual's heart attack risk by just 2 per cent–while only 0.04 per cent of the population use cocaine, which raises an individual's heart attack risk by 2,400 per cent. "The important message here is that while an individual's risk from air pollution is moderate or small, each of us is exposed, making the amount of risk intolerable for the entire community, " says Dr. Andrea Baccarelli, an associate professor of environmental health at the Harvard School of Public Health, who wrote an editorial that accompanied the study.

The Lancet study found that traffic exposure accounted for the highest percentage of heart attacks–more than 7 per cent–followed by extreme physical exertion, excess alcohol use, coffee, and a depressed mood. The Belgian researchers, however, couldn't tease out whether it was the frustration of sitting in traffic that contributed to heart attack risk or the exposure to air pollutants. Air pollutants may contribute to heart attack risk by promoting inflammation and increasing blood clotting, says Baccarelli.

These are both processes involved in the formation of unstable artery plaque, often involved in heart attacks. Air pollutants also appear to attach to lung cell receptors, sending a signal to the heart that the body is in trouble. "Studies have shown that people experience changes in heart rate on days when air quality is low," Baccarelli adds.

Those particularly susceptible to the heart-damaging effects of pollution include the elderly, who have weaker hearts, and people who are obese, because they have higher levels of inflammation as well as more strain on their hearts. To minimize your exposure to air pollutants, Baccarelli recommends trying the following:

**Stay Indoors as much as Possible on Poor Air Quality Days**

It's a good idea to check the government's air quality index map on hot summer days and to avoid strenuous outdoor activity on "unhealthy" or "hazardous" days.

**Keep Windows Closed if you live on a Busy Street or near a Highway**

In a 2009 study published in the journal Circulation, Baccarelli and his colleagues found that those who lived on a major traffic road had nearly a 50 per cent higher risk of developing a vascular condition called deep vein thrombosis than those who lived several blocks away. Previous studies have shown that concentrations of exhaust pollutants decline to insignificant levels beyond 1,200 to 1,500 feet from a clogged roadway.

If you do live in a high-traffic area, your best bet is to keep your windows closed as much as possible, especially on poor air quality days, and use the air conditioner, which filters out pollutants pretty effectively, says Baccarelli. You can also consider investing in an air purifier, which also removes contaminants from the air.

**Use Recirculated Air in your Car**

Keep your car windows closed and push the recirculated air button or switch on your dashboard–especially when you're driving in heavy traffic.

**Avoid Exercising Outdoors During the Morning and Evening Rush Hour**

That's, of course, when car emission levels reach their peak. If you're biking to and from work during rush hour, try to find a bike path that's a bit off the busy thoroughfares, if possible. Ditto if you're walking. You're still better off being active, even in high traffic areas, than skipping exercise altogether, says Baccarelli, since the health benefits of exercise outweigh the risks of pollution.

**PSA Testing Declines Little Despite Limited Value**

Two studies published last week in the Journal of the National Cancer Institute underscore the controversies over screening for prostate cancer by measuring a blood marker called prostate specific antigen or PSA. Fewer men, at least in one Northwest health network, are getting screened, and the test was found to be less useful than previously thought for catching early cancer.

In the first study, Seattle researchers found that PSA testing declined moderately in a local VA network since 2008 when the US Preventive Services Task Force–an independent group that makes screening recommendations–advised against routine screening for men over 75. The task force in 2002 stopped recommending that all men be tested and suggested they speak to their doctors about the risks and benefits before making a decision. The decline was 3 percentage points among men ages 40 to 54; 2.7 percentage points among those ages 55 to 74; and 2.2 percentage points among men ages 75 years and older. That's pretty modest considering that some research suggests that the PSA test not only doesn't save lives but also increases the risk of being treated for cancers that aren't life-threatening.

**Should you Worry about Getting Measles**

Measles has been big news in Boston recently, with several cases either confirmed or suspected in the Back Bay area. Health officials have taken the precaution of holding free measles vaccination clinics for those in the vicinity of St. James Street near the Park Square Building. Should you worry, though, if you weren't in that area? Unless you've come in contact with an infected person, there's little reason to be alarmed, says Sagar Nigwekar, an internist at Brigham and Women's Hospital and author of "5 Questions to Ask Your Doctor." "It's not enough to just be in the same building," he adds. "You have to come within a few feet of a person who has measles to catch the virus."

You also probably don't need to worry if you were vaccinated against measles as a child. The combination measles-mumps-rubella vaccine is given twice at ages 12 to 15 months and then again at 4 to 6 years of age. After childhood vaccination, the vast majority of adults will retain their immunity over the rest of their lives, says Nigwekar. And those born before 1957 probably have natural immunity to the virus which was in wide circulation before that time.

**Diabetes Exhibit at Museum of Science**

We've all heard of diabetes–maybe even have it ourselves–but how does it occur and what does it really do to our body? A temporary exhibit that recently opened at the Museum of Science will help you visualize what's happening beneath your skin and what to do to avoid or better manage diabetes.

Produced by the Detroit Science Center with support from the Joslin Diabetes Center, the interactive exhibit takes you through a 40-foot-long blood vessel to see how excess blood sugar does its damage. Video games show just how tough it is to get a new diabetes treatment through the testing pipeline. And, you can even exercise while you're there on some gym equipment. But don't expect to get any food from the giant refrigerator on display. It just raps about what you should eat to keep the disease at bay.

## AIR POLLUTION AND HUMAN HEALTH

People have no choice but to breathe the air around them regardless of its quality. When they breathe ground-level ozone or air laden with particulates, they do so at risk to their health. Symptoms of air pollution exposure include irritation to one's eyes, throat, and lungs. Wheezing, coughing, burning eyes, chest tightness, headaches, and difficulty breathing are all commonly reported when air quality is poor. Increased doctor visits, hospitalization, and school absences also frequently occur at such times. People's symptoms often disappear once air quality improves, but air pollution exposure can also result in tragedy.

The first major air pollution event to be studied occurred in 1930 in the Meuse Valley of Belgium. A dense blanket of smog hung over the valley for five days, which killed 63 people and caused 6000 others to become ill. A similar event occurred in Donora, Pennsylvania in 1948. Almost 6,000 of the 14,000 people in the town became ill, and 20 died. The most notorious of these tragedies occurred in London in 1952 as a result of dense smog caused by the burning of coal fires. More than 4000 people died over a five-day period. But tragic consequences are not always immediate. Slow and subtle health effects from long-term air pollution exposure are also of great concern and may culminate in life-threatening illnesses such as cancer.

Certain people appear to be more vulnerable to air pollution, namely the elderly, the young, and those with cardiopulmonary disease such as asthma or severe bronchitis. Ozone and air particles appear to be especially harmful to children's health. This is because their lungs are still growing, they are often outside for long periods, and they are usually quite active. As a result, pound for pound they inhale more polluted outdoor air than adults typically do.

Although people have no choice but to breath the air around them, they do have choices that can help them stay healthy. Paying attention to air quality forecasts, avoiding outside exercise or spending more time inside when ozone levels are high, and supporting measures to improve air quality and reduce sources of pollution are all commendable and health-saving measures. Such actions are a positive response to a problem that can literally steal one's breath away.

## SOURCES

Air pollution comes from many different sources. Natural processes that affect air quality include volcanoes, which produce sulfur, chlorine, and ash particulates. Wildfires produce smoke and carbon monoxide. Cattle and other animals emit methane as part of their digestive process. Even pine trees emit volatile organic compounds (VOCs). Many forms of air pollution are human-made. Industrial plants, power plants and vehicles with internal combustion engines produce nitrogen oxides, VOCs, carbon monoxide, carbon dioxide, sulfur dioxide and particulates.

In most megacities, such as Mexico City and Los Angeles, cars are the main source of these pollutants. Stoves, incinerators, and farmers burning their crop waste produce carbon monoxide, carbon dioxide, as well as particulates. Other human-made sources include aerosol sprays and leaky refrigerators, as well as fumes from paint, varnish, and other solvents. One important thing to remember about air pollution is that it doesn't say in one place. Winds and weather play an important part in transport of pollution locally, regionally, and even around the world, where it affects everything it comes in contact with. There are many different chemical substances that contribute to air pollution. These chemicals come from a variety of sources.

Among the many types of air pollutants are nitrogen oxides, carbon monoxides, and organic compounds that can evaporate and enter the atmosphere. Air pollutants have sources that are both natural and human. Now, humans contribute substantially more to the air pollution problem. Forest fires, volcanic eruptions, wind erosion, pollen dispersal, evaporation of organic compounds, and natural radioactivity are all among the natural causes of air pollution. Usually, natural air pollution does not occur in abundance in particular locations. The pollution is spread around throughout the world, and as a result, poses little threat to the health of people and ecosystems.

Though some pollution comes from these natural sources, most pollution is the result of human activity. The biggest causes are the operation of fossil fuel-burning power plants and automobiles that combust fuel. Combined, these two sources are responsible for about 90 per cent of all air pollution in the

United States. Some cities suffer severely because of heavy industrial use of chemicals that cause air pollution. Places like Mexico City and Sao Paulo have some of the most deadly pollution levels in the world.

## AIR POLLUTION CAUSES AND EFFECTS

Humans probably first experienced harm from air pollution when they built fires in poorly ventilated caves. Since then we have gone on to pollute more of the earth's surface. Until recently, environmental pollution problems have been local and minor because of the Earth's own ability to absorb and purify minor quantities of pollutants. The industrialization of society, the introduction of motorized vehicles, and the explosion of the population, are factors contributing towards the growing air pollution problem. At this time it is urgent that we find methods to clean up the air. The primary air pollutants found in most urban areas are carbon monoxide, nitrogen oxides, sulfur oxides, hydrocarbons, and particulate matter (both solid and liquid). These pollutants are dispersed throughout the world's atmosphere in concentrations high enough to gradually cause serious health problems. Serious health problems can occur quickly when air pollutants are concentrated, such as when massive injections of sulfur dioxide and suspended particulate matter are emitted by a large volcanic eruption.

### Air Pollution in the Home

You cannot escape air pollution, not even in your own home. "In 1985 the Environmental Protection Agency (EPA) reported that toxic chemicals found in the air of almost every American home are three times more likely to cause some type of cancer than outdoor air pollutants". The health problems in these buildings are called "sick building syndrome". "An estimated one-fifth to one-third of all U.S. buildings are now considered "sick". The EPA has found that the air in some office buildings is 100 times more polluted than the air outside. Poor ventilation causes about half of the indoor air pollution problems.

The rest come from specific sources such as copying machines, electrical and telephone cables, mold and microbe-harbouring air conditioning systems and ducts, cleaning fluids, cigarette smoke, carpet, latex caulk and paint, vinyl molding, linoleum tile, and building materials and furniture that emit air pollutants such as formaldehyde. A major indoor air pollutant is radon-222, a colourless, odourless, tasteless, naturally occurring radioactive gas produced by the radioactive decay of uranium-238. "According to studies by the EPA and the National Research Council, exposure to radon is second only to smoking as a cause of lung cancer". Radon enters through pores and cracks in concrete when indoor air pressure is less than the pressure of gasses in the soil. Indoor air will be healthier than outdoor air if you use an energy recovery ventilator to provide a consistent supply of fresh filtered air and then seal air leaks in the shell of your home.

The two main sources of pollutants in urban areas are transportation (predominantly automobiles) and fuel combustion in stationary sources, including residential, commercial, and industrial heating and cooling and coal-burning power plants. Motor vehicles produce high levels of carbon monoxides (CO) and a major source of hydrocarbons (HC) and nitrogen oxides (NOx). Whereas, fuel combustion in stationary sources is the dominant source of sulfur dioxide ($SO_2$).

**Carbon Dioxide**

Carbon dioxide ($CO_2$) is one of the major pollutants in the atmosphere. Major sources of $CO_2$ are fossil fuels burning and deforestation. "The concentrations of $CO_2$ in the air around 1860 before the effects of industrialization were felt, is assumed to have been about 290 parts per million (ppm). In the hundred years and more since then, the concentration has increased by about 30 to 35 ppm that is by 10 per cent". Industrial countries account for 65 per cent of $CO_2$ emissions with the United States and Soviet Union responsible for 50 per cent.

Less developed countries (LDCs), with 80 per cent of the world's people, are responsible for 35 per cent of $CO_2$ emissions but may contribute 50 per cent. "Carbon dioxide emissions are increasing by 4 per cent a year". In thousand million tons of carbon dioxide (equivalent to 5 thousand million tons of carbon) were released into the atmosphere, but the atmosphere showed an increase of only 8 billion tons (equivalent to 2.2 billion tons of carbon". The ocean waters contain about sixty times more $CO_2$ than the atmosphere. If the equilibrium is disturbed by externally increasing the concentration of $CO_2$ in the air, then the oceans would absorb more and more $CO_2$.If the oceans can no longer keep pace, then more $CO_2$ will remain into the atmosphere. As water warms, its ability to absorb $CO_2$ is reduced.

$CO_2$ is a good transmitter of sunlight, but partially restricts infrared radiation going back from the earth into space. This produces the so-called greenhouse effect that prevents a drastic cooling of the Earth during the night. Increasing the amount of $CO_2$ in the atmosphere reinforces this effect and is expected to result in a warming of the Earth's surface. Currently carbon dioxide is responsible for 57 per cent of the global warming trend. Nitrogen oxides contribute most of the atmospheric contaminants.

*NOX-nitric oxide (NO) and nitrogen dioxide ($NO_2$):*

- Natural component of the Earth's atmosphere.
- Important in the formation of both acid precipitation and photochemical smog (ozone), and causes nitrogen loading.
- Comes from the burning of biomass and fossil fuels.
- 30 to 50 million tons per year from human activities, and natural 10 to 20 million tons per year.

- Average residence time in the atmosphere is days.
- Has a role in reducing stratospheric ozone.

*$N_2O$-nitrous oxide*:

- Natural component of the Earth's atmosphere.
- Important in the greenhouse effect and causes nitrogen loading.
- Human inputs 6 million tons per year, and 19 million tons per year by nature.
- Residence time in the atmosphere about 170 years.
- 1700 (285 parts per billion), 1990 (310 parts per billion), 2030 (340 parts per billion).
- Comes from nitrogen based fertilizers, deforestation, and biomass burning.

Sulfur dioxide is produced by combustion of sulfur-containing fuels, such as coal and fuel oils. Also, in the process of producing sulfuric acid and in metallurgical process involving ores that contain sulfur. Sulfur oxides can injure man, plants and materials. At sufficiently high concentrations, sulfur dioxide irritates the upper respiratory tract of human beings because potential effect of sulfur dioxide is to make breathing more difficult by causing the finer air tubes of the lung to constrict. "Power plants and factories emit 90 per cent to 95 per cent of the sulfur dioxide and 57 per cent of the nitrogen oxides in the United States.

Almost 60 per cent of the $SO_2$ emissions are released by tall smoke stakes, enabling the emissions to travel long distances". As emissions of sulfur dioxide and nitric oxide from stationary sources are transported long distances by winds, they form secondary pollutants such as nitrogen dioxide, nitric acid vapor, and droplets containing solutions of sulfuric acid, sulfate, and nitrate salts. These chemicals descend to the earth's surface in wet form as rain or snow and in dry form as a gases fog, dew, or solid particles. This is known as acid deposition or acid rain.

**Chlorofluorocarbons (CFCs)**

CFCs are lowering the average concentration of ozone in the stratosphere. "Since 1978 the use of CFCs in aerosol cans has been banned in the United States, Canada, and most Scandinavian countries. Aerosols are still the largest use, accounting for 25 per cent of global CFC use".

Spray cans, discarded or leaking refrigeration and air conditioning equipment, and the burning plastic foam products release the CFCs into the atmosphere. Depending on the type, CFCs stay in the atmosphere from 22 to 111 years. Chlorofluorocarbons move up to the stratosphere gradually over several decades. Under high energy ultra violet (UV) radiation, they break down and release chlorine atoms, which speed up the breakdown of ozone ($O_3$) into oxygen gas ($O_2$). Chlorofluorocarbons, also known as Freons, are greenhouse gases that contribute to global warming. Photochemical air pollution is commonly referred to as "smog". Smog, a contraction of the words smoke and fog, has been caused throughout recorded history by water condensing on smoke particles, usually from burning coal. With the introduction of petroleum to replace coal economies in countries, photochemical smog has become predominant in many cities, which are located in sunny, warm, and dry climates with many motor vehicles. The worst episodes of photochemical smog tend to occur in summer.

Photochemical smog is also appearing in regions of the tropics and subtropics where savanna grasses are periodically burned. Smog's unpleasant properties result from the irradiation by sunlight of hydrocarbons caused primarily by unburned gasoline emitted by automobiles and other combustion sources.

The products of photochemical reactions includes organic particles, ozone, aldehydes, ketones, peroxyacetyl nitrate, organic acids, and other oxidants. Ozone is a gas created by nitrogen dioxide or nitric oxide when exposed to sunlight. Ozone causes eye irritation, impaired lung function, and damage to trees and crops. Another form of smog is called industrial smog.

This smog is created by burning coal and heavy oil that contain sulfur impurities in power plants, industrial plants, etc. . . The smog consists mostly of a mixture of sulfur dioxide and fog. Suspended droplets of sulfuric acid are formed from some of the sulfur dioxide, and a variety of suspended solid particles. This smog is common during the winter in cities such as London, Chicago, Pittsburgh. When these cities burned large amounts of coal and heavy oil without control of the output, large-scale problems were witnessed. In 1952 London, England, 4,000 people died as a result of this form of fog. Today coal and heavy oil are burned only in large boilers and with reasonably good control or tall smokestacks so that industrial smog is less of a problem. However, some countries such as China, Poland, Czechoslovakia, and some other eastern European countries, still burn large quantities of coal without using adequate controls.

**Pollution Damage to Plants**

With the destruction and burning of the rain forests more and more $CO_2$ is being released into the atmosphere. Trees play an important role in producing oxygen from carbon dioxide. "A 115 year old Beech tree exposes about 200,000 leaves with a total surface to 1200 square meters. During the course of one sunny day such a tree inhales 9,400 litres of carbon dioxide to produce 12 kilograms of carbohydrate, thus liberating 9,400 litres of oxygen. Through this mechanism about 45,000 litres of air are regenerated which is sufficient for the respiration of 2 to 3 people". This process is called photosynthesis which all plants go though but some yield more and some less oxygen. As long as no more wood is burnt than is reproduced by the forests, no change in atmospheric $CO_2$ concentration will result. Pollutants such as sulfur dioxide, nitrogen oxides, ozone and peroxyacl nitrates (PANs), cause direct damage to leaves of crop plants and trees when they enter leaf pores (stomates). Chronic exposure of leaves and needles to air pollutants can also break down the waxy coating that helps prevent excessive water loss and damage from diseases, pests, drought and frost. "In the midwestern United States crop losses of wheat, corn, soybeans, and peanuts from damage by ozone and acid deposition amount to about $5 billion a year".

**Reducing Pollution**

You can help to reduce global air pollution and climate change by driving a car that gets at least 35 miles a gallon, walking, bicycling, and using mass transit when possible. Replace incandescent light bulbs with compact fluorescent bulbs, make your home more energy efficient, and buy only energy efficient appliances. Recycle newspapers, aluminum, and other materials. Plant trees and avoid purchasing products such as Styrofoam that contain CFCs. Support much stricter clean air laws and enforcement of international treaties to reduce ozone depletion and slow global warming. Earth is everybody's home and nobody likes living in a dirty home. Together, we can make the earth a cleaner, healthier and more pleasant place to live.

## POLLUTANTS

### WHAT ARE POLLUTANTS

These are defined as the substances which cause pollution. They can be physical or chemical. They involve the smoke, gases and heat etc. They are mostly the waste product or their by products. Sometimes it is necessary to add them for the benefit. For example, the soil needs a phosphate and nitrates for its fertility. They may cause the water pollution too.

The pollutants cause an adverse effect on the environment. The increase in concentration of carbon dioxide and the decrease in the concentration of oxygen also cause the pollution. The pollutants are classified into different types and they depend on the form, existence and the natural disposal.

On the basis of the form they can be primary or secondary pollutant. The primary pollutants are those which maintain their form in the environment. It includes the DDT.

The secondary pollutants are those which are not able to maintain their form in the environment. It includes the peroxyacyl nitrate which is formed due to the reaction between the primary pollutants like the hydrocarbons and nitrogen oxides. On the basis of the existence they can be qualitative or quantitative. The qualitative pollutants are those which do not occur in the environment normally but occur due to the human activities. It includes the DDT, fungicides etc. The quantitative pollutants are those which occur in the environment normally but do occur due to the human activities also. It includes the CO, $CO_2$. On the basis of the natural disposal they can be biodegradable and non-degradable. The biodegradable pollutants are those in which the waste products are degraded slowly by the microbes.

When the production is more than the capacity it results in the pollution of environment. The non biodegradable pollutants are those in which the waste products are not degraded slowly by the microbes. It includes the DDT, glass, plastics, phenols, radioactive substances and metal containers. There are different types of pollution. This includes the air, water, soil, radioactive and noise pollution.

## DISPERSION CHARACTERISTICS OF STACK PLUMES

Dispersion is the process of spreading out pollution emission over a large area and thus reducing their concentration. Wind speed and environmental lapse rates directly influence the dispersion pattern. Five classifications of plume behaviour, which may occur under some commonly encountered metrological conditions.

### Coning

A coning plume, occurs under essentially neutral stability, when environmental lapse rate is equal to adiabatic lapse rate, and moderate to strong winds occur. The plume enlarges in the shape of a cone. A major part of pollution may be carried fairly far downwind before reaching ground.

### Looping

Under super-adiabatic condition, both upward and downward movement of the plume is possible. Large eddies of a strong wind cause a looping pattern. Although the large eddies tend to disperse pollutants over a wide region, high ground level concentrations may occur close to the stack.

### Fanning

A fanning plume occurs in the presence of a negative lapse rate when vertical dispersion is restricted. The pollutants disperse at the stack height, horizontally in the from of a fanning plume.

### Fumigation

As when the emission from the stack is under an inversion layer, the movement of the pollutants in the upward direction is restricted. The pollutants move downwards. The resulting fumigation can lead to a high ground level concentration downwind of the stack.

### Lofting

When the stack is sufficiently high and the emission is above an inversion layer, mixing in the upward direction is uninhibited, but downward motion is restricted. Such lofting plumes do not result in any significant concentration at ground level. However, the pollutants are carried hundreds of kilometers from the source.

## STABILITY CLASSIFICATION

*For the purpose of calculation of concentration of pollutants downwind of a source the stability of the atmosphere is classified as:*

| | | | | | |
|---|---|---|---|---|---|
| A | = | Extremely unstable, unstable, | B | = | Moderately |
| C | = | Slightly unstable, | D | = | Neutral, |
| E | = | Slightly stable, | F | = | Stable |

These classifications are arrived at from metrological condition of wind speed, solar insulation and cloudiness, given in Table. a-at 10 m above ground, b-sun higher than 60° and clear sky, c-sun 35°– 60° and few broken clouds or clear sky, d-sun 15°- 35°, cloudy. The above described stability classification is known as Pasquill's stability classification which is the most popular one because of its simplicity.

## GAUSSIAN PLUME MODEL

Gaussian plume model is used to calculate the concentration of a pollutant downwind of a point source. Figure shows the three dimensional coordinates system and the Gaussian plume equation for the case when the value of z is zero, for calculating ground level concentration: Where,

| | | |
|---|---|---|
| C (x, y) | = | Concentration at ground level |
| Q | = | Emission rate of pollutants, μg/s |
| H | = | Effective stack height, m |
| U | = | Average wind speed at effective stack height |
| s y and s z | = | Horizontal and vertical dispersion coefficients, m, respectively |

The dispersion coefficient depends on the atmospheric stability class and increase with the downwind distance from the source. Figure gives values of the dispersion coefficients as a function of distance for various stability classes. The Gaussian plume equation can be used to predict ground level pollutant concentrations under different 25 stability conditions for a given pollution source.

Figure shows the downwind ground level concentrations due to emissions from a thermal power plant calculated for different stability classes. Note that the turbulence in an unstable atmosphere results in a high concentration near the stack. Downwind, however, concentrations drop off very quickly. The stable atmosphere, on the other hand, has a much lower peak. However, it continues to be appreciable for a considerable distance downwind.

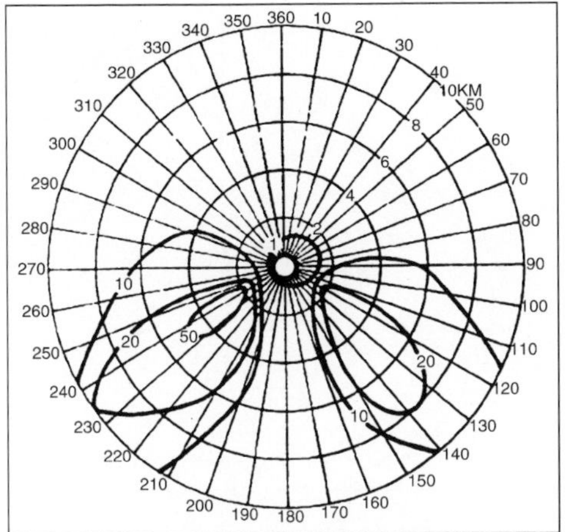


As shown in Figure, the ground level concentration is also quite sensitive to stack height. In order to disperse pollutants over a larger area, stack heights in the range of 200 to 250 m are quite common. The stack heights shown in Figure are effective stack heights, which are higher than the actual stack heights because of the buoyancy of the plume at the emission point, Figure. It is possible to calculate average ground level concentrations over a specified period by integrating values occurring under different stability conditions and wind speed and direction. Such data can be presented in the form of isopleths as shown in Figure.

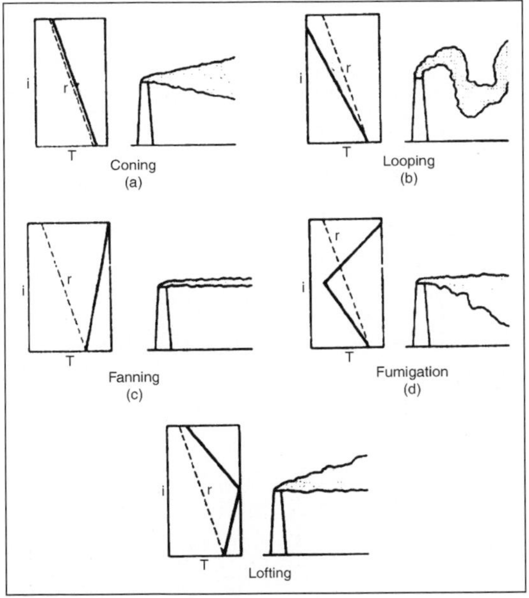

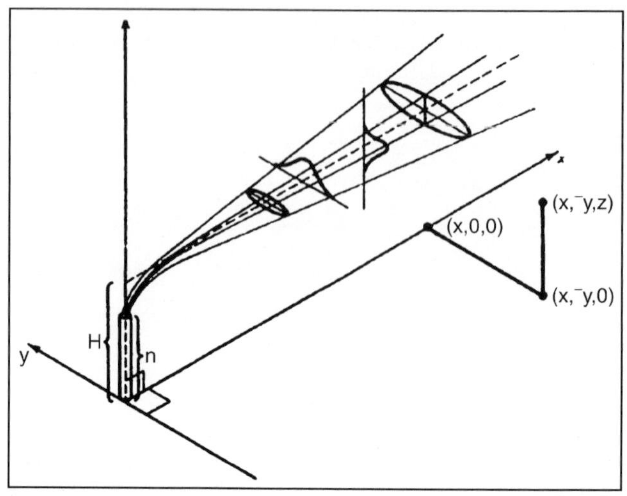


$$C(x,y) = \frac{Q}{\pi u \sigma_y \sigma_z}\left(\exp\frac{-H^2}{2\sigma_z^2}\right)\left(\exp\frac{-y^2}{2\sigma_y^2}\right)$$

Where,

C(x, y) = Concentration at Ground-level at the Point (x, y), $mg/m^3$

x = Distance Directly Downwind, m

y = Horizontal distance from the Plume Centerline, m

Q = Emission rate of Pollutants, mg/s

H = Effective stack height, m, (H=h+gh, where h = actual stack height, and gh = plume rise

u = Average wind Speed at the effective height of the stack, m/s

$s_y$ = Horizontal Dispersion Coefficient (standard deviation), m

$s_z$ = Vertical Dispersion Coefficient (standard deviation), m

## ORGANIC AND INORGANIC POLLUTANTS

### ORGANIC POLLUTION

Organic chemical pollutants are both naturally and synthetically (man-made) produced and consist of a combination of only a few basic element, primarily carbon, hydrogen, oxygen, nitrogen, sulfur, phosphorous, and halogens such as fluorine, chlorine, bromine, and iodine. Carbon is the foundation of all organic chemicals and because of the ability of carbon to form carbon-to-carbon bonds an almost unlimited number of stable organic chemical can be formed.

The primary organic pollutants of concern in urban stormwater are most often synthetic organic chemicals. These chemicals are used in every aspect of human activity and are often highly toxic to humans and the environment. As well, many are also very difficult to remove once introduced to the environment. This *Species Selection Tool* categories the organic pollutants of primary concern in urban stormwater into three categories: pesticides, petroleum hydrocarbons, and other toxic organic which include chlorinated solvents, explosives, PCB's, and other organic pollutants. The below chemical structures highlight just a few of the organic pollutants potentially found in urban stormwater that have potential to be removed using phytoremediation in constructed wetlands.

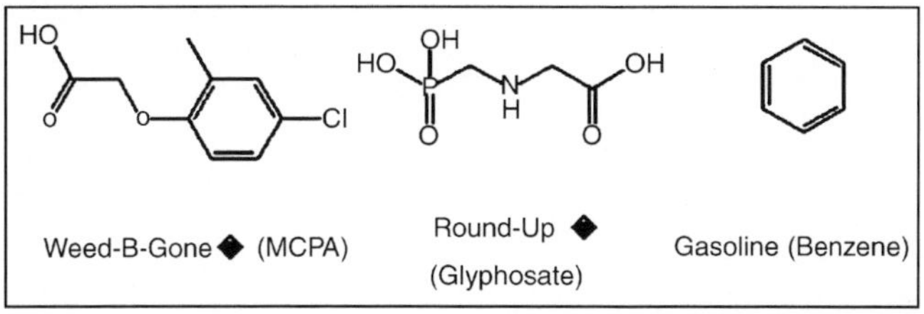

Weed-B-Gone◆ (MCPA)

Round-Up ◆ (Glyphosate)

Gasoline (Benzene)

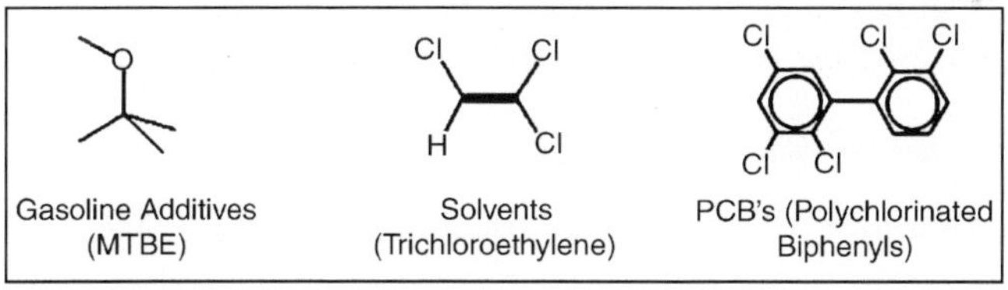

Gasoline Additives (MTBE)

Solvents (Trichloroethylene)

PCB's (Polychlorinated Biphenyls)

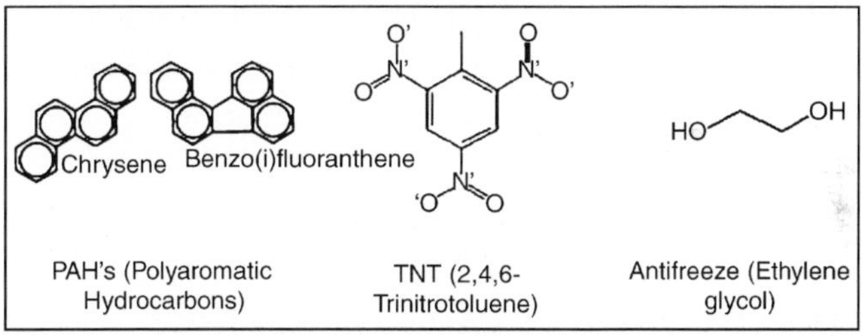

Chrysene Benzo(i)fluoranthene

PAH's (Polyaromatic Hydrocarbons)

TNT (2,4,6-Trinitrotoluene)

Antifreeze (Ethylene glycol)

Organic pollution has been on the rise in the Danube over the last century, as human activities have resulted in increasing loads of wastewater rich in organic matter. The most serious organic pollution problems occur in tributaries that regularly receive untreated or inadequately treated wastewater from industrial plants and municipalities. Organic pollution has been increasing in parts of the Danube Basin up till the end of the last century, as industrial production and household consumption have resulted in increasing loads of wastewater rich in organic matter. Other organic matter of natural origin includes natural soil erosion with a high organic content and the decomposition of dead plants and animals.

In terms of organic pollution, water quality in the Danube ranges between Class II (moderate pollution) and Class II-III (moderate to critical pollution) of the interim ICPDR classification. The most serious organic pollution problems occur in tributaries that regularly receive untreated or inadequately treated wastewater from industrial plants and municipalities, such as on the Arges river.

## Organic Compounds in Freshwater

Organic pollution occurs when large quantities of organic compounds, which act as substrates for microorganisms, are released into watercources. During the decomposition process the dissolved oxygen in the receiving water may be used up at a greater rate than it can be replenished, causing oxygen depletion and having severe consequences for the stream biota. Organic effluents also frequently contain large quantities of suspendid solids which reduce the light available to photosynthetic organisms and, on settling out, alter the characteristics of the river bed, rendering it an unsuitable habitat for many invertebrates. Toxic ammonia is often present.

Organic pollutants consist of proteins, carbohydrates, fats and nucleic acids in a multiplicity of combinations. Raw sewage is 99,9 per cent water, and of the 0,1 per cent solids, 70 per cent is organic (65 per cent proteins, 25 per cent carbohydrates, 10 per cent fats). Organic wastes from people and their animals may also be rich in disease-causing (pathogenic) organisms.

### *What are the Origins of Organic Pollutants*

Organic pollutants originate from domestic sewage (raw or treated), urban run-off, industrial (trade) effluents and farm wastes. Sewage effluents is the greatest source of organic materials discharged to freshwaters. In England and Wales there are almost 9000 discharges releasing treated sewage effluent to rivers and canals and several hundred more discharges of crude sewage, the great majority of them tot the lower, tidal reaches of rivers or, via long outfalls, to the open sea. It has been assumed, certainly incorrectly, that the sea has an almost unlimited capacity for purifying biodegradable matter.

### *The Effects of Organic Effluents on Receiving Waters*

When an organic polluting load id discharged into a river it is gradually eliminated by the activities of micro organisms in a way very similar to the processes in the sewage treatment works. This self-purification requires sufficient concentrations of oxygen, and involves the breakdown of complex organic molecules into simple in organic molecules.

Dilution, sedimentation and sunlight also play a part in the process. Attached micro organisms in streams play a greater role than suspended organisms in self-purification. Their importance increases as the quality of the effluent increases since attached microorganisms are already present in the stream, whereas suspended ones are mainly supplied with the discharge.

### *Effects on the Biota*

Organic pollution affects the organisms living in a stream by lowering the available oxygen in the water. This causes reduced fitness, or, when severe, asphyxiation. The increased turbidity of the water reduces the light available to photosynthetic organisms. Organic wastes also settle out on the bottom of the stream, altering the characteristics of the substratum. The general effects of fairly severe organic pollution are illustrated in

## INORGANIC POLLUTANTS

Inorganic chemical pollutants are naturally found in the environment but due to human development these pollutants are often concentrated and released into the environment in urban stormwater. The primary inorganic pollutants of concern in urban stormwater are cadmium, copper, lead, zinc, nitrogen, nitrate, nitrite, ammonia, phosphorous, and phosphate. These chemicals are used in every aspect of human activity and are often highly toxic to humans and the environment.

This *Species Selection Tool* categories the inorganic pollutants of primary concern in urban stormwater into two categories: heavy metals and nutrients. The below chemical structures highlight just a few of the inorganic pollutants potentially found in urban stormwater that have potential to be removed using phytoremediation in constructed wetlands.

| Cadmium (Cd) | Copper (Cu) | Lead (Pb) | Zinc (Zn) | Nitrogen (N) |
|---|---|---|---|---|
| Phosphorous (P) | Nitrate | Nitrite | Ammonia | Phosphate |

### Common Inorganic Pollutants of Water Pollution

Industrial, agricultural, and domestic wastes can contribute to the pollution of twater, and water pollutants can damage human and animal health. One of the most important categories of water pollutants is inorganic chemical. These inorganic chemicals are usually substances of mineral origin. Metals, salt, and minerals are examples of inorganic chemicals. The chemicals are discussed below, which are the most common inorganic pollutants of of water pollution.

### Alkalinity

Alkalinity is a measure of the ability of water to neutralize acids. It is a commonly measured water characteristic that has little meaning or importance to the typical homeowner. Calcium (Ca) is a major component of alkalinity, as it is with hardness. High alkalinity of water, that means it is probably hard, too. There is no drinking water standard for alkalinity.

### Arsenic (As)

Arsenic occurs in groundwater from both natural sources and human activities. It is odourless and tasteless in drinking water. Arsenic has a primary drinking water standard because it can cause skin lesions, circulatory problems, and nervous system disorders. Prolonged exposure also can cause various forms of cancer. The present arsenic drinking water standard (0.05 mg/L) is being studied and will likely be lowered to 0.005 mg/L in the near future.

**Barium (Ba)**

Like arsenic (As), barium occurs naturally in small concentrations in many groundwater supplies. Barium has a primary drinking water standard of 2.0 mg/L because it causes nervous and circulatory system problems, especially high blood pressure. Standard water softeners are effective in removing barium.

**Hardness**

It is a general term used to refer to the calcium carbonate ($CaCO_3$) content of water. Hardness does not pose a health threat, but it does cause aesthetic problems. you can ruin hot water heater elements, reduce soap lathering, and make laundry difficult to clean. Moderate levels of hardness are beneficial because they inhibit plumbing system corrosion. Removal of hardness using a water softener is necessary only if the water is causing aesthetic problems. Use of water softeners may result in undesirable levels of sodium in drinking water and may increase plumbing system corrosion.

**Chloride ($Cl^-$)**

Chloride occurs naturally in most groundwater but may become elevated due to leaching from salt storage areas around highways or from brines produced during gas well drilling. Other possible sources of chloride are sewage effluent, animal manure, and industrial waste. Chloride has a secondary drinking water standard of 250 mg/L because it may cause a salty taste in the water. Groundwater in Pennsylvania usually contains less than 25 mg/L of chloride.

**Copper (Cu)**

Copper usually originates from corrosion of copper plumbing in the home. Copper has a secondary drinking water standard of 1.0 mg/L because it causes a bitter, metallic taste in water and a blue-green stain in sinks and bathtubs. Copper levels above 1.3 mg/L are a health concern because they may cause severe stomach cramps and intestinal illnesses.

**Iron (Fe)**

Iron is a common natural problem in groundwater that may be worsened by mining activities. Iron does not occur in drinking water in concentrations of health concern to humans. The secondary drinking water standard for iron is 0.3 mg/L because it causes a metallic taste and orange-brown stains that make water unsuitable for drinking and clothes washing.

**Lead (Pb)**

If lead is detected in drinking water, it probably originated from corrosion of the plumbing system. If your water supply is corrosive, then any lead present in the plumbing system may be dissolved into the drinking water. Lead concentrations are usually highest in the first water out of the tap (known as "first-draw" water), since this water has been in contact with the plumbing for a longer time. Lead concentrations typically decrease as water is flushed through the plumbing system. Lead poses a serious health threat to the safety of drinking water. Long-term exposure to lead concentrations in excess of the drinking water standard has been linked to many health effects in adults including cancer, stroke, and high blood pressure.

At even greater risk are the fetus and infants up to four years of age, whose rapidly growing bodies absorb lead more quickly and efficiently. Lead can cause premature birth, reduced birth weight, seizures, behavioural disorders, brain damage, and lowered IQ in children.

In rare cases, the source of lead in drinking water might be from groundwater pollution rather than corrosion of the plumbing system. Such pollution may be the result of industrial or landfill contamination of an aquifer. The source of the lead usually can be determined by comparing water test results from a first-draw sample versus a sample collected after the water runs for several minutes. If the lead concentration is high in both samples, then the source of the lead is likely from groundwater contamination.

**Hydrogen Sulfide ($H_2S$)**

It is a naturally occurring gas that is common in groundwater. Very small concentrations of hydrogen sulfide in water are offensive to most individuals. Although hydrogen sulfide is a highly toxic gas, only under the most unusual conditions would it reach levels toxic to humans as a result of its occurrence in drinking water. More often, it is simply an aesthetic odour problem that can be removed using several treatment processes.

**Manganese (Mn)**

As like iron, manganese is a naturally occurring metal that can be worsened by mining activities. Manganese at concentrations normally found in drinking water does not constitute a health hazard; however, even small amounts of manganese may impart objectionable tastes or blackish stains to water. For this reason, manganese has a recommended drinking water standard of 0.02 mg/L.

**Nitrate or Nitrate Nitrogen**

Nitrate in drinking water usually originates from fertilizers or from animal or human wastes. Nitrate concentrations in water tend to be highest in areas of intensive agriculture or where there is a high density of septic systems. Nitrate has a secondary drinking water standard that was established to protect the most sensitive individuals in the population (infants under 6 months of age and a small component of the adult population with abnormal stomach enzymes). These segments of the population are prone to methomoglobinemia (blue baby disease) when consuming water with high nitrates. Nitrate may be reported on the water test report as either nitrate or nitrate-nitrogen. Look carefully at your report to determine which form of nitrate is being reported. The primary drinking water standard or MCL is 10 mg/L as nitrate-nitrogen ($NO_3$-N) but it is 45 mg/L as nitrate (($NO_3$).

### Sulfate ($SO_4$)

Sulfates normally are present at some level in all private water systems. Sulfates occur naturally as a result of leaching from sulfur deposits in the earth. It has a secondary drinking water standard of 250 mg/L because it may impart a bitter taste to the water at this level.

A proposal also exists to make sulfate a primary contaminant with an MCL of 500 mg/L, because it may have a laxative effect and cause other gastrointestinal upset above this concentration.

## EFFECTS OF POLLUTION ON PLANTS

- *Learning Objective*: The students will be able to identify how air, water and soil pollution affects vegetation, and cite specific effects on agricultural crops and forests. Key Question: How does pollution affect the chemical reactions in plant populations?
- *Motivation*: Plants are being assaulted on several levels by human activities which threatens our own survival.

### DEVELOPMENT WITH PIVOTAL QUESTIONS

Life exists on Earth largely because of the ability of plants (producers) to capture and use radiant energy from the sun to create glucose from carbon dioxide and water during photosythesis:

$$6CO_2 + 6H_2O = C_6H_{12}0_6 + 60_2$$

This reaction produces oxygen which provided for the development of air-breathing consumers like humans who use the chemical energy in plant carbohydrates as fuel in aerobic respiration (the reverse reaction of photosynthesis.) In Biology you learned about food chains and webs, the pyramid of biomass and energy, and nutrient cycles. You know about how the extinction of dinosaurs may be related to an astronomical event, but do you realise that human beings may turn out to be more deadly than any comet ever could be? Many scientists believe we are currently in a period of unprecedented mass extinctions and habitat destruction generated by humans.

*Loss of biological diversity and important producers like the rain forest may be the death knell for our own species as well*:

- Air Pollution is harmful to plants. Solfur dioxide damages barley, alfalfa, cotton and wheat. Particulate matter affects climate in that those with a diameter greater than 1 micrometer scatter incoming light away from the Earth's surface. How would this affect weather and plant production? Ozone is very toxic to leafy vegetation such as tomatoes and tobacco. In California, crop damage caused by ozone and other photochemical pollutants costs the state millions of dollars a year. You can review recent research funded by the US Department of Agriculture at its Agricultural Research Services which specializes in the relationship between air quality and plant growth and development.
- Water Pollution that affects plants would include acid rain. Do you recall from the Air Pollution lesson which pollutants form acid rain in the atmosphere? Sulfuric and nitric acid rain washes nutrients out of the soil, damages the bark and leaves of trees and harms the fine root hairs of many plants which are needed to absorb water. Since the acid concentration increases near the base of clouds by density, high altitude trees and vegetation may be exposed to pH levels as low as 3. Unique areas such as the Black Forest in Germany and Vermont's sugar maples are particularly threatened.
- Soil Pollution problems affecting plants include the overuse of pesticides which selects for resistant strains of insects and weeds, the loss of topsoil due to overdevelopment and soil erosion, and the uptake of toxic chemicals through the mismanagement of solid waste. Not only are agricultural crops affected, but natural vegetation in special habitats like in Hawaii are threatened. Although occupying less than 0.2 per cent of the US landmass, Hawaii contains 27 per cent of its endangered birds and plants. Seventy two per cent of all the US species that have become extinct in the past years did so on the Hawaiian Islands.
- The Greenhouse Effect is the name given to the process by which gases in the atmosphere trap solar radiation and warmth like in the interior of a greenhouse. Greenhouse gases include carbon dioxide ($CO_2$), methane ($CH_4$) and nitrous oxide ($N_2O$). The delicate balance that maintained a steady level of $CO_2$ in the atmosphere for plants and animals has been increasingly disturbed by human activity such as the burning of fossil fuels and the clearcutting of $CO_2$ sinks (a site where $CO_2$ is absorbed) like the rain forest. Today the atmosphere holds over 27 per cent more $CO_2$ than 100 years ago and this contributes to global warming. Doubling of $CO_2$ is thought to warm

the Earth by 2 to 4 degrees celsius. Why is this a problem? Polar ice may melt, raising the sea level and submerging coastal areas like Long Island and Florida. Weather patterns may change such that we in the Eastern US experience more intense hurricanes while there is less rain in the midwest which may lose its status as the breadbasket of the world due to droughts. We may lose the plants we need to survive.

- Agricultural Crops are directly affected by air pollution and acid rain. Locally, research is conducted in fields at Brookhaven National Laboratory as well as in North Carolina. From the Pacific coast, pictorial evidence of the effects of ozone on grape leaves is shown on the right:
- Forest ecosystems are also affected by pollution. Acid rain is killing high altitude forests and photochemical smog and ozone damage sensitive redwoods and other old growth forests closer to sea level. Many scientists believe a wealth of medicinal substances, possibly a cure for cancer, is being destroyed along with the precious rain forest habitats, the most productive areas in the world.

**WHAT CAN BE DONE TO SAVE PLANT POPULATIONS**

A normal reaction is that one person cannot make a difference. However if YOU eat less beef, fewer rain forest acres will be cleared to graze cattle. If You drive less and use less electricity, $CO_2$ and NOx and SOx emissions will be lower and that will reduce both acid rain and global warming. If YOU reduce, reuse and recycle, You can help prevent vegetation extinctions

**OXIDANTS**

Ozone is the main pollutant in the oxidant smog complex. Its effect on plants was first observed in the Los Angeles area in 1944. Since then, ozone injury to vegetation has been reported and documented in many areas throughout North America, including the southwestern and central regions of Ontario.

Throughout the growing season, particularly July and August, ozone levels vary significantly. Periods of high ozone are associated with regional southerly air flows that are carried across the lower Great Lakes after passing over many urban and industrialised areas of the United States.

Localized, domestic ozone levels also contribute to the already high background levels. Injury levels vary annually and white bean, which are particularly sensitive, are often used as an indicator of damage. Other sensitive species include cucumber, grape, green bean, lettuce, onion, potato, radish, rutabagas, spinach, sweet corn, tobacco and tomato. Resistant species include endive, pear and apricot.

**Fig.** Ozone Injury to Soybean Foliage.

Ozone symptoms characteristically occur on the upper surface of affected leaves and appear as a flecking, bronzing or bleaching of the leaf tissues. Although yield reductions are usually with visible foliar injury, crop loss can also occur without any sign of pollutant stress. Conversely, some crops can sustain visible foliar injury without any adverse effect on yield. Susceptibility to ozone injury is influenced by many environmental and plant growth factors. High relative humidity, optimum soil-nitrogen levels and water availability increase susceptibility. Injury development on broad leaves also is influenced by the stage of maturity. The youngest leaves are resistant. With expansion, they become successively susceptible at middle and basal portions. The leaves become resistant again at complete maturation.

**SULFUR DIOXIDE**

Major sources of sulfur dioxide are coal-burning operations, especially those providing electric power and space heating. Sulfur dioxide emissions can also result from the burning of petroleum and the smelting of sulfur containing ores. Sulfur dioxide enters the leaves mainly through the stomata (microscopic openings) and the resultant injury is classified as either acute or chronic. Acute injury is caused by absorption of high concentrations of sulfur dioxide in a relatively short time.

The symptoms appear as 2-sided (bifacial) lesions that usually occur between the veins and occasionally along the margins of the leaves. The colour of the necrotic area can vary from a light tan or near white to an orange-red or brown depending on the time of year, the plant species affected and weather conditions. Recently expanded leaves usually are the most sensitive to acute sulfur dioxide injury, the very youngest and oldest being somewhat more resistant.

Chronic injury is caused by long-term absorption of sulfur dioxide at sub-lethal concentrations. The symptoms appear as a yellowing or chlorosis of the leaf, and occasionally as a bronzing on the under surface of the leaves. Different plant species and varieties and even individuals of the same species may vary considerably in their sensitivity to sulfur dioxide.

**Fig**. Acute Sulfur Dioxide Injury to Raspberry. Note that the Injury occurs between the Veins and that the Tissue Nearest the Vein Remains Healthy.

These variations occur because of the differences in geographical location, climate, stage of growth and maturation. The following crop plants are generally considered susceptible to sulfur dioxide: alfalfa, barley, buckwheat, clover, oats, pumpkin, radish, rhubarb, spinach, squash, Swiss chard and tobacco. Resistant crop plants include asparagus, cabbage, celery, corn, onion and potato.

## FLUORIDE

Fluorides are discharged into the atmosphere from the combustion of coal; the production of brick, tile, enamel frit, ceramics, and glass; the manufacture of aluminium and steel; and the production of hydrofluoric acid, phosphate chemicals and fertilizers. Fluorides absorbed by leaves are conducted towards the margins of broad leaves (grapes) and to the tips of monocotyledonous leaves (gladiolus). Little injury takes place at the site of absorption, whereas the margins or the tips of the leaves build up injurious concentrations. The injury starts as a gray or light-green water-soaked lesion, which turns tan to reddish-brown. With continued exposure the necrotic areas increase in size, spreading inward to the midrib on broad leaves and downward on monocotyledonous leaves.

**Fig**. Fluoride Injury to Plum Foliage. The Fluoride Enters the Leaf through the Stomata and is Moved to the Margins where it Accumulates and causes Tissue Injury.

Studies of susceptibility of plant species to fluorides show that apricot, barley (young), blueberry, peach (fruit), gladiolus, grape, plum, prune, sweet corn and tulip are most sensitive. Resistant plants include alfalfa, asparagus, bean (snap), cabbage, carrot, cauliflower, celery, cucumber, eggplant, pea, pear, pepper, potato, squash, tobacco and wheat.

## AMMONIA

Ammonia injury to vegetation has been observed frequently in Ontario in recent years following accidents involving the storage, transportation or application of anhydrous and aqua ammonia fertilizers. These episodes usually release large quantities of ammonia into the atmosphere for brief periods of time and cause severe injury to vegetation in the immediate vicinity.

Complete system expression on affected vegetation usually takes several days to develop, and appears as irregular, bleached, bifacial, necrotic lesions. Grasses often show reddish, interveinal necrotic streaking or dark upper surface discolouration. Flowers, fruit and woody tissues usually are not affected, and in the case of severe injury to fruit trees, recovery through the production of new leaves can occur. Sensitive species include apple, barley, beans, clover, radish, raspberry and soybean. Resistant species include alfalfa, beet, carrot, corn, cucumber, eggplant, onion, peach, rhubarb and tomato.

**Fig.** Severe Ammonia Injury to Apple Foliage and Subsequent Recovery through the Production of new Leaves following the Fumigation.

## PARTICULATE MATTER

Particulate matter such as cement dust, magnesium-lime dust and carbon soot deposited on vegetation can inhibit the normal respiration and photosynthesis mechanisms within the leaf. Cement dust may cause chlorosis and death of leaf tissue by the combination of a thick crust and alkaline toxicity produced in wet weather. The dust coating also may affect the normal action of pesticides and other agricultural chemicals applied as sprays to foliage. In addition, accumulation of alkaline dusts in the soil can increase soil pH to levels adverse to crop growth.

**Fig.** Cement-dust coating on apple leaves and fruit. The dust had no injurious effect on the foliage, but inhibited the action of a pre-harvest crop spray.

## INVESTIGATION OF AIR-POLLUTION INJURY TO VEGETATION

The Ministry of the Environment monitors air quality at 33 stations across the province. The sites are set in both urban and rural settings and monitor the 6 most common air pollutants: sulfur dioxide, ozone, nitrogen dioxide, total reduced sulfur compounds, carbon monoxide and suspended particles. The sites are monitored in real-time on an hourly basis. Pollutant concentrations are converted into an Air Quality Index (AQI) with a lower AQI translating into cleaner air. AQI values above 50 can cause crop injury.

# 4

# Air Pollution: Prevention and Control

## AIR POLLUTION

### WHAT IS AIR POLLUTION

'I'll go out for a breath of fresh air' is an often-heard phrase. But how many of us realise that this has become irrelevant in today's world, because the quality of air in our cities is anything but fresh. The moment you step out of the house and are on the road you can actually see the air getting polluted; a cloud of smoke from the exhaust of a bus, car, or a scooter; smoke billowing from a factory chimney, flyash generated by thermal power plants, and speeding cars causing dust to rise from the roads. Natural phenomena such as the eruption of a volcano and even someone smoking a cigarette can also cause air pollution.

**Table. The Gaseous Composition of Unpolluted Air**

| The Gases (vol) | Parts per million (vol) | The Gases | Parts per million |
|---|---|---|---|
| Nitrogen | 756,500 | Oxygen | 202,900 |
| Water | 31,200 | Argon | 9,000 |
| Carbon Dioxide | 305 | Neon | 17.4 |
| Helium | 5.0 | Methane | 0.97-1.16 |
| Krypton | 0.97 | Nitrous oxide | 0.49 |
| Hydrogen | 0.49 | Xenon | 0.08 |
| Organic vapours | ca. 0.02 | | |

Air pollution is aggravated because of four developments: increasing traffic, growing cities, rapid economic development, and industrialization. The Industrial Revolution in Europe in the 19th century saw the beginning of air pollution as we know it today, which has gradually become a global problem.

Learn more about indoor air pollution, the air and its major pollutants, what you can do, and the problem of acid rain and smog. Here are many types of air pollution. Some are visible. Others are invisible. But all of them can make people sick, harm plants, pollute waterways, or even cause global warming. Some of these dangerous particles get into the atmosphere naturally from forest fires, lightning strikes, and erupting volcanoes, but most of the stuff that we call air pollution is put into the atmosphere by humans.

*Here's a list of a few of the common air pollutants and where they come from*:

- Carbon monoxide, is an invisible air pollutant and is released into the air when fossil fuels such as coal, oil and gas are burned.
- Nitrogen dioxide is a brown gas that comes mostly from cars and trucks that burn fossil fuels, plants that make electricity, and burning wood. Breathing nitrogen dioxide in the air makes people's lungs irritated. A similar type of pollution, nitric oxide, creates haze. It damages plants and makes materials like fabric and metal wear apart easily.
- Ground-level ozone, a harmful type of ozone, forms in the air when sunlight interacts with nitrogen oxides, which are released from factories, and organic compounds, which are released naturally from plants. Ozone near the ground is not beneficial like the ozone layer in the stratosphere, and is a major part of smog.
- Ultrafine aerosols are tiny particles in the air. They are released from cars, trucks, and power plants. When people breath aerosols, the particles travel deep into their lungs and the tissue of the lungs can become inflamed.

## AIR AND ITS MAJOR POLLUTANTS

- One of the formal definitions of air pollution is as follows–'The presence in the atmosphere of one or more contaminants in such quality and for such duration as is injurious, or tends to be injurious, to human health or welfare, animal or plant life. ' It is the contamination of air by the discharge of harmful substances. Air pollution can cause health problems and it can also damage the environment and property. It has caused thinning of the protective ozone layer of the atmosphere, which is leading to climate change.
- Modernisation and progress have led to air getting more and more polluted over the years. Industries, vehicles, increase in the population, and urbanization are some of the major factors responsible for air

pollution. The following industries are among those that emit a great deal of pollutants into the air: thermal power plants, cement, steel, refineries, petro chemicals, and mines.

- Air pollution results from a variety of causes, not all of which are within human control. Dust storms in desert areas and smoke from forest fires and grass fires contribute to chemical and particulate pollution of the air. The source of pollution may be in one country but the impact of pollution may be felt elsewhere. The discovery of pesticides in Antarctica, where they have never been used, suggests the extent to which aerial transport can carry pollutants from one place to another. Probably the most important natural source of air pollution is volcanic activity, which at times pours great amounts of ash and toxic fumes into the atmosphere. The eruptions of such volcanoes as Krakatoa in Indonesia, Mt. St. Helens in Washington, USA and Katmai in Alaska, USA, have been related to measurable climatic changes.

**Table. National Ambient Air Quality Standards**

| Pollutants | Average Time | Concentration |
|---|---|---|
| Sulphur dioxide ($SO_2$) | Annual average<br>24 hour | 60 $\mu g/m^3$<br>80 $\mu g/m^3$ |
| Oxides of Nitrogen ($NO_2$) | A. A<br>24H | 60 $\mu g/m^3$<br>80 $\mu g/m^3$ |
| Suspended Particulate Matter (SPM) | A. A<br>24H | 140 $\mu g/m^3$<br>200 $\mu g/m^3$ |
| Lead | A. A<br>24H | 0.75 $\mu g/m^3$<br>1.0 $\mu g/m^3$ |
| Carbon Monoxide | A. A<br>24H | 2.0 $\mu g/m^3$<br>4.0 $\mu g/m^3$ |
| Respirable Particulate Matter (RPM) | A. A<br>24H | 60 $\mu g/m^3$<br>100 $\mu g/m^3$ |

- Listed below are the major air pollutants and their sources.
- Carbon monoxide (CO) is a colourless, odourless gas that is produced by the incomplete burning of carbon-based fuels including petrol, diesel, and wood. It is also produced from the combustion of natural and synthetic products such as cigarettes. It lowers the amount of oxygen that enters our blood. It can slow our reflexes and make us confused and sleepy.
- Carbon dioxide ($CO_2$) is the principle greenhouse gas emitted as a result of human activities such as the burning of coal, oil, and natural gases.
- Chloroflorocarbons (CFC) are gases that are released mainly from air-conditioning systems and refrigeration. When released into the

air, CFCs rise to the stratosphere, where they come in contact with few other gases, which leads to a reduction of the ozone layer that protects the earth from the harmful ultraviolet rays of the sun.

- Leadis present in petrol, diesel, lead batteries, paints, hair dye products, etc. Lead affects children in particular. It can cause nervous system damage and digestive problems and, in some cases, cause cancer.
- Ozone occur naturally in the upper layers of the atmosphere. This important gas shields the earth from the harmful ultraviolet rays of the sun. However, at the ground level, it is a pollutant with highly toxic effects. Vehicles and industries are the major source of ground-level ozone emissions. Ozone makes our eyes itch, burn, and water. It lowers our resistance to colds and pneumonia.
- Nitrogen oxide (Nox) causes smog and acid rain. It is produced from burning fuels including petrol, diesel, and coal. Nitrogen oxides can make children susceptible to respiratory diseases in winters.
- Suspended particulate matter (SPM) consists of solids in the air in the form of smoke, dust, and vapour that can remain suspended for extended periods and is also the main source of haze which reduces visibility. The finer of these particles, when breathed in can lodge in our lungs and cause lung damage and respiratory problems.
- Sulphur dioxide ($SO_2$) is a gas produced from burning coal, mainly in thermal power plants. Some industrial processes, such as production of paper and smelting of metals, produce sulphur dioxide. It is a major contributor to smog and acid rain. Sulfur dioxide can lead to lung diseases.

## SMOG

The term smog was first used in 1905 by Dr H. A. Des Voeux to describe the conditions of fog that had soot or smoke in it. Smog is a combination of various gases with water vapour and dust. A large part of the gases that form smog is produced when fuels are burnt. Smog forms when heat and sunlight react with these gases and fine particles in the air. Smog can affect outlying suburbs and rural areas as well as big cities. Its occurrences are often linked to heavy traffic, high temperatures, and calm winds. During the winter, wind speeds are low and cause the smoke and fog to stagnate; hence pollution levels can increase near ground level. This keeps the pollution close to the ground, right where people are breathing.

It hampers visibility and harms the environment. Heavy smog is greatly decreases ultraviolet radiation. In fact, in the early part of the 20th century, heavy smog in some parts of Europe resulted in a decrease in the production of natural vitamin D leading to a rise in the cases of rickets. Smog causes a misty haze similar to fog, but very different in composition. In fact the word smog has been coined from a combination of the words fog and smoke. Smog refers to hazy air that causes difficult breathing conditions.

The most harmful components of smog are ground-level ozone and fine airborne particles. Ground-level ozone forms when pollutants released from gasoline and diesel-powered vehicles and oil-based solvents react with heat and sunlight. It is harmful to humans, animals, and plants. The industrial revolution in the 19th century saw the beginning of air pollution in Europe on a large scale and the presence of smog mainly in Britain. The industries and the households relied heavily on coal for heating and cooking. Due to the burning of coal for heat during the winter months, emissions of smoke and sulphur dioxide were much greater in urban areas than they were during the summer months.

Smoke particles trapped in the fog gave it a yellow/black colour and this smog often settled over cities for many days. The effects of smog on human health were evident, particularly when smog persisted for several days. Many people suffered respiratory problems and increased deaths were recorded, notably those relating to bronchial causes. A haze of dense harmful smog would often cover the city of London. The first smog-related deaths were recorded in London in 1873, when it killed 500 people. In 1880, the toll was 2000.

London had one of its worst experiences with smog in December 1892. It lasted for three days and resulted in about 1000 deaths. London became quite notorious for its smog. By the end of the 19th century, many people visited London to see the fog. Despite gradual improvements in air quality during the 20th century, another major smog occurred in London in December 1952. The Great London Smog lasted for five days and resulted in about 4000 more deaths than usual. In response to the Great London Smog, the government passed its first Clean Air Act in 1956, which aimed to control domestic sources of smoke pollution by introducing smokeless zones.

In addition, the introduction of cleaner coals led to a reduction in sulphur dioxide pollution. In the 1940s, severe smog began covering the city of Los Angeles in the USA. Relatively little was done to control any type of pollution or to promote environmental protection until the middle of the 20th century. Today, smoke and sulphur dioxide pollution in cities is much lower than in the past, as a result of legislation to control pollution emissions and cleaner emission technology.

## ACID RAIN

Another effect of air pollution is acid rain. The phenomenon occurs when sulphur dioxide and nitrogen oxides from the burning of fossil fuels such as, petrol, diesel, and coal combine with water vapour in the atmosphere and fall as rain, snow or fog. These gases can also be emitted from natural sources like volcanoes. Acid rain causes extensive damage to water, forest, soil resources and even human health. Many lakes and streams have been contaminated and this has led to the disappearance of some species of fish in Europe, USA and Canada as also extensive damage to forests and other forms of life. It is said that it can corrode buildings and be hazardous to human health.

Because the contaminants are carried long distances, the sources of acid rain are difficult to pinpoint and hence difficult to control. For example, the acid rain that may have damaged some forest in Canada could have originated in the industrial areas of USA. In fact, this has created disagreements between Canada and the United States and among European countries over the causes of and solutions to the problem of acid rain. The international scope of the problem has led to the signing of international agreements on the limitation of sulphur and nitrogen oxide emissions.

## FLYASH

With the boom in population and industrial growth, the need for power has increased manifold. Nearly 73 per cent of India's total installed power generation capacity is thermal, of which 90 per cent is coal-based generation, with diesel, wind, gas, and steam making up the rest. Thermal power generation through coal combustion produces minute particles of ash that causes serious environmental problems. Commonly known as fly ash, these ash particles consist of silica, alumina, oxides of iron, calcium, and magnesium and toxic heavy metals like lead, arsenic, cobalt, and copper.

The 80-odd utility thermal power stations in India use bituminous coal and produce large quantities of fly ash. According to one estimate, up to 150 million tonnes of fly ash will be produced in India in the year 2000, primarily by thermal power plants and, to a lesser extent, by cement and steel plants and railways. This poses problems in the form of land use, health hazards, and environmental dangers. Both in disposal and in utilization utmost care has to be taken to safeguard the interest of human life, wild life, and such other considerations. The prevalent practice is to dump fly ash on wastelands, and this has lain to waste thousands of hectares all over the country.

To prevent the fly ash from getting airborne, the dumping sites have to be constantly kept wet by sprinkling water over the area. The coal industry in USA spends millions of dollars on lining fly ash dumping grounds. But in India, these sites are not lined and it leads to seepage, contaminating groundwater and soil. It lowers soil fertility and contaminates surface and ground water as it can leach into the subsoil. When fly ash gets into the natural draining system, it results in siltation and clogs the system. It also reduces the pH balance and portability of water. Fly ash interferes with the process of photosynthesis of aquatic plants and thus disturbs the food chain. Besides, fly ash corrodes exposed metallic structures in its vicinity.

In Delhi, the problem of fly ash is particularly severe as three power stations are located here. Being very minute, fly ash tends to remain airborne for a very long period leading to serious health problems as the airborne ash can enter the body. It causes irritation to eyes, skin, and nose, throat, and respiratory tract. Repeated inhalation of fly ash dust containing crystalline silica can cause bronchitis and lung cancer.

**Tackling the Problem of Fly Ash**

Fly ash management has taken considerable strides over the past few years. Researches have been attempting to convert this waste into wealth by exploring viable avenues for fly ash management. Fly ash is oxide-rich and can be used as the raw material for different industries. Today, fly ash bricks can be used as a building material. The American Embassy in India has used fly ash bricks in some of its recent construction. Use of fly ash as a part replacement of cement in mortar and concrete has started with the Indian Institute of Technology, Delhi taking the lead. Use of fly ash in the construction of roads and embankments has been successfully demonstrated in the country and it is gaining acceptance. The NTPC (National Thermal Power Corporation) is setting up two fly ash brick manufacturing plants at Badarpur and Dadri near Delhi.

At TERI, researchers have proven that fly ash dumps can be reclaimed by suitable addition of organic matter and symbiotic fungi, making it commercially viable for activities like floriculture and silviculture. TERI researchers have successfully reclaimed a part of an ash pond at the Badarpur Thermal Power Station by introducing a mycorrhizal fungi-based organic bio-fertilizer. As the fungus germinates, it sustains on the partner plant and quickly spreads to the roots and beyond.

It improves the plant's water and nutrient uptake, helps in the development of roots and soil-binding, stores carbohydrates and oils for use when needed, protects the plants from soil-borne diseases, and detoxifies contaminated soils. This helps in keeping both air and water pollution under control. It also helps revive wastelands and saves millions of litres of precious water from going down the fly ash slurries. Marigold, tuberose, gladiolus, carnation, sunflower, poplar, sheesham, and eucalyptus now grow at the demonstration site of the power station.

Use of fly ash in agricultural applications has been well demonstrated and has been accepted by a large number of farmers. The National Capital Power Station of the NTPC has come up with an innovative technology for commercial utilization of this by-product. Known as the dry ash technology, it is considered environment-friendly. Under the dry ash technology, the fly ash is collected in huge mounds with a filter bed provided at the bottom of the mound. Grass is planted on the slopes of the fly ash mounds and polymer layering is also done to prevent the ash from being blown by the wind. Fly ash treated by this method develops certain physical properties that make it more suitable for commercial purposes.

## INDOOR AIR POLLUTION

It refers to the physical, chemical, and biological characteristics of air in the indoor environment within a home, building, or an institution or commercial facility. Indoor air pollution is a concern in the developed countries, where energy efficiency improvements sometimes make houses relatively airtight, reducing ventilation and raising pollutant levels. Indoor air problems can be subtle and do not always produce easily recognized impacts on health. Different conditions are responsible for indoor air pollution in the rural areas and the urban areas. In the developing countries, it is the rural areas that face the greatest threat from indoor pollution, where some 3.5 billion people continue to rely on traditional fuels such as firewood, charcoal, and cowdung for cooking and heating. Concentrations of indoor pollutants in households that burn traditional fuels are alarming. Burning such fuels produces large amount of smoke and other air pollutants in the confined space of the home, resulting in high exposure. Women and children are the groups most vulnerable as they spend more time indoors and are exposed to the smoke. In 1992, the World Bank designated indoor air pollution in the developing countries as one of the four most critical global environmental problems. Daily averages of pollutant level emitted indoors often exceed current WHO guidelines and acceptable levels. Although many hundreds of separate chemical agents have been identified in the smoke from biofuels, the four most serious pollutants are particulates, carbon monoxide, polycyclic organic matter, and formaldehyde.

Unfortunately, little monitoring has been done in rural and poor urban indoor environments in a manner that is statistically rigourous. In urban areas, exposure to indoor air pollution has increased due to a variety of reasons, including the construction of more tightly sealed buildings, reduced ventilation, the use of synthetic materials for building and furnishing and the use of chemical products, pesticides, and household care products. Indoor air pollution can begin within the building or be drawn in from outdoors.

*Other than nitrogen dioxide, carbon monoxide, and lead, there are a number of other pollutants that affect the air quality in an enclosed space*:

- Volatile organic compounds originate mainly from solvents and chemicals. The main indoor sources are perfumes, hair sprays, furniture polish, glues, air fresheners, moth repellents, wood preservatives, and many other products used in the house. The main health effect is the imitation of the eye, nose and throat. In more severe cases there may be headaches, nausea and loss of coordination. In the long term, some of the pollutants are suspected to damage to the liver and other parts of the body.

- Tobacco smoke generates a wide range of harmful chemicals and is known to cause cancer. It is well known that passive smoking causes a wide range of problems to the passive smoker (the person who is in the same room with a smoker and is not himself/herself a smoker) ranging from burning eyes, nose, and throat irritation to cancer, bronchitis, severe asthma, and a decrease in lung function.
- Pesticides, if used carefully and the manufacturers, instructions followed carefully they do not cause too much harm to the indoor air.
- Biological pollutants include pollen from plants, mite, hair from pets, fungi, parasites, and some bacteria. Most of them are allergens and can cause asthma, hay fever, and other allergic diseases.
- Formaldehyde is a gas that comes mainly from carpets, particle boards, and insulation foam. It causes irritation to the eyes and nose and may cause allergies in some people.
- Asbestos is mainly a concern because it is suspected to cause cancer.
- Radon is a gas that is emitted naturally by the soil. Due to modern houses having poor ventilation, it is confined inside the house causing harm to the dwellers.

## GASEOUS POLLUTANTS

### INTRODUCTION

Generally, gaseous pollution is caused by the burning of fuels. Pollutants such as sulphur dioxide, hydrogen sulphide, and nitrogen dioxide combine with moisture in the air to form acids that attack and damage library material. Ozone is a powerful oxidant which severely damages all organic materials. It is a product of the combination of sunlight and nitrogen dioxide from car exhaust; it may also be produced by electrostatic filtering systems used in some air conditioners, as well as by electrostatic photocopy machines.

Smoking, cooking, and off-gassing from unstable materials (cellulose nitrate film, paint finishes, fire-retardant coatings, and adhesives) may also produce harmful gaseous pollutants. Wood, particularly oak, birch and beech, emit acetic and other acids, and vulcanized rubber releases volatile sulphides that are especially damaging to photographs.

The composition of all equipment, materials, and finishes used for the storage, transport, and display of objects should be tested by recognised methods to ascertain whether they are likely to produce harmful emissions. Bryophytes are generally easier to identify and are as equally susceptible to air pollution as lichens, yet less attention has been paid to their use in gaseous air pollution monitoring. A reason for this may be the larger number of lichen species (particularly epiphytic species) available for air pollution monitoring.

Most studies are associated with regional and urban sulphur dioxide ($SO_2$) contamination. On a national and multi-national monitoring scale, bryophytes are used more as bioaccumulative indicators of aerial metal contamination than as bioindicators of gaseous air pollution. Most literature regarding gaseous pollutants and bryophytes is concerned with measuring and assessing impacts of pollution on bryophyte communities as a valuable ecological group in their own right. Hallingback and Tan discussed the IUCN/IAB Bryophyte Committees' commitment to endangered bryophytes and presented a skeleton action plan for these species. The conservation and protection of these 'key-stone' species was emphasised and air pollution was listed as one of the major threats to species. Biomonitoring of air pollution impacts on bryophytes therefore plays a role in biodiversity and conservation.

Most of the published literature on research with regard to bryophytes and gaseous air pollutants concentrates on pollutant effects and harm and little is available in terms of their role as biomonitors and bioindicators. However, such work may still be used to ascertain pollution problems and indicate potential threats to other biological systems including humans. Standard practices such as sampling, analysis and species selection in bryophyte monitoring are underdeveloped in comparison to lichen monitoring.

## SULPHUR DIOXIDE ($SO_2$)

Adams and Preston presented comprehensive evidence of long-term effects of gaseous pollutants, particularly $SO_2$, on bryophyte distribution in the UK. The methods, patterns and correlations discussed highlight air pollution trends and much of the information is highly applicable to gaseous pollutant biomonitoring using bryophytes.

In order to draw their conclusions, the authors collated evidence from various sources. Analysis of the quaternary sub-fossil record demonstrated that the reduction in *Sphagnum* cover in peat profiles corresponded with the presence of soot deposits due to the advent of the industrial revolution. Bryophyte distributions determined from the British Bryological Society's mapping scheme from 1960 onwards and herbarium collections at national and local scales appeared to be correlated with long-term direct air monitoring data. The national recording scheme revealed bryophyte species which had been adversely affected by atmospheric pollution on a national scale.

**Table. Bryophyte Species which Appear to have been most Adversely Affected by Atmospheric Pollution in terms of Distribution at a National Scale**

| Epiphytes | Epiliths | Species which can grow as epiphytes or epiliths |
|---|---|---|
| Cryphaea heteromalla | Grimmia affinis | Antitrichia curtipendula |
| Frullania dilatata | G. decipiens | Leucodon sciuroides |

| | |
|---|---|
| Neckera pumila | G. laevigata |
| Orthotrichum lyellii | G. orbicularis |
| O. obtusifolium | G. ovalis |
| O. schimperi | |
| O. speciosum | |
| O. stramineum | |
| O striatum | |
| O. tenellum | |
| Tortula laevipila | |
| Ulota crispa var. Crispa | |
| U. crispa var. Norvegica | |

More intensive distribution studies in the heavily polluted London and Essex areas enabled the development of an epiphytic bryophyte sensitivity scale (Table), similar to the highly recognised lichen zonation scale drawn up by Hawksworth and Rose. It is important to note that the position of a species on the scale will vary with humidity in the area and acidity of the bark substrate. Epiphytic species may be indirectly affected by the soil through influences on the bark chemistry, as demonstrated by Gustafsson and Eriksson in aspen *Populus tremula* in Sweden. Some species may prefer limestone to trees.

In addition, it would be difficult for this scale to consider the effects of events such as short seasonal elevations in $SO_2$associated with prevailing winds. It is difficult to distinguish between the response of moss communities subjected to a few severe air pollution events or continued chronic exposure. It would prove difficult to develop a scale to include mosses in polluted environments, which may be protected against the effects of air pollutants if located in sheltered areas such as deep valleys. Although not as detailed or as accurate as the lichen scale, the proposed bryophyte scale lends itself as a positive biomonitoring tool.

**Table. Epiphytic Bryophytes that Showed Poor or Restricted Growth During the Maximum Phase of $SO_2$ Pollution, or were Exterminated, with their Approximate Equivalent $SO_2$ thresholds**

| **Mean winter $SO_2$($\mu g\ m^{-3}$)** | | **Approximate pollution zone*** | **Species** | **Last or only date recorded** |
|---|---|---|---|---|
| | | | **Epping forest** | **Outer Essex** |
| Pure | 9-10 | #Antitrichia curtipendula | c 1800 | 1874 |
| | 9 | Orthotrichum sprucei | - | 1866 |
| <30 | 9 | O. schimperi | - | 1873 |
| | 9 | O. tenellum | - | 1870 |
| | 9 | Ulota crispa var. crispa | c 1800 | 1874 |
| | 9-8 | Orthotrichum. striatum | c 1800 | 1870 |

| **Mean winter $SO_2$(μg $m^{-3}$)** | **Approximate pollution zone*** | **Species** | **Last or only date recorded** | |
|---|---|---|---|---|
| c 35 | 8 | Zygodon conoideus | 1885 | 1886 |
| | 8 | Neckera pumila | 1890 | 1874 |
| | 8-7 | Tortula papillosa | - | 1874 |
| | 7 | Ulota crispa var. norvegica | - | |
| c 40 | 7 | #Anomodon viticulosus | 1932 | |
| | 7 | #Radula complanata | c 1890 | |
| | 7-6 | Leucodon sciuroides | 1885 | |
| | 6 | Orthotrichum lyellii | 1898 | |
| | 6 | Cryphaea heteromalla | c 1800 | |
| c 50 | 6 | Frullania dilatata | 1923 | |
| | 6 | #Homalia trichomanoides | 1973 | |
| | 6 | Porella platyphylla | c 1890 | |
| | 6-5 | Isothecium myurum | 1885 | |
| | 5 | Tortula laevipila | c 1980 | |
| | 5 | Neckera complanata | | |
| c 60 | 5 | Zygodon viridissimus | | |
| | 5 | Orthotrichum affine | | |
| | 5 | O. diaphanum | | |
| | 5-4 | Homalothecium sericeum | | |
| | 4 | Hypnum mammillatum | | |
| c 70 | 4 | Hypnum cupressiforme var. cupressiforme | | |
| | 4 | Dicranum scoparium | | extant |
| | 4 | Isothecium myosuroides | | |
| | 4 | Bryum capillare | | |
| | 4-3 | Dicranoweisia cirrata | | |
| c 125 | 3 | Hypnum cupressiforme var. resupinatum | | |
| | 3 | Lophocolea heterophylla | | |
| c 150 | 2-3 | Ceratodon purpureus | | |

# Species which may be limited by factors other than specific sensitivity to $SO_2$* Hawksworth and Rose

National and local lists showed relatively good correlation. Disparities may be due to factors other than $SO_2$, which may cause local variation. Furthermore, local species decline due to localized high $SO_2$ levels may not be reflected on the national scale. The same authors also reported similar reductions in the distribution of terrestrial and saxicolous bryophytes in relation to high $SO_2$ levels. As is the case with lichens, recent declines in $SO_2$ levels are reflected in bryophyte recolonisation to formerly polluted areas.

*However, the sequence of decline in species associated with $SO_2$ pollution may not be as marked in bryophyte recolonisation for several reasons*:

- Inconspicuous species may have simply been missed in previous studies.
- Bryophytes have shown a slower response to reduced $SO_2$ levels than lichens.
- Species may colonise areas where they were not previously found.
- Species may not recolonise areas where they were previously found because of changes in habitat. In a study of epiphytic bryophytes in Groningen in the Netherlands certain species (*e.g. Leucodon sciuroides* and *Orthotrichum lyelli*) became scarce in areas of relatively good air quality due to Dutch Elm Disease. These species rely on vulnerable *Ulmus* sp. for survival, illustrating the importance of available habitats even in improved air pollution conditions.
- Increasing levels of other atmospheric pollutants such as nitrogen oxides and ozone may reach threatening levels.
- Recolonisation processes depend on the mobility of the species and the proximity of source populations.

Changes in bryophyte cover in forest ecosystems in response to air pollution have been reported. In Estonia, two groups of moss in Scots pine forests were observed: moss whose distribution depended on the level of air pollution and those whose distribution was more dependent on other factors such as climate and soil. Bryophyte cover also increased in thickness along a pollution gradient from the polluted north-east to less polluted south-west. In a study of bryophyte communities in two Atlantic forests in Brazil, higher percentage cover and biomass of bryophytes were observed in the least polluted forest.

## URBAN AND INDUSTRIAL

In a Finnish study of effects of metal, chemical and fertilizer plants on forest floor vegetation in the vicinity of the plants, the common forest bryophytes were more sensitive than lichens. Further away from the factories *Ceratodon purpureus* and *Pohlia nutans* became frequent. Huber found that bryophyte species distribution in the Swiss Canton of Basel-Stadt was similar to that in the proximity of a cellulose factory. Furthermore, the area of suspected good air quality in the Canton contained similar species distribution to an area 200 m above the cellulose factory.

In Romania, the zone of influence on bryoflora of the industrial Alba district extended to a distance of 20 km. The main atmospheric burden was from $SO_2$. Winner reviewed the responses of bryophytes to air pollution by presenting an overview of North American studies in this field. Types of studies include determination of impact zones, Indices of Atmospheric Purity (IAP) methods, bioaccumulation and physiological responses to gaseous air pollution. IAP methods have been applied to moss communities in much the same way as to lichens. IAP values have been calculated and mapped with respect to urban areas and point emission sources. Zones have been determined representing changes in bryophyte communities relative to the location of an air pollution source.

*According to Winner, IAP values should be regarded tentatively and cannot be compared between different sites for the following reasons*:

- All changes in moss communities are too often attributed to air pollution alone, whereas other environmental factors such as elevation, humidity and temperature may have an effect.
- Cities generally have higher temperatures and fewer appropriate substrates than rural areas.
- IAP may not be an accurate measure of the complexity of bryophyte communities.

## TRANSPLANTS AND EFFECTS

The biochemical and physiological effects of $SO_2$ on bryophytes will not be discussed here. Recent reviews in this area are presented by Brown, Kovács and Winner. A recent review of transplantation exercises using bryophytes can be found in Brown. These studies are concerned with assessing effects of air pollutants on bryophytes rather than their use as biomonitors.

A technique that was not mentioned by Burton was bryometers. Bryometers were developed in Japan to measure phytotoxic air pollution. Mosses were placed in small, transparent plant chambers and two chambers were placed at a study site. One chamber was filled with ambient air and the other exposed to filtered, pollution free air. In this way the presence or absence of ambient air pollutants could be detected.

## NITROGEN AND ITS COMPOUNDS (N, $NO_X$ AND $NH_3$)

Much of the published literature in relation to inorganic nitrogen (N), acid rain and bryophytes is concerned with the effects of this type of deposition on bryoflora. This research is gaining profile in response to the coincidence of reduced atmospheric $SO_2$ levels and increases in nitrogen deposition and acid rain. However relatively few attempts have been made to use bryophytes to directly monitor nitrogen deposition. The following paragraphs attempt to highlight aspects which may be suitable for monitoring purposes in the future. Due to their dependence on atmospheric inputs of nutrients, total N content in bryophyte tissues can reflect atmospheric inputs of N.

Press *et al.* chose two ombrotrophic mires dominated by bryophytes to study the ecological significance of increased atmospheric nitrate deposition in the UK *Sphagnum* sp. transplanted to a relatively polluted site from a non-polluted site showed an increase in tissue nitrogen concentration and reduced growth in the polluted site. The authors concluded that elevated nitrogen deposition in the polluted area might be affecting the growth and metabolism of ombrotrophic *Sphagnum* sp.

Comparison of bryophyte tissue N content in herbarium samples and field samples collected in 1989 illustrated temporal N deposition trends in the UK. The study demonstrated increasing N deposition levels apparent throughout the UK and potential harm to the ecosystems themselves. Percentage increase in tissue N content at a range of sites of varying pollution climates corresponded with atmospheric increases in N levels. Certain sites remained unpolluted over the thirty-year period and showed insignificant elevations in bryophyte tissue concentrations. A 62 per cent increase in tissue N content of ombrotrophic Sphagna at Moor House, Cumbria, is of concern.

The relationship between tissue N content with atmospheric N inputs was expressed as: $N_{Bry} = 0.62 + 0.022\ N_{Dep}$ *i.e.* tissue N increases at 0.022 mg $g^{-1}$dry weight per kg total N deposited. The authors recommended the application of this equation as a rough measure of atmospheric N deposition in areas where monitoring equipment is impractical. Mäkipää proposed changes of forest floor moss biomass as an early warning indicator of atmospheric N and S deposition in boreal forests.

In a study of acidic deposition, experimental plots of boreal forest in Finland were exposed to annual treatments of ammonium sulphate over a four year period. The forest floor mosses, dominated by *Pleurozium schreberi* and *Dicranum polysetum*, were more sensitive to N and S deposition than vascular plants. Biomass of bryophytes decreased by 60 per cent over the study period and N content in moss tissues was greater than in the control areas. Although this experiment was not undertaken under natural conditions, such experiments are valuable in enhancing our knowledge of the response of bryophytes to N deposition if they are to be used as biomonitoring tools.

Woolgrove and Woodin analysed tissue-N content in the late snowbed bryophyte, *Kiaeria starkei*, collected from sites in the Scottish Highlands. Results corresponded with the atmospheric N loads to which the bryophyte was exposed, which due to acid flushes during snowmelt and the sensitivity of the species pose a threat to these snowbed species. Urban and industrial point source studies in relation to nitrogen oxide pollution gradients are limited.

A study of the distribution of epiphytic bryophytes in Naha City in Japan showed that few species were found in the city centre where $NO_2$ levels were $> 0.02$ ppm. An interesting observation of this study was the lack of correlation between IAP values and $NO_2$ concentration. If bioindication of $NO_2$ pollution using bryophytes and lichens is to develop, additional research into the applicability of indices and methods used in $SO_2$ monitoring should be encouraged.

## OTHER GASES

In contrast to Burton's review, little published literature was found in relation to fluoro-compounds and bryophytes. With regard to ozone ($O_3$), literature is concerned with the effects of this gas on bryophytes since they are regarded as an important ecological group. An open chamber fumigation experiment of the effects of elevated $O_3$ concentrations on forest floor moss cover was undertaken by Stanosz *et al.*

Significant negative correlations between moss cover (predominantly *Ditrichum pusillum*) and $O_3$ concentrations were demonstrated. In the UK, *Sphagnum recurvum*and *Polytrichum commune* were exposed to long-term chronic $O_3$ concentrations. Both species showed a reduction in growth when exposed to $O_3$ in comparison to control experiments. *Sphagnum* appeared to be more sensitive to fumigation than *Polytrichum*. Fumigation studies will enable effects to be determined which could possibly be used as bioindication responses, since bryophytes may be more sensitive to $O_3$ than higher, economically important plants.

## NITROGEN OXIDES

**Table. The Oxiths of Nitzogen**

| Name | Formula | Structure | Description |
|---|---|---|---|
| Dinitrogen Monoxide (nitrous oxide) | $N_2 0$ | N–N–0 linear | Colourless Gas (bp–88.5°C) |
| (Mono)natrogen monoxide (nitric oxide) | NO | N–0 | Colourless Paramagneac Gas (bp -151.8° C): Liquid and Solid are also Colourless when Pure |
| Dinitrogen Trioxide | $N_2 0_3$ | 0\N–N/0 \0 planar | Blue Solid (mp —100.7° C): Dissociates Reversibly NO and $NO_2$ |
| Nitrogen Dioxide | $NO_2$ | 0/N\0 planar | Brown Paramagnetic Gas; Dimerizes Reversibly to $N_2O_4$ |
| Dinitrogen Tetroxide | $N_2O_4$ | 0\N–N/0 0/ \0 planar | Colourless Liquid (mp —11.2° C); Dissociates Reversibly in in Gas Phase to $NO_2$ |
| Dimtrogen Pentoxide | $N_2O_5$ $[NO_2]^*[NO_2]$ | 0\N–0–N/0 0/ \0 planar | Colourless Ionic Solid; Sublimes at 32.4° C to Unstable Molecular Gas (N– 0 – N angle ~ 180°) |
| Nitrogen Trioxide | $No_2$ | 0–N/0 \0 planar | Unstable Paramagnetic Radical |

Nitrogen (N) forms oxides in which nitrogen exhibits each of its positive oxidation numbers from +1 to +5. These molecules are all well characterized and are shown in Table. Nitrous oxide (dinitrogen oxide), $N_2O$, is formed when ammonium nitrate, $NH_4NO_3$, is heated. This oxide, which is a colourless gas with a mild, pleasing odour and sweet taste, is used as an anesthetic for minor operations, especially in dentistry. It is called laughing gas because of its intoxicating effect.

It is also widely used as a propellant in aerosol cans of whipped cream. Nitric oxide, NO, can be prepared in several ways. The lightning that occurs during thunderstorms brings about the direct union of nitrogen and oxygen in the air to produce small amounts of nitric oxide, as does heating the two elements together. Commercially, nitric oxide is produced by burning ammonia ($NH_3$), while in the laboratory it can be produced by the reduction of dilute nitric acid ($HNO_3$) with, for example, copper (Cu).

$$3Cu + 8HNO_3 \rightarrow 2NO + 3Cu(NO_3)^2 + 4H_2O$$

Gaseous nitric oxide is the most thermally stable oxide of nitrogen and is also the simplest known thermally stable paramagnetic molecule–*i.e.,* a molecule with an unpaired electron. It is one of the environmental pollutants generated by internal-combustion engines, resulting from the reaction of nitrogen and oxygen in the air during the combustion process. At room temperature nitric oxide is a colourless gas consisting of diatomic molecules. However, because of the unpaired electron, two molecules can combine to form a dimer by coupling their unpaired electrons.

$$2NO \rightarrow N_2O_2$$

Thus, liquid nitric oxide is partially dimerized, and the solid consists solely of dimers. When a mixture of equal parts of nitric oxide and nitrogen dioxide, $NO_2$, is cooled to-21 C (-6 F), the gases form dinitrogen trioxide, a blue liquid consisting of $N_2O_3$ molecules. This molecule exists only in the liquid and solid states. When heated, it forms a mixture of NO and $NO_2$.

Nitrogen dioxide is prepared commercially by oxidizing NO with air, but it can be prepared in the laboratory by heating the nitrate of a heavy metal, as in the following equation, or by adding copper metal to concentrated nitric acid. Like nitric oxide, the nitrogen dioxide molecule is paramagnetic. Its unpaired electron is responsible for its colour and its dimerization. At low pressures or at high temperatures, $NO_2$ has a deep brown colour, but at low temperatures the colour almost completely disappears as $NO_2$ dimerizes to form dinitrogen tetroxide, $N_2O_4$. At room temperature an equilibrium between the two molecules exists.

$$2NO_2 \rightarrow N_2O_4$$

Dinitrogen pentoxide, $N_2O_5$, is a white solid formed by the dehydration of nitric acid by phosphorus(V) oxide.

$$P_4O_{10} + 4HNO_3 \rightarrow 4HPO_3 + 2N_2O_5$$

Above room temperature $N_2O_5$ is unstable and decomposes to $N_2O_4$ and $O_2$. Two oxides of nitrogen are acid anhydrides; that is, they react with water to form nitrogen-containingoxyacids. Dinitrogen trioxide is the anhydride of nitrous acid, $HNO_2$, and dinitrogen pentoxide is the anhydride of nitric acid, $HNO_3$.

$$N_2O_3 + H_2O \rightarrow 2HNO_2$$

$$N_2O_5 + H_2O \rightarrow 2HNO_3$$

There are no stable oxyacids containing nitrogen with an oxidation number of +4. Nitrogen dioxide reacts with water in one of two ways. In cold water $NO_2$ disproportionates to form a mixture of $HNO_2$ and $HNO_3$, while at higher temperatures $HNO_3$ and NO are formed. In their chemical activity, the nitrogen oxides undergo extensive oxidation-reduction reactions. Nitrous oxide resembles oxygen in its behaviour when heated with combustible materials.

It is a strong oxidizing agent that decomposes upon heating to form nitrogen and oxygen. Because one-third of the gas liberated is oxygen, nitrous oxide supports combustion better than air. All the nitrogen oxides are in fact good oxidizing agents. Dinitrogen pentoxide reacts violently with metals, non-metals, and organic materials, as in the following reactions with potassium (K) and iodine gas ($I_2$).

$$N_2O_5 + K \rightarrow KNO_3 + NO_2$$

$$N_2O_5 + I_2 \rightarrow I_2O_5 + N_2$$

## PARTICULATE POLLUTANTS

### PARTICULATE MATTER

The term Particulate Matter (PM) includes both solid particles and liquid droplets found in air. Many man-made and natural sources emit PM directly or emit other pollutants that react in the atmosphere to form PM. These solid and liquid particles come in a wide range of sizes. Particles less than 10 micrometers in diameter tend to pose the greatest health concern because they can be inhaled into and accumulate in the respiratory system. Particles less than 2.5 micrometers in diameter are referred to as "fine" particles. Sources of fine particles include all types of combustion (motor vehicles, power plants, wood burning, etc.) and some industrial processes. Particles with diameters between 2.5 and 10 micrometers are referred to as "coarse". Sources of coarse particles include crushing or grinding operations, and dust from paved or unpaved roads. The Environmental Protection Agency uses its Air Quality Index to provide general information to the public about air quality and associated health effects. An Air Quality Index (AQI) of 100 for any pollutant corresponds to the level needed to violate the federal health standard for that pollutant. For $PM_{2.5}$, an AQI of 100 corresponds to 40 micrograms per cubic meter (averaged over 24 hours)–the current federal standard. An AQI of 100 for $PM_{10}$ corresponds to a $PM_{10}$ level of 150 micrograms per cubic meter (averaged over 24 hours).

**PARTICULATE MATTER STANDARDS**

In 1997 the Environmental Protection Agency (EPA) proposed a new standard of ozone and particulate matter levels in the atmosphere. The particulate matter levels of up to 10 microns in diameter($PM_{10}$) at each monitor within an area must not exceed 150g/m3, in one hour more than once per year, averaged over 3 years. The particulate matter levels of up to 2.5 microns in diameter ($PM_{2.5}$), must not exceed 15 micrograms per cubic meter (g/m3) and 65g/m3, respectively, each year and 24-hour period.

However, a coalition of business and industry interests sued to have those standards blocked, claiming they were too expensive and ill-conceived. In 1999 a federal court agreed, issuing a ruling blocking implementation of the tougher standards. Changes were made again in February 2001, when the U.S. Supreme Court unanimously upheld the constitutionality of the Clean Air Act as EPA had interpreted it in setting health-protective air quality standards for ground-level ozone and particles. The Supreme Court also reaffirmed EPA's long-standing interpretation that it must set these standards based solely on public health considerations without consideration of costs.

However, the Supreme Court did find that the EPA's plans for implementing the rules were unreasonable, and it ordered the agency to develop new implementation policies. Industry opponents immediately promised to use this aspect of the ruling as the basis for new legal challenges to weaken implementation of the new standards. It remains to be seen if the new standards will truly take effect as legislated.

*According to the EPA, the new particulate matter and ozone standards will have the following effects*:

- Reduced risk of significant decreases in children's lung functions. The new standards should provide approximately 1 million fewer incidences of difficulty of breathing or shortness of breath in children each year. These problems can limit a healthy child's activities or result in increased medication use, or medical treatment for children with asthma.
- Reduced risk of moderate to severe respiratory symptoms in children. The new standards should result in hundreds of thousands of fewer incidences each year of symptoms such as aggravated coughing and difficult or painful breathing.

- Reduced risk of hospital admissions and emergency room visits for respiratory causes. The new standards should result in thousands fewer admissions and visits for individuals with asthma.
- Reduced risks of more frequent childhood illnesses and more subtle effects such as repeated inflammation of the lung, impairment of the lung's natural defence mechanisms, increased susceptibility to respiratory infection, and irreversible changes in lung structure. Such risks can lead to chronic respiratory illnesses such emphysema and chronic bronchitis later in life and/or premature aging of the lungs.
- Reduce the yield loss of major agricultural crops, such as soybeans and wheat, and commercial forests by almost $500,000,000.

## WHAT ARE THE HEALTH EFFECTS FROM PARTICULATE MATTER

The Environmental Protection Agency has found that numerous health effects arise from both fine and coarse particles when they accumulate in the respiratory system. Coarse particles can aggravate respiratory conditions such as asthma. Exposure to fine particles is associated with several serious health effects, including premature death.

*Health effects have been associated with exposures to PM over both short (such as a day) and longer periods (a year or more)*:

- When exposed to even small levels of PM, people with existing heart or lung diseases-such as asthma, chronic obstructive pulmonary disease, congestive heart disease, or ischemic heart disease-are at increased risk of premature death and or admission to hospitals or emergency rooms.
- The elderly are very sensitive to PM exposure. They are at increased risk of admission to hospitals or emergency rooms and premature death from heart or lung diseases.
- Children and people with existing lung disease may not be able to breathe as deeply or vigourously as they normally would, and they may experience symptoms such as coughing and shortness of breath when exposed to levels of PM.
- PM can increase the susceptibility to respiratory infections and can aggravate existing respiratory diseases, such as asthma and chronic bronchitis, causing more use of medication and more doctor visits.

## SIZES AND SOURCES OF PARTICULATE MATTER

Particulate matter comes in variety of sizes–coarse, fine, ultrafine–and from a variety of sources. These different sizes affect the aerodynamic properties of each particle, meaning it affects *how* they travel through the air and *how* they are removed from it. More importantly, the size also helps dictate how each particulate enters our respiratory system. Coarse particles are largely the result of a "mechanical break-up" of larger particles–think combustion of engines, fires, dust, pollen, and plant and insect parts. Fine particles, on the other hand, are attributed more to the formation of gases, but also can come from mobile sources. Finally, ultrafine particulate matter are formed by "nucleation", or the first step in a gas becoming a liquid.

## THE DANGERS OF PARTICULATE MATTER

The dangers of these particles are real. Not only do some particulate matter contribute to allergies and other low-level respiratory problems, but they also contribute to a reduction in life expectancy by helping facilitate and promote the risk of cardio-pulmonary and lung cancer ailments. For example, fine particles are especially worrisome because they can reach the deepest regions of the lungs, according to the EPA. This usually results in "Health effects including asthma, difficult or painful breathing, and chronic bronchitis, especially in children and the elderly. "

In many cases, this is more often seen in long-term exposure to particulate matter. However, even relatively short exposure can have adverse health effects. In a number of cohort studies, researchers have found that both short and long-term exposure to particulate matter has led to poorer respiratory health, as well as the difficulty both impose of resolving previous exposure. Because of this, the World Health Organization has argued that "ambient [outdoor air] particulate matter *per se* is considered responsible for the health effects seen in the large multi-city epidemiological studies relating ambient PM to mortality and morbidity... In the Six Cities and ACS [American Cancer Society] cohort studies, PM but not gaseous pollutants with the exception of sulfur dioxide was associated with mortality.

That ambient PM is responsible *per se* for effects on health is substantiated by controlled human exposure studies, and to some extent by experimental findings in animals. " In other words, outdoor air pollution can have a serious effect on our health. Unsurprisingly, indoor air pollution can be just as threatening. The WHO stresses the major complications cooking and heating with solid fuels or open fires can facilitate, mainly citing that "indoor air pollution is responsible for 2.7 per cent of the global burden of disease. " The EPA stresses the importance of awareness concerning radon, a radioactive gas that is emitted through the ground and creeps indoors through cracks, second hand smoke, combustion pollutants, molds, and volatile organic compounds like those found in paint thinners and cleaning supplies, and other asthma triggers such as dander and dust mites.

## HOW TO AVOID THE BAD

Despite the seemingly ubiquitous nature of particulate matter and air pollution, you can still take measures to protect yourself and your family.

*Here are a few strategies to consider*:

- Undergo regular testing for chemicals such as radon
- Do not smoke indoors
- Have indoor plants and/or an indoor garden
- Keep your bathrooms and other areas exposed to water clean
- Use all natural cleaners in place of harmful chemical ones
- Install a carbon monoxide alarm
- Ventilate while cleaning your home
- Consider buying a HEPA air filter for your home
- Change your filters regularly
- Raise awareness around your house!

*I'd like to thank GreenFacts. org for their excellent and informative article on Air Pollution–Particulate Matter, as well as the EPA for additional resources.*

## PHOTOCHEMICAL OXIDANTS

Air pollutants formed by the action of sunlight on oxides of nitrogen and hydrocarbons. Any of the chemicals which enter into oxidation reactions in the presence of light or other radiant energy are also termed as photochmecial oxidant. Ozone, a natural constituent of the stratosphere formed by the photolysis of molecular oxygen, can be transported by atmospheric circulation into the lower atmosphere. Natural hydrocarbons including terpenes from trees and vegetation are also subject to photochemical reactions producing oxidants. These two processes are the natural sources of background ozone concentrations. Ozone and peroxyacylnitrates are formed in the lower atmosphere by reactions between oxides of nitrogen and an array of photochemically reactive hydrocarbons. The chemical structure and reactivity of each organic hydrocarbon determines its importance in the formation of oxidants. Motor vehicles, space heating, power plants, and industrial processes are major sources of these oxidant precursors. A balance between oxidizing and reducing agents would be maintained in the atmosphere, thus avoiding accumulation of ozone and other photochemical oxidants, if it were not for the photochemical degradation of hydrocarbons into peroxy radicals.

Peroxy radicals rapidly convert nitric oxide to nitrogen dioxide, thus shifting the equilibrium towards ozone production during daylight. At night, emissions of nitric oxide into the atmosphere serve as a sink for ozone. Welding and the manufacture of hydrogen peroxide are the main sources of occupational exposure to ozone. The use of ultraviolet lamps, electrostatic precipitators, or photocopying machines may also generate ozone.

### ENVIRONMENTAL CONCENTRATIONS AND EXPOSURES

Since significant concentrations of oxidants in urban areas are generally restricted to a period of 4-6 h per day, oxidant or ozone data are most often reported in terms of maximum 1-h concentrations or in terms of the number of hours or days recorded with hourly concentrations exceeding a specified value. In isolated places far removed from sources of pollution, maximum hourly ozone concentrations of 100 $\mu g/m^3$ (0.05 ppm) have been recorded.

Transport of oxidants from urban areas for distances of 100-700 kilometres appears to be a widespread phenomenon, and 1-h ozone concentrations of 120 $\mu g/m^3$ (0.06 ppm) or more have been observed in rural areas. In some large cities, maximum 1-h oxidant concentrations exceed 200 $\mu g/m^3$ (0.1 ppm) on 5-30 per cent of days, while in Los Angeles it is commonplace for maximum 1-h oxidant values to exceed 200 $\mu g/m^3$ (0.1 ppm) on most days of the month between May and October.

Diurnal patterns in oxidant levels are an important feature of the urban environment and result from hourly changes in solar radiation and pollutant emission intensity. Maximum hourly ozone levels frequently occur around noon and are often preceded by peak concentrations of nitrogen dioxide. Concentrations of peroxyacetylnitrate are typically between 1/50th and 1/100th of those of ozone and, in general, closely follow temporal variations in ozone levels.

On a seasonal basis, oxidant concentrations tend to increase during the high temperature season, and the frequency of days on which oxidant concentrations exceed 200 $\mu g/m^3$ (0.1 ppm) is greatest during this period. Oxidant concentrations indoors tend to be lower than those outdoors, and are reduced by destructive reaction on material surfaces. They are also reduced by activities that generate nitric oxide such as smoking and cooking.

## IMPACTS OF PHOTOCHEMICAL OXIDANTS (OZONE)

Photochemical oxidants are the products of reactions between NOx and a wide variety of volatile organic compounds (VOCs). The most well known 'oxidants' are ozone ($O_3$), peroxyacetyle nitrate (PAN) and hydrogen peroxide ($H_2O_2$). The main impact on the natural environment is mostly due to elevated $O_3$. Excessive concentrations of tropospheric $O_3$ have toxic effects on both plants and human health. Effects on vegetation include visible injury, early senescence of leaves, and reduction of crop yield. Experiments with open-top chambers in various parts of Europe (including the UK) show that exposure of plants to concentrations above 40 ppb for several weeks can reduce growth and the yield of sensitive crops species. However, it is difficult to translate this kind of information into effects on crops growing in the field and on natural communities. The most comprehensive information is available for wheat, and the evidence indicates that yields are probably reduced in some parts of Britain in high ozone years. With regard to the effects on natural vegetation, it has proved difficult to determine whether there are important ecological effects under UK conditions and more technically-challenging research is needed. For a discussion of the problems and challenges of assessing the effects on natural vegetation see Davison and Barns.

In 1988 the UNECE accepted the critical levels/loads approach as the basis for abatement of transboundary air pollutants. The ozone critical level defines the maximum exposure, over a threshold of 40 ppb, below which there is no negative effect according to present knowledge. Ozone critical levels for Europe were initially agreed at a UNECE workshop, revised at the Kuopio workshop and they have just been subject to further debate (UNECE meeting Goteborg, November 2002). The levels are defined using the AOT40 index which is the Accumulated Ozone concentration over a Threshold of 40 ppb, *i.e.*, the sum of the difference between hourly mean concentrations and 40 ppb when the concentration exceeds 40 ppb, hence the units are ppb hours.

*The latest critical levels are*:

- *Crops and semi-natural vegetation*: 3, 000 ppb hours during May to July daylight hours
- *Forests*: 5,000 ppb hours during April to September daylight hours (where: daylight hours are when solar radiation exceeds 50 Wm-2,exceedance is assessed on a five year mean)

An important weakness of the present critical levels is that they refer to the exposure of the plant to the ozone (concentration x time) but the effects depend on how much is taken in through the stomata. Because stomatal conductance is affected by several environmental factors, notably humidity and soil water deficit, the critical level over estimates the amount of ozone absorbed. For example, in warm sunny weather ozone tends to be high, but if there is also a lack of moisture the stomata close, reducing ozone uptake. In that situation the critical level may be exceeded but the plants may show no effect because it did not take up enough ozone to be toxic.

Furthermore, most studies of ozone have used in open-top chambers where the ozone concentration is easily maintained at a specific concentration. These were originally adopted because it was thought that they produce an environment that is similar to the open air, but it is now recognised that the turbulence, temperature and humidity, do not accurately reflect field conditions so the plants do not respond to ozone in the same manner as they would in a natural environment. In open-top chamber experiments a well defined threshold concentration for damage to the vegetation is usually found, but in field experiments plants often show damage appearing over a wide range of concentrations. This inconsistency between chamber and field experiments is probably due to differences in ozone uptake.

Therefore, although the AOT40 can indicate the potential for effects on vegetation it does to correlate well with the magnitude of effects, making it difficult to make economic assessments of ozone damage. The scientific community and the UNECE are well aware of these shortcomings and research is in progress that aims to provide a critical level that is expressed in terms of the absorbed dose. Some experiments are being done in the open air, without the use of chambers.

## PHOTOCHEMICAL SMOG

Smog is a type of air pollution. Smog is a mixture of smoke and fog. Smog usually forms when smoke from pollution mixes with fog. For example, London, England, is often very foggy. Most people in London used to heat their homes by burning coal. The coal made lots of smoke, which mixed with fog to form smog. London used to have a lot of smog. There is a special kind of smog called photochemical smog. It forms when photons of sunlight hit molecules of different kinds of pollutants in the atmosphere. The photons make chemical reactions happen. The pollution molecules turn into other kinds of nasty chemicals. That mixture of bad chemicals is called photochemical smog.

The chemicals in photochemical smog include nitrogen oxides, Volatile Organic Compounds (VOCs), ozone, and PAN (peroxyacytyl nitrate). Nitrogen oxides mostly come from the engines of cars and trucks. VOCs are given off by paint, gasoline, and pesticides. Ozone is a form of oxygen that is harmful. PAN is a type of pollution that is made by chemical reactions between other kinds of pollution. Smog smells bad and makes it hard for people to breath. It can also damage materials. Smog is a very harmful kind of air pollution.

Photochemical smog is a unique type of air pollution which is caused by reactions between sunlight and pollutants like hydrocarbons and nitrogen dioxide. Although photochemical smog is often invisible, it can be extremely harmful, leading to irritations of the respiratory tract and eyes. In regions of the world with high concentrations of photochemical smog, elevated rates of death and respiratory illnesses have been observed.

Smog itself is simply airborne pollution which may obscure vision and cause various health conditions. It is caused by small particles of material which become concentrated in the air for a variety of reasons. Commonly, smog is caused by an inversion, in which cool air presses down on a column of warm air, forcing the air to remain stationary. Inversions are notorious in Southern California, where smog can sometimes get so severe that people are warned to stay indoors.

Some of the particulate matter in the air can oxidize very readily when exposed to the UV spectrum. Nitrogen dioxide and various hydrocarbons produced through combustion will interact with sunlight to break down into hazardous chemicals. It doesn't have to be sunny for photochemical smog to form; UV light can also penetrate clouds. The pollutants released through human activity in this situation are known as "primary pollutants, " and they include sulfur dioxide, carbon monoxide, and other volatile organic compounds. When these compounds interact with the sun, they form "secondary pollutants" like ozone and additional hydrocarbons. While ozone is an excellent thing in the upper atmosphere, since it protects the delicate environment of the Earth, it is not desired at ground level. Ozone can be extremely irritating to the respiratory tract, leading to fits of coughing and various medical conditions if exposure is prolonged. The mixture of hazardous pollutants formed by the reaction between UV rays and smog can travel on the wind to rural areas, meaning the photochemical smog does not just impact big cities.

Some measures have been taken around the world to reduce photochemical smog. Tight emissions regulations on vehicles and factories are one such step; many factories must use scrubbers and treatment systems before releasing air from their manufacturing facilities, for example. The use of harmful chemicals is also restricted in some regions of the world, since these chemicals can create photochemical smog. Government agencies also monitor air quality through testing, citing companies which violate the law and issuing warnings when smog levels are dangerous.

## MOVEMENTS AND EFFORTS

Up until the time of the first civil court decision in 1973, the movement involving the disease victims and their supporters was mainly oriented towards a protest against the Chisso Company, with the company trying to negotiate by means of a third-party system established through governmental intervention. As a result of these reverberating interactions and interrelationships, the disease victims' movement was divided, in response to company and administrative obstructionism, into smaller factions employing differing tactics that were broadly characterized by direct negotiation and court settlements. after the end of Japan's high-economic-growth period in 1974, systematic pollution policies created by various governmental organs produced more problems for the disease victims than the company ever did. This government-induced oppression of victims' movements, such as was seen in the attempted indictment of Kawamoto mentioned earlier, and in many other lawsuits and legal manoeuvrings, had the effect of heightening public debate but did not bring about a meaningful resolution of outstanding problems and contentions.

The most salient problem for the victims' movement was the system established for designating official disease patients, for, given the legal circumstances of the times, lawyers came to take a leading role in determining which of the disease victims should give up their legal struggle for patient designation, and which should go on to a second- or even third-level court. In some of these court battles, the judicial system was able to achieve, at least for some of the victims, legal designation as verified Minamata disease patients, but for all the effort involved in this process, the results have been minuscule in every respect. The indictment brought against the company by the victims in 1960 found the company to be in error, but the court struggle continues to this day without any clear end in sight. Although, from the beginning, a great deal of the fault in relation to the continuation of the Minamata problem rests with government administrative error and ineptitude, as represented by the continued obstruction of justice and limiting of due process, there exists no effective means of designation and punishing governmental duplicity. The administrative organs of government were asked by the Minamata disease patients to provide thorough examinations and observations of disease victims in order to compile a complete epidemiological picture of the disease, but the fact remains that such work, even on the most fundamental level, has yet to begin. Within the context of the present political climate, there is very little hope of change in the Minamata situation. There are no records of the exact number of disease victims, nor is there any likelihood that primary policy orientations will change in such a manner as to respond more adequately to the plight of the many victims that probably exist.

However, even in these difficult circumstances, the disease victims have continued with their rehabilitation under their own auspices. Kawamoto once said: "Because there was no Minamata disease patients" movement, and because the said patients are weak, the truth about the Minamata disease is still unknown." Although the victims and the various supporters view the problem from varying perspectives, the words of Kawamoto are recognized by all as portraying the truth of the situation. It goes without saying that there have been efforts at disease-victim rehabilitation, but these have been sporadic and limited.

The results of the United Nations Conference on the Human Environment have already been mentioned, but this conference provided the first instance of physically handicapped disease victims being sent abroad through the support of a coalition of supporting citizens' groups.

In 1975, as soon as it became known that the same kind of disease was being discovered among certain Canadian Indians, the Minamata disease victims invited the Indians to Minamata and Niigata. In 1976, some of the disease victims from Japan participated as delegates in the Habitat Conference in Vancouver, and visited Indian reservations in Ontario and Quebec. In 1982, delegates from Japan went to the Environment Conference in Nairobi, Kenya, and spoke on problems and developments during the ten-year period since the last environment conference in Stockholm. At that time warnings were sounded in relation to the spread of environmental diseases in certain other Asian countries.

The Minamata Research Group, which was formed at the time of the first civil court struggle, was able to provide reports and survey results of high academic quality. They provided Kumamoto University with well-documented Minamata disease sources, and this work still continues. In this regard, the contribution of Seirinsha must be remembered.

Seirinsha was the creation of Noriaki Tsuchimoto, who led Japan in the production of documentary films on the Minamata disease and the supporters' movements. These films were shown in many places and this contributed greatly to spreading knowledge of the disease. His works have also been recognized internationally.

In 1978, over a six-month period, the Seirinsha group took their movies to 133 locations in 65 villages within a radius of 30 km of Minamata City, in and around the Shiranui Sea area. In 1975, Tsuchimoto produced a three-part series on the Minamata disease from a medical perspective, the results being a compilation of the many differing aspects of the disease. The narrative was written in collaboration with the Minamata Disease Research Group using all available material, including that of Dr. Hosokawa and a number of other co-operating medical practitioners.

Dr. Masazumi Harada and the group of medical doctors who participated in the Minamata Disease Research Group tried to understand the total picture through research centring on the patients themselves. It is a well-known fact that Minamata disease patients exhibit symptoms characterized by high blood pressure, diabetic-type responses, and liver ailments, all of which are complicated by Hunter-Russel-type responses. These factors have been identified from pathological investigations that have traced the distribution of mercury poisons in various organs of the human body.

Long-term observations in real-life situations have shown that the absorption of even small amounts of mercury over an extended period result in undeniable dangers to human health. However, the combination of clinical symptoms characteristic of the Minamata disease is similar to that seen in the pathological processes of ageing, and as a result it is difficult to differentiate the Minamata disease from geriatric problems in cases where the degree of mercury poisoning is limited. Various treatments aimed at ameliorating the sustained degenerative effects of the Minamata disease have been tried. Certain rehabilitation exercises may be of some use in regaining lost motor functions, but there is no hope of recovery from the pathological processes brought on by mercury poisoning.

1976 saw the formation of an academic research group incorporating both the natural and social sciences; subsequently a report, *Shiranui-kai sogo chosadan* was published. Several young people who were involved in disease-victim support movements now live in Minamata and continue their various activities. One of their projects is the building of the Soshisha (Mutual Concern) Centre with contributions sent from all over Japan. The centre functions to provide work for patients whose handicaps are not severe, as well as offering training activities for young people; to this end it gives one-year internships to young people from urban situations. This project is similar to programmes in India in which urban youths participate in rural community camps.

Akira Sunada, a professional actor, lives in Minamata with the patients, earning his livelihood from organic farming. At the same time he has continued his mission, and through the presentation of his unique but traditional plays has told the story of Minamata and the disease patients to a large and varied audience. Through these activities people are reminded that this problem is still very much with us, and that there is a need for continued financial and moral support. A network that provides sales outlets for the organic agricultural products has been established with the help of various consumer organizations. Treatment of patients with oriental medicine has been tried, and many other projects have been launched. All activities are sustained on a voluntary basis and are not supported by any established funding organizations a fact that gives the community a certain feeling of autonomy.

In front of the Minamata City railway station there is a clinic to serve the needs of the Minamata disease patients. This clinic was set up by the Japan Association of Democratic Medical Organizations to provide a broad range of medical services to disease victims. It has a great deal of meaning for this community which once revolved around the lordship of an industrial complex, and great strength is derived from the knowledge that adequate care is available at the hands of professionals. Attempts to renew the local community have just begun, but the problems of the Minamata disease are far from over.

Minamata disease came into being as a result of one chemical complex that was, at a certain point in time, positioned at the heart of a new and rapidly growing industry. Because of the company's pride in its own technological prowess, it was blinded to the dangers of the waste effluents that it allowed to enter the human environment. The industry and various governmental organizations understood pollution problems only in terms of economic viability, and these same sectors of society tried to evade and cover up these problems through an initially successful series of oppressive measures. However, the problem reared its ugly head again, and this time the company was forced into a situation in which it could no longer continue operations. The way in which these problems were dealt with is beyond the comprehension of the present age. When industry and government circles were faced with this crisis, instead of attempting to deal in a straightforward manner with the realities involved, they simply initiated a cover-up in order to maintain traditional social structures and relationships for the sake of profit.

Because of this obstructionism and perfidy, the destruction of the environment was worsened to a such a degree that all attempts at a solution have failed. Within the context of this crisis, the Minamata disease victims themselves, their supporters, and citizens' movements have worked to arouse public opinion, and to encourage the struggle of all disease victims. The majority of victims endured their suffering in silence in the hope that someday people would forget their existence. Through the application of universal democratic social action, a small minority of disease victims, refusing to be beaten, brought about a new start for the many victims of the disease.

Where there are no people's movements, no progress is made towards meaningful solutions and the complexities are finally forgotten. Through active citizens' movements, new methods are discovered and new ways opened up. Most Minamata-disease-related citizens' movements employed non-violent direct action. In this way, the disease victims, even though they were socially weak, did not lose the battle, but were able to turn their deeply shared mutual experiences into powerful weapons for the fight. When such movements aim at returning basic human rights to the people, and when the appeal is directed at the full humanity of all persons, success is assured and progress can be made on a firm basis of human support.

One issue that has not yet been fully dealt with is that of the potential contribution of the disease victims to a full assessment of the damage done to the human environment by methyl mercury poisoning. The magnitude of the problem is such that, on the basis of present knowledge, it cannot be left to experts in the environmental sciences alone. It is obvious that Japanese political circles, and the related administrative organs of government, are totally incapable of coming to terms with the multitude of problems centring around this massive pandemic. Thus the adequacy of the work ahead depends upon how positively the socially weak disease victims are allowed, and able, to participate in the much needed overview of the problem, and in the process of finding solutions. These experiences are ones that we should all learn from, as we look ahead to other potential environmental disasters brought about by human greed.

## TRADE AND THE ENVIRONMENT

One contentious issue in international trade policy discussions concerns the connection between international trade and the environment. Many environmental groups claim that freer trade as implemented through the WTO agreements, or in free trade agreements such as NAFTA, results in negative environmental outcomes. For example the Sierra Club argues that,

"Economic globalization ties the world together as never before. But it also poses serious new threats to our health and the environment. Trade agreements promote international commerce by limiting governments' ability to act in the public interest. Already food safety, wildlife and pollution control laws have been challenged and weakened under trade rules as illegal "barriers to trade."

In contrast, the WTO, a frequent target for criticism by environmental groups, points to the WTO agreement which states, (WTO member) relations in the field of trade and economic endeavour should be conducted with a view to raising standards of living... while allowing for the optimal use of the world's resources in accordance with the objective of sustainable development, seeking both to protect and preserve the environment and to enhance the means for doing so in a manner consistent with their respective needs and concerns at different levels of economic development".Arguably, the stated goals of free trade-oriented groups and environmental groups are very similar at least as highlighted in the documents produced by both sides.

What differs are the methods used to achieve the objectives. For reasons to be elucidated below, the WTO has argued that environmental concerns are not directly within the purview of the WTO agreement, but despite that, environmental policies and international environmental agreements are not prohibited by, nor are they inconsistent with, the WTO accords. In essence, the argument by some has been that the WTO agreement, and free trade agreements more generally, are intended to be about trade and are not intended to solve tangential problems related to the environment. On the other hand, environmental groups have pointed out that sometimes WTO and free trade agreement decisions have a negative effect upon environmental outcomes and thus these agreements should be revised to account for these negative effects.Below we will consider these issues with respect to one type of environmental concern, pollution caused by consumption of an imported good. Although we will not consider many of the other contested environmental and trade issues, this one example will suffice to establish some important and generalizable conclusions.

## TRADE LIBERALIZATION WITH ENVIRONMENTAL POLLUTION

Consider a small country importing gasoline with a tariff in place initially such that the domestic tariff- inclusive price is $P_1$. At this price domestic supply is $S_1$, domestic demand (or consumption) is $D_1$, and the level of imports is ($D_1$ - $S_1$), shown on the adjoining diagram. Suppose that domestic consumption of gasoline causes air pollution. This means that consumption has a negative external effect on all users of air, *i.e.*, there is a negative consumption externality.

Let's assume that the cost to society (in dollar terms) of the air pollution is an increasing function of domestic consumption. In other words, the greater the consumption of gasoline, the greater is the pollution and the greater is the subsequent harm caused to people in the country. For simplicity, assume the environmental cost, EC(D), is a linear function of total domestic demand, D. The height of EC at any level of demand represents the additional dollar cost of an additional gallon of gasoline consumption. This implies the total environmental cost of a consumption level, say $D_2$, is the area under the EC curve between the origin and $D_2$. With the initial tariff in place, domestic demand is $D_1$, which implies that the total societal cost of pollution is given by area (h

+ i + j). Note that despite the cost of pollution it does make sense to produce and consume this good if the objective is national welfare. Consumer surplus is given by area (a + b) and producer surplus is (c + g). The sum of these two clearly exceeds the social cost of pollution, (h + i + j). *Note*: these statements are true for the particular graph shown above, it is not true in general. By drawing the EC curve very steep, corresponding to a much higher cost of pollution, it might not make sense to produce and consume the good in the market equilibrium. Next suppose that the country agrees to remove the tariff on imported gasoline after signing a trade liberalization agreement. The question we ask is, can trade liberalization have such a negative effect on the environment that it makes a country worse off? The answer, as we'll see is yes.

Suppose the tariff is removed and the price of gasoline falls to $P_2$. The lower price causes a reduction in production to $S_2$, an increase in consumption to $D_2$, and an increase in imports from the blue line segment ($D_1$–$S_1$) to the red line segment ($D_2$–$S_2$). Since domestic consumption of gasoline rises there is also an increase in pollution. The welfare effects of the tariff elimination are summarized in the Table below. The letters refer to the area in the previous graph. Red letters indicate losses while black letters indicate gains.

**Table. Welfare Effects of a Tariff Elimination with a Negative Environmental Consumption Externality**

| | **Importing Country** |
|---|---|
| Consumer Surplus | + (c + d + e + f) |
| Producer Surplus | –c |
| Govt. Revenue | – e |
| Pollution Effect | – k |
| National Welfare | (d + f) – k |

Consumers of gasoline benefit by areas (c + d + e + f) from the lower free trade price. Domestic producers lose (c) with a reduction in producer surplus. The government also loses tariff revenue (e). The net total efficiency gains from trade are given by areas (d + f). However the presence of the environmental consumption externality means there is an additional cost (k) caused by the pollution from higher domestic consumption of gasoline.

The national welfare effect of the tariff elimination is given by (d + f – k). For a particular level of efficiency gains, the total national effect will depend on the size of the pollution cost. In the graph, the curves are drawn such that area k is slightly larger than d + f. Thus, trade liberalization can cause a reduction in national welfare. The cost of additional pollution may be greater than the efficiency improvements from free trade. However, if the environmental cost of consumption were lower, the EC(D) line would be flatter and area k would become smaller. Thus for lower environmental costs trade liberalization might raise national welfare. The net effect, positive or negative, will depend on the magnitude of the pollution costs relative to the efficiency benefits.

## Trade Policy vs. Domestic Policy

In general, the Theory of the Second-Best suggests that, in the presence of a market imperfection or distortion, a properly chosen trade policy might be found that will raise a small country's national welfare. However, for most imperfections, trade policy will be a second-best policy. A better policy, a first-best policy, will always be that policy which attacks the imperfection or distortion most directly. In most instances, the first-best policy will be a domestic policy rather than a trade policy. In this case, environmental pollution caused by the consumption of gasoline is a market imperfection because gasoline consumption has a negative external effect (via pollution) on others within the society. Economists call this a negative consumption externality. This problem can be corrected with any policy that reduces the negative effect at a cost that is less then the benefit. A tariff on imports is one such policy that could work. However, the most direct policy option, hence the first-best policy choice, is a consumption tax. Below we'll show the welfare effects of a tariff and a domestic consumption tax and compare the results to demonstrate why a consumption tax is first-best while a tariff is second-best.

## A COMPARISON: TRADE POLICY VERSUS DOMESTIC POLICY

More interesting is the comparison between the welfare effects of a tariff with that of a consumption tax. Since the two policies are set at identical levels it is easy to compare the effects. The distributional effects, that is, who wins and who loses are slightly different in the two cases. First, the effects on consumers is the same since both policies raise the price to the same level. However, domestic producers suffer a loss in producer surplus with a tariff, whereas they are unaffected by the consumption tax. To some, this may look like a bad effect since domestic production of the polluting good is not reduced with the consumption tax. However, it is the net effect that matters. Next, the government collects more revenue with the domestic tax than with the tariff since both taxes are set at the same rate and consumption is greater than imports. Finally, the environmental effect is the same for both since consumption is reduced to the same level.

The net welfare effect of the consumption tax ($NW^C = k - f$) clearly must exceed the net welfare effect from a tariff ($NW^T = k - d - f$), that is $NW^C > NW^T$. The reason is that the tariff incurs two separate costs on society to receive the environmental benefit whereas the consumption tax incurs only one cost for the same benefit. Specifically, the tariff causes a loss in both consumption and production efficiency (d and f) while the consumption tax only causes a consumption efficiency loss (f). For this reason, we say it is more efficient (that is, less costly) to use a domestic consumption tax to correct for a negative consumption externality such as pollution, than to use a trade policy, even thouth the trade policy may improve national welfare.

### A Source of Controversy

For many environmental advocates, trade liberalization, or globalization more generally, clearly has the potential to cause environmental damage to many ecosystems. Concerns include pollution from industrial production, pollution from consumption, clear-cutting of tropical forests, extinction of plant and animal species and global warming, among others. Although only one type of environmental problem is addressed above, the principles of the Theory of the Second-Best will generally apply to all of these concerns.

The analysis above accepts the possibility that consumption causes pollution and that pollution is bad for society. The model shows that under these assumptions, trade policy can potentially be used to improve environmental outcomes and can even be in society's overall interest. However, trade policy is not the most efficient means to achieve the end. Instead, resources will be better allocated if a domestic policy, such as a consumption tax, is used instead. Since the domestic policy attacks the distortion most directly, it minimizes the economic cost. For this reason, a properly chosen consumption tax will always do better than any tariff. With respect to other types of environmental problems, a similar conclusion can be reached. The best way to correct for most pollution and other environmental problems will be to use a domestic policy intervention such as a production tax, consumption tax, factor-use tax or other types of domestic regulations.

Trade policies, although potentially beneficial, are not the most efficient policy instruments to use. It is worth emphasizing that the goal of most economic analysis should in many instances be aligned with the goal of environmentalists. It is the extraction and use of natural resources that contributes to environmental damage. At the same time, it is the extraction and use of natural resources that is necessary to produce the goods and services needed to raise human standards of living to acceptable levels. Thus, if we minimize the use of resources to produce a particular level of output we can achieve both the economist's goal of maximizing efficiency and the environmentalist's goal of minimizing damage to the environment.

## AIR (PREVENTION AND CONTROL OF POLLUTION) ACT, 1981

To counter the problems associated with air pollution, ambient air quality standards were established, under the 1981 Act. The Act provides means for the control and abatement of air pollution. The Act seeks to combat air pollution by prohibiting the use of polluting fuels and substances, as well as by regulating appliances that give rise to air pollution. Under the Act establishing or operating of any industrial plant in the pollution control area requires consent from state boards. The boards are also expected to test the air in air pollution control areas, inspect pollution control equipment, and manufacturing processes. National Ambient Air Quality Standards (NAAQS) for major pollutants were notified by the CPCB in April 1994. These are deemed to be levels of air quality necessary with an adequate margin of safety, to protect public health, vegetation and property (CPCB 1995 cited in Gupta, 1999). The NAAQS prescribe specific standards for industrial, residential, rural and other sensitive areas. Industry-specific emission standards have also been developed for iron and steel plants, cement plants, fertilizer plants, oil refineries and the aluminium industry. The ambient quality standards prescribed in India are similar to those prevailing in many developed and developing countries.

To empower the central and state pollution boards to meet grave emergencies, the *Air (Prevention and Control of Pollution) Amendment Act*, 1987, was enacted. The boards were authorized to take immediate measures to tackle such emergencies and recover the expenses incurred from the offenders. The power to cancel consent for non-fulfilment of the conditions prescribed has also been emphasized in the Air Act Amendment.

### The Air (Prevention and Control of Pollution) Rules formulated in 1982

Defined the procedures for conducting meetings of the boards, the powers of the presiding officers, decision-making, the quorum; manner in which the records of the meeting were to be set etc. They also prescribed the manner and the purpose of seeking assistance from specialists and the fee to be paid to them. Complementing the above Acts is the *Atomic Energy Act* of 1982, which was introduced to deal with radioactive waste. In 1988, the *Motor Vehicles Act,* was enacted to regulate vehicular traffic, besides ensuring proper packaging, labelling and transportation of the hazardous wastes. Various aspects of vehicular pollution have also been notified under the EPA of 1986. Mass emission standards were notified in 1990, which were made more stringent in 1996. In 2000 these standards were revised yet again and for the first time separate obligations for vehicle owners, manufacturers and enforcing agencies were stipulated. In addition, fairly stringent Euro I and II emission norms were notified by the Supreme Court on April 29, 1999 for the city of Delhi. The notification made it mandatory for car manufacturers to conform to the Euro I and Euro II norms by May 1999 and April 2000, respectively, for new non-commercial vehicle sold in Delhi.

## AIR POLLUTION

As serious as water pollution is to the health and welfare of man, in many parts of the world air pollution represents an even more serious threat to human existence. Most of the major cities of the world and many rural areas as well now have serious air quality problems. The relative small (500,000 people) and modern capital of Malaysia, Kuala Lumpur, a beautiful city surrounded by luxuriant green hills, now has an emerging air pollution problem as its economy begins to prosper and expand. It arises from rapidly increasing automobile, truck and bus traffic, and burgeoning industrial development. Even the remote capital of Nepal, Kathmandu, which is nestled in the Himalayan mountains north of India, began to show symptoms of air pollution in 1971. Famous for its clear mountain air and startling views of the high Himalaya, the valley of Kathmandu is often hazy with exhaust smoke from rapidly increasing automobile, truck, bus, and airplane traffic. The valley and city of Kathmandu are vulnerable to such pollution because of a natural air inversion a layer of warm tropical air over cold mountain air resting in an enclosed and densely populated valley. Unless control measures are taken now, the magnificent scenery which brings tourists to Nepal, will be viewed through a polluted haze in future years.

Ten years ago many engineers and environmental scientists insisted that there were few if any demonstrable ill effects on human health from the levels of air pollution then existing in most cities. They admitted exceptions to this, of course, but these were primarily confined to critical situations such as the famous London smog of 1952 in which 4,000 to 5,000 people died from respiratory distress in a persistent smog, or the crisis in Donora, Pennsylvania in 1948, in which hundreds of people suffered similar fatalities during and after a severe smog condition which hung over the city for several weeks. Other than these rare circumstances, there was little evidence that chronic levels of urban air pollution represented a public health problem.

Now there is abundant evidence that the levels of air pollution in many cities do represent a major medical problem. This evidence comes from both experimental research with animals and clinical studies in human health. In fact, the health hazards of air pollution in the United States, Japan and parts of Europe are now more clearly documented than are those of water pollution. This is not true, of course, throughout much of Latin America, Africa, and Asia, where impure water still represents a major health problem.

Some of the first clues to the health hazards of air pollution came from clinical observations of increasing rates of emphysema, chronic bronchitis, and respiratory distress in city dwellers, and from experimental studies on laboratory animals exposed to air pollutants. In the latter category of experimental studies, it was noted in Los Angeles several years ago that laboratory mice exposed to ambient air pollution (*i.e.*, normally existing air pollution) developed significant pathology compared to control mice in clean air. Aging inbred mice showed increased frequency of pulmonary adenoma, and one strain of rice showed increased mortality of young adult males. Severe smog episodes in Los Angeles caused basic changes in the cellular structure of lung tissue in these animals. Guinea pigs and rabbits developed altered hormone excretion patterns and differential enzyme levels in blood serum in contrast to clean air controls.

Pathologic effects of air pollution have also been clearly demonstrated in people. In one study in New York City, children under 8 years of age showed a prevalence of respiratory symptoms directly related to levels of particulate matter and carbon monoxide. In many cities along the Eastern seabord, increasing evidence of dyspnea, bronchitis, cough, sputum production, wheezes, eye irritations and general malaise was elicited as air pollution levels increased. In Los Angeles, a correlation was shown between carbon monoxide pollution levels and case fatality rates in patients with heart trouble.

Sulfur dioxide is one of the common gaseous air pollutants which is most injurious to human health. It irritates respiratory epithelium and impairs normal beathing. Most cities have $SO_2$ levels less than 0.5 ppm, and human effects are not prominent until 0.8 or 1.0 ppm are attained. Frequently, ambient levels in cities exceed 0.8 ppm when stagnant air remains for several days and gaseous pollutants accumulate.

Other investigators have shown correlations between photochemical air pollutants and respiratory distress, emphysema and susceptibility to respiratory infection. Chronic as well as acute effects have been documented. In the industrial complex of Bayonne and Elizabeth, New Jersey, the death rate from respiratory cancer in males was 35 per cent higher in an area of high air pollution, compared to a similar population living in a lower air pollution environment only a few miles away. On Staten Island, the rate of lung cancer in women in polluted areas was shown to be twice that of women in clear areas. Other studies have shown that urban air pollution increases the rate of lung cancer in men three times above that of rural men. This type of evidence could be cited in considerable detail to leave little doubt that air pollution is detrimental to human health. In fact, respiratory ailments related to air pollution—emphysema, chronic bronchitis, lung ancer, and severe asthma—are among the most rapidly increasing health problems in industrialized nations. Various disease problems associated with air pollution were reviewed by Lave and Seskin who also evaluated the economic costs of respiratory disease attributable the air pollution. They estimated that a 50 per cent reduction in US urban air pollution would save the nation $2,080 million dollars per year in health and medical costs.

Air pollution is as complex in origin and type as water pollution. It has been estimated that 164 million metric tons of pollutants enter the United States air every year. This pollutant load is composed of a wide range of particulate matter, suspended particles, and gases. In New York City alone, the daily emission of air pollutants includes 3,200 tons of $SO_2$, 4,200 tons of CO, and 280 tons of particulate dirt. In Philadelphia in 1959, daily releases of air pollutants amounted to 830 tons of $SO_2$, 300 tons of $NO_2$, 1,350 tons of hydrocarbons, and 470 tons of particulates.

Particulate matter fall-out affords a dramatic example of dirty air. In many of the world's cities, the average daily air pollution fall-out is in the order 0.5 to 3.0 tons per square mile per day. Some cities have over 4 tons per square mile per day. In Pittsburgh, fall-out was reduced from over 5.5 tons per square mile per day to less than 0.9 tons by a vigourous clean air and smoke abatement programme. Up to one per cent of urban dust may be lead, which is toxic to humans in fairly low concentrations. Yet particulate fall-out, while it may be one of the most dramatic forms of air pollution, is certainly not necessarily the most harmful. Most particles which fall out are over 100 micra in diameter, and these seldom if ever reach the alveolar tissue of the lung where irritation occurs. Fine suspended particles and gaseous substances are the primary agents of respiratory distress. Some of the specific components of air pollution known to have pathologic effects in humans are listed in Table 5A.

One of the most persistent forms of air pollution which has not responded well to control measures is automobile exhaust. In the late 1960's, over 90 million motor vehicles in the United States produced 66 million tons of carbon monoxide, and 20 million tons of other air pollutants per year. Although various devices have been developed to reduce exhaust emission, these devices are not consistently well maintained by the public. Los Angeles, which has mounted an outstanding campaign of industrial air pollution control, still faces a major smog problem from automobile exhaust. The city has 4,000,000 automobiles for 3,000,000 people.

Two-thirds of the area of the central city has been taken over by the automobile in the form of parking space, streets and freeways. One study in Los Angeles showed a direct correlation between air pollution and the frequency of motor vehicle accidents. In the air on crowded freeways, carbon monoxide levels may reach 400 parts per million. Automobile drivers thought to be responsible for accidents showed elevated carbon monoxide levels in their blood. Many experts predict that the internal combustion engine must be either outlawed or significantly altered within the next few years if we are to avert air pollution tragedies in our cities.

There is no doubt that air pollution is a very significant and increasing factor in environmental deterioration. Airplane pilots who have flown for twenty or thirty years report a great increase in "ground haze" and air pollution domes encapsulating cities. Whereas cities were often seen from aerial distances of 30 to 40 miles some years ago, they are now usually enshrouded by air pollution and not visible from more than 5 or 10 miles. In many areas, air pollution has caused dramatic injury to plants, both agricultural crops and natural plant communities. Citrus groves, truck garden crops of lettuce, tomatoes, onions and celery, field crops of alfalfa, sweet corn, and tobacco, and even forests of pine, spruce, and deciduous trees have all fallen victims to air pollution in various parts of the world.

Air pollution also takes its toll of buildings and other man-made objects. When moisture accumulates in polluted air, the oxides of sulfur, carbon and nitrogen form weak sulfuric acid, carbonic acid and nitrous acid which are corrosive to metals, stone, paint, rubber, textiles and even some plastics. One study estimated that air pollution in 1960 cost the average American family $800 per year in property damage. Current projections of this figure would certainly put these costs well over $1000 per family per year.

Throughout Europe, many famous buildings, monuments and art treasures of former centuries are deteriorating at an alarming rate due to the erosional effects of air pollution. In Athens, the President of the Greek Academy of Sciences, estimated in 1971 that the Parthenon on top of the Acropolis has deteriorated more in the last 50 years than in the previous 2000 years. Athens is normally blessed with clean fresh air, but this depends entirely on sea breezes. On still, windless days, a noxious air pollution haze quickly forms, even shrouding out the Aegean Sea near Pathens.

A serious possibility of world-wide air pollution that was first detected in the late 1960s is the occurrence of stratospheric pollution and a global air pollution veil; that is, a ring of air pollution circling the globe in the northern hemisphere around the latitudes of the US, Europe and Japan. Such a global veil has been detected by satellite photos, and confirmed by both Russian and US scientists. The formations have been temporary, and apparently occurred during periods of unusually stable currents of air circulation around the world. In other words, pollutants were being carried intercontinenally, so that air entering North America contained Japanese contaminants, and air entering Europe contained North American pollutants. Normally, the oceanic mixing of air would break up pollution bands. This is not a major problem now, but it is a serious portent of things to come.

Certainly, many types of air pollution can be controlled by modern technology, but the costs must ultimately be borne by the public through higher prices for industrial goods, higher taxes, reduced profit margins in industry, and more careful monitoring of automobile exhaust control systems. In many cases of environmental quality control, the ultimate responsibility rests with the citizenry at large through established routes of political process. The present cost of air pollution in terms of ill health, agricultural damage, and accelerated deterioration of construction materials and personal goods is very great, but it is so diffuse and indirect that we are not aware of the great price we already are paying for it. When we reach the point where the cost of air pollution is greater than the cost of controlling it, the public must demand appropriate governmental action and be willing to support it.

## ANTIQUITY OF AIR POLLUTION

Air pollution, particularly in cities, is certainly not a new problem. Back in the Middle Ages the use of coal in cities such as London was beginning to escalate. The problems of poor urban air quality even as early as the end of the $16_{th}$ century are well documented. In the UK the Industrial Revolution during the $18_{th}$ and $19_{th}$ centuries was based on the use of coal. Industries were often located in towns and cities, and together with the burning of coal in homes for domestic heat, urban air pollution levels often reached very high levels. During foggy conditions, pollution levels escalated and urban smogs were formed. These often brought cities to a halt, disrupting traffic but more dangerously causing death rates to dramatically rise. The effects of this pollution on buildings and vegetation also became obvious. The 1875 Public Health Act contained a smoke abatement area to try and reduce smoke pollution in urban areas. During the first part of the $20_{th}$ century, tighter industrial controls lead to a reduction in smog pollution in urban areas. The 1926 Smoke Abatement Act was aimed at reducing smoke emissions from industrial sources, but despite the declining importance of coal as a domestic fuel, pollution from domestic sources remained significant.

The Great London Smog of 1952, which resulted in around 4,000 extra deaths in the city, led to the introduction of the Clean Air Acts of 1956 and 1968. These introduced smokeless zones in urban areas, with a tall chimney policy to help disperse industrial air pollutants away from built up areas into the atmosphere. Following the Clean Air Acts, air quality improvements continued throughout the 1970s. Further regulations were introduced through the 1974 Control of Air Pollution Act. This included regulations for the composition of motor fuel and limits for the sulphur content of industrial fuel oil. However, during the 1980s the number of motor vehicles in urban areas steadily increased and air quality problems associated with motor vehicles became more prevalent. In the early 1980s, the main interest was the effects of lead pollution on human health, but by the late 1980s and early 1990s, the effects of other motor vehicle pollutants became a major concern. The 1990s have seen the occurrence of wintertime and summertime smogs. These are not caused by smoke and sulphur dioxide pollution but by chemical reactions occurring between motor vehicle pollutants and sunlight.

These are known as'photochemical smogs'. In 1995, the Government passed its Environment Act, requiring the publication of a National Air Quality Strategy to set standards for the regulation of the most common air pollutants. Published in 1997, the National Air Quality Strategy has set commitments for local authorities to achieve new air quality objectives throughout the UK by 2005. It is reviewed periodically.

**HUMAN HEALTH**

Healthy people do not normally notice any effects from air pollution, except when the pollution is very high. However, people sensitive to pollution, such as asthmatics, and those with heart conditions or lung diseases, may experience distress and other health effects, even at lower levels of pollution. The Government uses an air pollution banding system to describe the potential health impacts of poor air quality.

Particulates may be seen as the most critical of all pollutants, and some estimates have suggested that particulates are responsible for up to 10,000 premature deaths in the UK each year. The extent to which particulates are considered harmful depends largely on their composition. Sea salt, for example, is believed to have a positive effect on health. Man-made sources of particulates, however, are rarely harmless. In towns and cities, these are extensively from diesel vehicle exhausts. The effects of particulate emissions are considered detrimental due to their composition, containing mainly unburned fuel oil and hydrocarbons that are known to be carcinogenic among laboratory animals. Very fine particulates can penetrate deep into the lung and cause more damage, as opposed to larger particles that may be filtered out through the airways' natural mechanisms. Ozone differs from most pollutants in that it is created as a secondary pollutant by the action of sunlight on volatile organic compounds (VOCs) and oxides of nitrogen. Ozone is a toxic gas that can bring irreversible damage to the respiratory tract and lung tissue if delivered in high quantities. Asthmatics are known to adopt these symptoms more easily.

Nitrogen oxides consist mainly of nitrogen dioxide ($NO_2$) and nitric oxide (NO). Nitric oxide is more readily emitted to the atmosphere as a primary pollutant, from traffic and power stations, and is often oxidised to nitrogen dioxide following dispersal. The amount of nitrogen dioxide emitted directly to the atmosphere is relatively small. Nitric oxide is relatively non-toxic, but at high concentrations the health effects include changes to lung function. Nitrogen dioxide, however, is damaging to health, due to its toxicity. Health effects of exposure to nitrogen dioxide include shortness of breath and chest pains. Transport, tobacco smoke and gas appliances are the major sources of carbon monoxide. Its link with haemoglobin, the oxygen carrying component of the blood stream, forms carboxyhaemaglobin (COHb) which can be life-threatening in high doses.

The effects of carbon monoxide pollution are more damaging to pregnant women and their foetus. Research into smoking and pregnancy shows that concentrations within the blood stream of unborn infants is as high as 12 per cent, causing retardation of the unborn child's growth and mental development. A significant proportion of atmospheric lead comes from traffic emissions, due to the lead content in petrol. This has been significantly reduced in recent years but lead is still a serious air pollutant especially to those living near to areas of dense traffic. Damage to the central nervous system, kidneys and brain can result from high concentrations in the blood. Children, however, display vulnerability to the toxic effects of lead at much lower concentrations than for adults. It has been shown that there is a strong link between high lead exposures and impaired intelligence.

Even moderate concentrations of sulphur dioxide may result in a fall in lung function in asthmatics. Tightness in the chest and coughing may also result at higher levels. Sulphur dioxide pollution is considered more harmful when particulate and other pollution concentrations are high. This is known as the"cocktail effect." Some VOCs are quite harmful. Benzene, for example, has been linked with an increase susceptibility to leukaemia, if exposure is maintained over a period of time.

## INDOOR AIR POLLUTION

We spend a large part of our lives indoors at home. Keeping the air which we breathe at home clean is therefore of necessary importance, particularly for certain vulnerable members, including babies, children, pregnant women and the unborn babies, the elderly, and those suffering from respiratory or allergic diseases, such as asthma. In most homes the level of indoor air pollution is very low, because there are controls on the design and construction of buildings. However, if ventilation of rooms is poor, or household appliances are faulty, pollution can build up to levels which may be detrimental to human health. There are many possible sources of air pollutants in the home and indoor air quality can vary widely. DIY work may lead to a temporary increase in indoor pollutants such as volatile organic compounds (VOCs), during painting or stripping in enclosed spaces, or laying loft insulation. Another significant source of indoor pollution is the burning of fuels in flueless appliances, such as paraffin stoves, portable gas heaters, gas stoves and ovens. If the appliance is faulty, incomplete combustion may result in the release of carbon monoxide, a highly poisonous gas.

Carbon monoxide also builds up when people smoke cigarettes indoors. Dirty homes or houses in disrepair may be a source of dustmite and mould spores. In some parts of the UK, and in other parts of the world, the radioactive gas radon can seep into the house from the underlying geology, and accumulate indoors if ventilation is poor. Housing and public health legislation exists to help prevent air quality problems arising indoors in the first place. In the majority of homes there is no need for concern over existing levels of pollutants.

## INDUSTRY AND POWER GENERATION

Industry and power generation are main sources of sulphur dioxide emissions, a common air pollutant and the precursor for sulphuric acid in acid rain. In the UK power stations and all other types of industry account for 90 per cent of all sulphur dioxide pollution. During the Industrial Revolution industries were often located in urban areas. Following the UK Clean Air Acts in the 1950s and 1960s, and with the decline in heavy industry, few large industries and power stations are located in towns and cities today. Many large industries are now located in the more rural areas of the UK. Consequently, sulphur dioxide pollution in urban areas has been significantly reduced. The requirement of industries and power stations to disperse waste gases at elevated levels via a stack or chimney has also helped to reduce ground level concentrations of sulphur dioxide. However, this has significantly expanded the area of pollution dispersal, such that acid deposition is now the main pollution concern attributable to industry and power generation.

Power stations contribute significantly to the total emissions of nitrogen oxides in the UK. In 1999, 21 per cent of nitrogen oxides came from this source and a further 13 per cent from other industries, iron and steel and refineries. The major source of nitrogen oxides pollution in the UK is now road transport (44 per cent). Like sulphur dioxide, nitrogen oxides are also converted into acidic compounds when combined with water in the atmosphere, and contribute to acid rain.

## LEGISLATION

Ever since air pollution was recognised as a problem, legislators, regulators and governments have tried to control it. As early as 1273 the use of coal was prohibited in London as being "prejudicial to health". In 1306 the Royal Proclamation prohibited craftsmen from using sea-coal (a soft coal) in their furnaces. Since the beginning of the Industrial Revolution in Britain, numerous Acts have been passed in an attempt to reduce air pollution. These have included the Railway Clauses Consolidated Act of 1845 (requiring railway engines to consume their own smoke), the Improvement Clauses Act of 1847 (to reduce factory smoke), the Sanitary Act of 1866 (empowering sanitary authorities to take action in cases of smoke nuisances), the Public Health Act of 1875 (containing smoke abatement legislation that has been used to the present day), and the Smoke Abatement Act of 1926.

In the aftermath of the Great London Smog of 1952, the Government pass the two Clean Air Acts of 1956 and 1968, which aimed to control domestic sources of smoke pollution by introducing smokeless zones, and control industrial sources of pollution by the use of tall chimneys for waste gas dispersal. Since the 1970s when the UK joined the European Union, European legislation has been used to control the amount of pollution being emitted by industry, and now increasingly by transport. Over the last 30 years the European Commission has passed a number of EC directives to limit emissions of carbon monoxide, lead, hydrocarbons and smoke emissions from road vehicles, and to set health limits for the common air pollutants, including sulphur dioxide, particulate matter, lead and nitrogen dioxide.

In response to European legislation, the UK Government passed the Environmental Protection Act (1990) and the Environment Act (1995), bringing many smaller emission sources under air pollution control by local authorities for the first time, and providing a new statutory framework for local air quality management. In 1997 the National Air Quality Strategy was published which sets air quality standards and targets for the pollutants of most concern. Many of these targets will need to be met by 2005.

## LICHENS

Lichens are mutualistic associations of a fungus and an alga or cyanobacterium and occur as crusty patches or bushy growths on trees, rocks and bare ground. The names given to lichens strictly refer to the fungal partner; the algae have separate names. Lichens are very sensitive to sulphur dioxide pollution in the air. Since industrialisation, many lichen species have become extinct in large areas of lowland Britain, one example being the beard moss *Usnea articulata*. This is mainly due to sulphur dioxide pollution, but the loss of habitat, particularly ancient woodland, has also led to reductions in some species. Lichens are sensitive to sulphur dioxide because their efficient absorption systems result in rapid accumulation of sulphur when exposed to high levels of sulphur dioxide pollution. The algal partner seems to be most affected by the sulphur dioxide; chlorophyll is destroyed and photosynthesis is inhibited. Lichens also absorb sulphur dioxide dissolved in water.

Lichens are widely used as environmental indicators or bio-indicators. If air is very badly polluted with sulphur dioxide there may be no lichens present, just green algae may be found. If the air is clean, shrubby, hairy and leafy lichens become abundant. A few lichen species can tolerate quite high levels of pollution and are commonly found on pavements, walls and tree bark in urban areas. The most sensitive lichens are shrubby and leafy while the most tolerant lichens are all crusty in appearance. Since industrialisation many of the shrubby and leafy lichens such as *Ramalina, Usnea* and *Lobaria* species have had very limited ranges, often being confined to the parts of Britain with the purest air such as northern and western Scotland and Devon and Cornwall.

A lichen zone pattern may be observed in large towns and cities or around industrial complexes which corresponds to the mean levels of sulphur dioxide experienced. The most commonly used zonal index is the Hawksworth and Rose index, first published in 1970, consisting of a scale of 1 (poorest air quality) to 10 (purest air). Particular species of lichen present on tree bark can indicate the typical sulphur dioxide levels experienced in that area. For example if there are no lichens present, the air quality is very poor, whilst generally only crusty lichens such as *Lecanora conizaeoides* or *Lepraria incana* can tolerate poor air quality. In moderate to good air, leafy lichens such as *Parmelia caperata* or *Evernia prunastri* can survive and in areas where the air is very clean, rare species such as 'the string of sausages' *Usnea articulata* or the golden wiry lichen *Teloschistes flavicans* may grow.

The Hawksworth and Rose index zonation index applies only to areas where sulphur dioxide levels are increasing. If sulphur dioxide conditions are falling, lichens rarely colonise in exactly the same sequence. Lichens are slow growing and may take a year or two to recolonise bark or other substrates following a reduction in air pollution levels, and tiny recolonising specimens can be difficult to spot and identify.

During the early and mid-twentieth century, air pollution levels were much greater than they are today in towns and cities of the UK. Sulphur dioxide levels were highest in the inner city areas becoming less polluted out towards the edges of the urban areas. At such times, the lichen zone scale would often highlight zone 1 as the inner city area, moving through the zones to the cleaner air at the edge of the city. From the 1970s onwards, sulphur dioxide levels have been falling markedly in the central and outer areas of cities, such that there may be no differentiation between levels in central and outer areas of many cities. The fall in sulphur dioxide levels between the 1970s and the present day has led to a number of lichens recolonising in areas from which they had already been eliminated.

## POLLUTION CONTROL LEGISLATIONS FOR AIR, WATER AND-LAND

Environmental pollution, a threat to life on earth is primarily attributable to anthropogenic and natural factors. Nature possesses an inherent quality of curing pollution by itself. However, anthropogenic factors, such as, population explosion, rapid industrialisation and growing urbanisation disturb the self-purification system of nature resulting in environmental pollution. Realising the gravity of the problem and with a view to combat and mitigate pollution problems, several legislations and their enforcement came into being worldwide.

In India, Orissa took the lead in the matter by enacting the Orissa River Pollution Prevention Act, 1953. Committed to a clean environment, Orissa Pollution Control Board was constituted with effect from 15.7.1983 vide notification number 1484 after the Orissa Legislative Assembly adopted the Water (Prevention and Control of Pollution) Act, 1974 and Air (Prevention and Control of Pollution) Act, 1981. The Board was primarily entrusted with the responsibilities to execute and ensure proper implementation of different provisions of both the Acts.

*The functions of the Board can be broadly classified into three main categories*:

1. Enforcement
2. Advisory
3. Monitoring, research and creation of public awareness.

Subsequently the Environment (Protection) Act. 1986, an umbrella act was enacted wherein, Orissa Pollution Control Board has been delegated with wide varieties of responsibilities.

*The main features are as follows*:

- Hazardous Waste (Management and Handling)Rules, 1989.
- Manufacture, use, import, export, storage of Hazardous Micro-organism, Genetically Engineered Organisms or Cells Rules, 1989.
- Manufacture, storage and import of Hazardous Chemical Rules, 1989.
- Environment Audit Notification.

- EIA and Notification for conducting public hearing prior to the issue of NOC (Consent to establish a statutory requirement under EP Act and rules).
- Coastal Regulation Zone Notification, 1991.
- Chemical Accidents (Emergency Planning, Preparedness and Response) Rules, 1996.
- Biomedical Waste (Management and Handling) Rules, 1998.

The Board continues to discharge its duties and responsibilities under various Environment Acts and Rules framed thereunder as described above. The recent rulings of the Apex Court have put enormous responsibilities on the State Pollution Control Boards beyond the statutory duties and responsibilities *i.e.* settlement of various public interest litigation cases. Besides these, the Board through various enforcements issues directive to industries to adopt clean technologies, install adequate emission control measures and effluent treatment plants.

The Board also grants consent to operate an industry. Without the consent of the Board, industry is not supposed to carry out its operation. Along with the consent application, the industry supplies information on nature and quantity of raw-materials, production process, quality of effluent, atmospheric emission in each stack, process emission, particulate analysis, details of water and Air Pollution control measures etc. to the Board.

On having the application from the heavily polluting Small Scale Industries the Board makes enquiry under Rule-33 of the Water (Prevention and Control of Pollution) Rules 1976 for granting consent. The categories of heavily polluting industries. For other categories of Small Scale Industries, the acknowledgement of the application form by the Board serves the purpose of consent. However, the Board may conduct random checks or call for information from any Small Scale unit and make a formal consent order prescribing conditions, etc. as required. Similarly the Board grants consent to establish (NOC) for facilitating the entrepreneurs to establish industries in the State.

Another important task of the Board is to prepare Zoning Atlas for siting of industries at the instance of Government of India. It is because of the fact that industrialisation in India is occurring in a fast pace and the environmental pollution due to emission and waste generated by the industries is increasing proportionately. Lack of planned development in the industrial sector can put the environment to high risk due to poor land use compatibility and over exploitation of resources.

Much of the problems of pollution and cost thereof can be avoided by preventing indiscriminate siting of industrial units. An important pre-requisite for judicious siting of industries is to consider the environmental profile of the proposed site. It is achieved through preparation of Zoning Atlas. While preparing the Zoning Atlas the environmental parameters and conditions are evaluated and quantified and the suitability of the site is determined basing on their sensitivity to air, water and land pollution.

The Central and State Pollution Control Boards which are statutorily involved in pollution control and regulatory activities have now taken a lead to introduce environmental planning in India.

This environmental planning programme more popularly known as "Zoning Atlas" is intended to help achieve development in an environmentally sound manner. The Zoning Atlas for siting of industries zones and classifies the environment in a District and presents the pollution receiving potential of various sites/zones in the District and the possible alternate sites for industries through easy to read maps.

*The objectives of preparing a Zoning Atlas for siting of industries are*:

- To zone and classify the environment in a Districts;
- To identify locations for siting of industries and
- To identify industries suitable to the identified sites.

The Zoning Atlas, in addition to streamlining the decision-making process has several benefits, such as;

- Provides a ready-reckoner for best suitable site and relevant environmental information;
- Makes decision-making process simpler, faster, realistic, transparent and reliable;
- Helps in planning cost-effective pollution control measures and programmes;
- Helps entrepreneur in readily finding out the location best suited to site an industry thereby saving time, efforts, investment and risk instead of heading for an unknown site, conducting environmental impact assessment and awaiting clearance by the regulatory authorities.
- Helps develop infrastructure facilities, such as roads, water supply, electricity etc. and provide common waste treatment and disposal facilities;
- Helps check additional pollution in the areas already over-stressed with pollution;
- Helps achieve sustainable development.

In the Zoning Atlas, the information on physical characteristics and environmental status of the districts are analysed to identify suitable areas for siting of various types of polluting industries. The outputs are presented through easy-to-read maps in 1,250,000 scale as well as in digital format in GIS environment. The advantage of creating digital database is that it can be upgraded periodically with latest information.

Zoning Atlas for siting of industries is a World Bank sponsored national project coordinated by Central Pollution Control Board, Delhi and executed in association with State Pollution Control Boards and other agencies. Orissa State Pollution Control Board has been entrusted to execute the preparation of Zoning Atlas for Orissa. The necessary technical support is being provided by Central Pollution Control Board in coordination with the German Technical Cooperation.

Orissa State Pollution Control Board, in the first phase, has prepared the Zoning Atlas for siting of Industries, based on Environmental Considerations for Sundargarh district.

In second phase, the Board has prepared Zoning Atlas for undivided Sambalpur (Sambalpur, Deogarh, Jharsuguda and Bargarh) and Cuttack (Cuttack, Jajpur, Kendrapara and Jagatsinghpur) districts. The Industrial Estate planning for Paradeep area has already been completed. The Board is now enganged in preparation of Zoning Atlas for Puri, Mayurbhanj and Keonjhar districts.

## AIR POLLUTION CONTROL

Air Pollution Control is dedicated to cleaning the environment in which we work and live. Air Pollution Control specializes in the design, installation and service of dust, smoke, fume and mist collection systems. We can further assist in servicing your existing system and provide aftermarket parts and filters.

## WATER

There are solid and definable techniques in existence that can help to control water pollution, but in order for them to work, we must be aware of what they are, why they are important and how to properly enact them.

### Instructions

Control water pollution in your home by using non-toxic soaps, detergents and cleaning products. Refrain from the use of chemical fertilizers and pesticides on your lawn and gardens.

*Always dispose of paints, motor oil, gasoline, antifreeze and other harmful chemicals in accordance with your local laws and safety regulations*:

- Protect groundwater, which is critical for drinking water, irrigation systems and natural ecosystems. If you are using chemicals that may be harmful to the environment, store them correctly. Improperly stored chemicals can slowly seep into the groundwater system, so keep them in tightly sealed containers, inside of structures with cement floors, to avoid groundwater contamination.

- Prevent polluted run-off and soil erosion. Polluted run-off is caused when rain washes toxic pollutants into surface waters from sources that include city streets, farms, or logging and mining sites. Plant bushes and trees along roads and natural water sources. The roots of trees and bushes can slow the speed of run-off and erosion, protecting surface water.
- Write letters to your state representative and congressman to express your concerns about water pollution. Ask them to promote more sustainable agricultural methods and mention that you feel it is time for Congress to create and enforce stricter mandatory laws regarding water pollution.

## POLLUTION CONTROL-CONTAMINATED LAND

*Some land in this country has been contaminated in the past by industries such as*:

- Gasworks
- Tanneries
- Chemical works
- Landfills

These are often called brownfield sites.

### The Problem

*Brownfield sites can be a problem for two reasons*:

1. There may be harmful substances in, on or under the land
2. Water pollution might be caused by substances at the site

However, brownfield sites do not generally cause a problem unless they are redeveloped for a different use.

### Pollutant linkage

*Land is only declared 'contaminated' if*:

- It contains a source of pollution-the source and
- Someone (or something) could be affected by the pollutant-the receptor and
- The pollution can get to the 'receptor'-the pathway

These three elements together are known as the pollutant linkage.

### Action Required

If you own or occupy contaminated land now, or you did in the past, you may be responsible for cleaning up the pollution. You may still be responsible for cleaning up the pollution after you have sold the land. Some contamination can be a hazard to current occupants or neighbours and the law says the problem must be put right immediately. The law follows the 'polluter pays' principle- the person or organisation that caused or permitted the contamination must pay to have it put right. If that person or organisation is not known, then the current owner of the land may become responsible. Owners and occupiers of domestic properties are not usually liable for these costs.

**Re-use of Brownfield Sites**

The approval of an application for redevelopment of these sites will only be granted on condition that the contamination is cleaned up to a standard that makes it suitable for the new use of the land. You should obtain specialist advice from an environmental consultant or a specialist lawyer before you buy or sell contaminated land. When you buy land in Hertsmere, the Land Charges department at the local council will tell you if a site has been declared 'contaminated land'.

**What the Local Council does about Contaminated Land**

The local council is responsible for enforcing the 'contaminated land' legislation.

- Publishes a Contaminated Land Strategy, which says how it will find contaminated sites in its area
- Carries out inspections of land that may be contaminated
- Finds out who is responsible for putting right the contamination and discusses the problem with them
- Formally declares land contaminated agrees the necessary action and makes sure it is done
- Keeps a Public Register of contaminated land sites, the action that was required to put the problem right and any legal action that has been taken.

In some cases the Environment Agency may take over the regulation of a site from the council, once it has been declared as 'contaminated land'.

**Planning**

If a piece of land is to be re-developed or have its use changed, and the land could potentially be contaminated, then this is usually dealt with under the Town and Country Planning Act. It is the responsibility of the owners and/or developers of the site to establish the extent and nature of any potentially harmful materials on the site. This should normally be done before the formal planning permission is granted for the development. However, if this is not possible, permission can be granted, but certain conditions regarding land contamination will be attached to the permission, which must be met. If potential risks are identified, the land will need to be dealt with before development begins, to minimise all risks posed. This process is known as remediation. Guidelines have been produced to help applicants, developers, land owners and consultants meet the planning requirements for a potentially contaminated site. It was prepared by Hertsmere Borough Council, in association with other local authorities in Hertfordshire and Bedfordshire and the Environment Agency and sets out the information that a Local Planning Authority will require.

**Environmental Information**

The Council can provide environmental information on sites within the Borough in accordance with the Environmental Information Regulations (EIR) 2004. We hold various information relating to contaminated land issues, including details of site investigations and remedial works, location of former landfill sites, details of authorised industrial processes, pollution incident records and details of other information held on public registers. We currently charge £75 to provide written responses to environmental information requests.

**Public Register of Contaminated Land Sites**

In accordance with its duty under Section 78R of Part IIA of the Environmental Protection Act 1990, the Council has a Contaminated Land Register. Whilst the register is a public document it should be noted that it is not a list of contaminated or potentially contaminated sites within the Borough. The register merely documents enforcement action, which has been taken by the authority in relation to the 'clean up' of contaminated land.

# 5

# Acid Rain

Acid rain is rain or any other form of precipitation that is unusually acidic. It has harmful effects on plants, aquatic animals, and infrastructure. Acid rain is mostly caused by human emissions of sulfur and nitrogen compounds which react in the atmosphere to produce acids. In recent years, many governments have introduced laws to reduce these emissions.

"Acid rain" is a popular term referring to the deposition of wet (rain, snow, sleet, fog and cloudwater, dew) and dry (acidifying particles and gases) acidic components. A more accurate term is "acid deposition".

Distilled water, which contains no carbon dioxide, has a neutral pH of 7. Liquids with a pH less than 7 are acidic, and those with a pH greater than 7 are bases. "Clean" or unpolluted rain has a slightly acidic pH of about 5.2, because carbon dioxide and water in the air react together to form carbonic acid, a weak acid but unpolluted rain also contains other chemicals.

$$H_2O\ (l) + CO_2\ (g) \rightarrow H_2CO_3\ (aq)$$

Carbonic acid then can ionize in water forming low concentrations of hydronium and carbonate ions:

$$2\ H_2O\ (l) + H_2CO_3\ (aq) \rightleftharpoons CO_3^{2-}\ (aq) + 2\ H_3O^+\ (aq)$$

## EMISSIONS OF CHEMICALS LEADING TO ACIDIFICATION

The most important gas which leads to acidification is sulfur dioxide. Emissions of nitrogen oxides which are oxidized to form nitric acid are of increasing importance due to stricter controls on emissions of sulfur containing compounds. 70 Tg(S) per year in the form of $SO_2$ comes from fossil fuel combustion and industry, 2.8 Tg(S) from wildfires and 7-8 Tg(S) per year from volcanoes.

### Natural Phenomena

The principal natural phenomena that contribute acid-producing gases to the atmosphere are emissions from volcanoes and those from biological processes that occur on the land, in wetlands, and in the oceans. The major biological source of sulfur containing compounds is dimethyl sulfide. Acidic deposits have been detected in glacial ice thousands of years old in remote parts of the globe.

### Human Activity

The principal cause of acid rain is sulfur and nitrogen compounds from human sources, such as electricity generation, factories, and motor vehicles. Coal power plants are one of the most polluting. The gases can be carried hundreds of kilometres in the atmosphere before they are converted to acids and deposited. In the past, factories had short funnels to let out smoke, but this caused many problems locally; thus, factories now have taller smoke funnels. However, dispersal from these taller stacks causes pollutants to be carried farther, causing widespread ecological damage.

## CHEMICAL PROCESSES

### Gas Phase Chemistry

In the gas phase sulfur dioxide is oxidized by reaction with the hydroxyl radical via an intermolecular reaction:

$$SO_2 + OH\cdot \rightarrow HOSO_2\cdot$$

*which is followed by*:

$$HOSO_2\cdot + O_2 \rightarrow HO_2\cdot + SO_3$$

*In the presence of water, sulfur trioxide ($SO_3$) is converted rapidly to sulfuric acid*:

$$SO_3\ (g) + H_2O\ (l) \rightarrow H_2SO_4\ (l)$$

*Nitric acid is formed by the reaction of OH with nitrogen dioxide*:

$$NO_2 + OH\cdot \rightarrow HNO_3$$

### Chemistry in Cloud Droplets

When clouds are present, the loss rate of $SO_2$ is faster than can be explained by gas phase chemistry alone. This is due to reactions in the liquid water droplets.

*Hydrolysis*

Sulfur dioxide dissolves in water and then, like carbon dioxide, hydrolyses in a series of equilibrium reactions.

$$SO_2\ (g) + H_2O \rightleftharpoons SO_2\cdot H_2O$$
$$SO_2\cdot H_2O \rightleftharpoons H^+ + HSO_3^-$$
$$HSO_3^- \rightleftharpoons H^+ + SO_3^{2-}$$

*Oxidation*

There are a large amount of aqueous reactions that oxidize sulfur from S(IV) to S(VI), leading to the formation of sulfuric acid. The most important oxidation reactions are with ozone, hydrogen peroxide and oxygen.

## ACID DEPOSITION

Processes involved in acid deposition (note that only $SO_2$ and $NO_x$ play a significant role in acid rain).

### Wet Deposition

Wet deposition of acids occurs when any form of precipitation (rain, snow, etc.) removes acids from the atmosphere and delivers it to the Earth's surface. This can result from the deposition of acids produced in the raindrops or by the precipitation removing the acids either in clouds or below clouds. Wet removal of both gases and aerosols are both of importance for wet deposition.

### Dry Deposition

Acid deposition also occurs via dry deposition in the absence of precipitation. This can be responsible for as much as 20 to 60 per cent of total acid deposition. This occurs when particles and gases stick to the ground, plants or other surfaces.

## ADVERSE EFFECTS

Acid rain has been shown to have adverse impacts on forests, freshwaters and soils, killing insect and aquatic life-forms as well as causing damage to buildings and having impacts on human health.

### Surface Waters and Aquatic Animals

Both the lower pH and higher aluminum concentrations in surface water that occur as a result of acid rain can cause damage to fish and other aquatic animals. At pHs lower than 5 most fish eggs will not hatch and lower pHs can kill adult fish. As lakes and rivers become more acidic biodiversity is reduced.

Acid rain has eliminated insect life and some fish species, including the brook trout in some lakes, streams, and creeks in geographically sensitive areas, such as the Adirondack Mountains of the United States. However, the extent to which acid rain contributes directly or indirectly via runoff from the catchment to lake and river acidity (*i.e.*, depending on characteristics of the surrounding watershed) is variable.

The United States Environmental Protection Agencyý's (EPA) web stie states: "Of the lakes and streams surveyed, acid rain caused acidity in 75 per cent of the acidic lakes and about 50 per cent of the acidic streams".

### Soils

Soil biology and chemistry can be seriously damaged by acid rain. Some microbes are unable to tolerate changes to low pHs and are killed. The enzymes of these microbes are denatured (changed in shape so they no longer function) by the acid.

The hydronium ions of acid rain also mobilize toxins such as aluminium, and leach away essential nutrients and minerals such as magnesium.

$$2\ H^+\ (aq) + Mg^{2+}\ (clay) \rightleftharpoons 2\ H^+\ (clay) + Mg^{2+}\ (aq)$$

Soil chemistry can be dramatically changed when base cations, such as calcium and magnesium, are leached by acid rain thereby affecting sensitive species, such as sugar maple (Acer saccharum).

### Forests and Other Vegetation

Adverse effects may be indirectly related to acid rain, like the acid's effects on soil or high concentration of gaseous precursors to acid rain. High altitude forests are especially vulnerable as they are often surrounded by clouds and fog which are more acidic than rain. Other plants can also be damaged by acid rain but the effect on food crops is minimized by the application of lime and fertilizers to replace lost nutrients.

In cultivated areas, limestone may also be added to increase the ability of the soil to keep the pH stable, but this tactic is largely unusable in the case of wilderness lands. When calcium is leached from the needles of red spruce, these trees become less cold tolerant and exhibit winter injury and even death.

### Human Health

Scientists have suggested direct links to human health. Fine particles, a large fraction of which are formed from the same gases as acid rain (sulfur dioxide and nitrogen dioxide), have been shown to cause illness and premature deaths such as cancer and other diseases.

### Other Adverse Effects

Acid rain can also cause damage to certain building materials and historical monuments. This results when the sulfuric acid in the rain chemically reacts with the calcium compounds in the stones (limestone, sandstone, marble and granite) to create gypsum, which then flakes off.

$$CaCO_3\ (s) + H_2SO_4\ (aq) \rightleftharpoons CaSO_4\ (aq) + CO_2\ (g) + H_2O\ (l)$$

This result is also commonly seen on old gravestones where the acid rain can cause the inscription to become completely illegible. Acid rain also causes an increased rate of oxidation for iron. Visibility is also reduced by sulfate and nitrate aerosols and particles in the atmosphere.

## AFFECTED AREAS

Particularly badly affected places around the globe include most of Europe (particularly Scandinavia with many lakes with acidic water containing no life and many trees dead) many parts of the United States (states like New York are very badly affected) and South Western Canada. Other affected areas include the South Eastern coast of China and Taiwan.

### Potential Problem Areas in the Future

Places like much of South Asia (Indonesia, Malaysia and Thailand), Western South Africa (the country), Southern India and Sri Lanka and even West Africa (countries like Ghana, Togo and Nigeria) could all be prone to acidic rainfall in the future.

## PREVENTION METHODS

### TECHNICAL SOLUTIONS

In the United States, many coal-burning power plants use Flue gas desulfurization (FGD) to remove sulfur-containing gases from their stack gases. An example of FGD is the wet scrubber which is commonly used in the U.S. and many other countries. A wet scrubber is basically a reaction tower equipped with a fan that extracts hot smoke stack gases from a power plant into the tower. Lime or limestone in slurry form is also injected into the tower to mix with the stack gases and combine with the sulfur dioxide present.

The calcium carbonate of the limestone produces pH-neutral calcium sulfate that is physically removed from the scrubber. That is, the scrubber turns sulfur pollution into industrial sulfates. In some areas the sulfates are sold to chemical companies as gypsum when the purity of calcium sulfate is high.

In others, they are placed in landfill. However, the effects of acid rain can last for generations, as the effects of pH level change can stimulate the continued leaching of undesirable chemicals into otherwise pristine water sources, killing off vulnerable insect and fish species and blocking efforts to restore native life. Automobile emissions control reduces emissions of nitrogen oxides from motor vehicles.

### International Treaties

A amount of international treaties on the long range transport of atmospheric pollutants have been agreed *e.g.* Sulphur Emissions Reduction Protocol under the Convention on Long-Range Transboundary Air Pollution.

### Emissions Trading

Every current polluting facility is given or may purchase on an open market an emissions allowance for each unit of a designated pollutant it emits. Operators can then install pollution control equipment, and sell portions of their emissions allowances they no longer need for their own operations, thereby recovering some of the capital cost of their investment in such equipment. The intention is to give operators economic incentives to install pollution controls. The first emissions trading market was established in the United States by enactment of the Clean Air Act Amendments of 1990. The overall goal of the Acid Rain Programme established by the Act is to achieve significant environmental and public health benefits through reductions in emissions of sulfur dioxide ($SO_2$) and nitrogen oxides ($NO_x$), the primary causes of acid rain. To achieve this goal at the lowest cost to society, the programme employs both regulatory and market based approaches for controlling air pollution.

## ACID RAIN AND ITS IMPACT

Acid rain is a widespread term used to describe all forms of acid precipitation (rain, snow, hail, fog, etc.). Atmospheric pollutants, particularly oxides of sulphur and nitrogen, can cause precipitation to become more acidic when converted to sulphuric and nitric acids, hence the term acid rain. Acid deposition, acid rain and acid precipitation all relate to the chemistry of air pollution and moisture in the atmosphere. Scientists generally use the term acid deposition but all three terms relate to the same issue.

The term acid rain was first used by Robert Angus Smith, a scientist working in Manchester in the 1870s. The problem of acid rain is hence not a new one but the nature of the problem has changed from being a local problem for towns and cities to being an international problem. In Smith's time, acid rain fell both in towns and cities whilst today pollutants can be transported thousands of kilometres due to the introduction of tall chimneys dispersing pollutants high into the atmosphere.

Precipitation is naturally acidic because of carbon dioxide in the atmosphere. The burning of fossil fuels (coal, oil and gas) produces sulphur dioxide and nitrogen oxides which can increase the acidity of rain or other precipitation. Sources of sulphur dioxide and oxides of nitrogen may be natural such as volcanoes, oceans, biological decay and forest fires, or may arise from combustion sources. The increasing demand for electricity and the rise in the number of motor vehicles in recent decades has meant that emissions of acidifying pollutants have increased dramatically from human sources, particularly since the 1950s. Emissions of such pollutants are heavily concentrated in the northern hemisphere, especially in Europe and North America. As a result, precipitation is generally acidic in these countries.

In the 1970s and 1980s, Scandinavian countries began to notice the effects of acid deposition on trees and freshwaters. Much of the pollution causing this damage was identified as being transported from other more polluting countries. Acid rain became an international concern.

The pH scale is used to measure the acidity of acid rain which is determined by the hydrogen ion content (H+). This scale was invented by a Danish scientist called Sorenson in 1909. The pH scale ranges from 0, which is strongly acid, to 14 which is strongly alkaline, the scale point 7 being neutral. The pH scale is logarithmic rather than linear, so there is a ten-fold increase in acidity with each pH unit, such that rainfall with pH 5 is ten times more acidic than pH 6, rainfall with pH 4 is 100 times more acidic than pH 6 and rainfall with pH 3 is

1000 times more acidic than pH 6. Acid rain became particularly prominent as a media issue during the 1980s. However, during the 1970s many countries started to notice changes in fish populations in lakes and damage to certain trees. By the late 1970s concern led to international efforts to identify the causes and effects of long-range (transboundary) transport of air pollutants, and thus during the 1980s much research was conducted in Europe and North America. International legislation during the 1980s and 1990s has led to reductions in sulphur dioxide emissions in many countries but reductions in emissions of nitrogen oxides have been much less. Although media attention has shifted towards other environmental issues such as global warming, acid rain continues to be a problem at the beginning of the 21$^{st}$ century.

## ACID DEPOSITION

Acid rain is a general name for many phenomena including acid fog, acid sleet, and acid snow. Although we associate the acid threat with rainy days, acid deposition occurs all the time, even on sunny days.

Acid Deposition is the scientific term used to describe"Acid Rain". When atmospheric pollutants such as sulphur dioxide and nitrogen oxides mix with water vapour in the air, they are converted to sulphuric and nitric acids. These acids make the rain acidic, hence the term"acid rain". Rain returns the sulphur and nitrogen acids to Earth, and in high concentrations, can cause damage to natural environments including forests and freshwater lakes. This form of acid deposition is known as wet deposition. A second method of acid deposition is known as dry deposition. Whilst wet deposition involves the precipitation of acids, dry deposition occurs when the acids are first transformed chemically into gases and salts, before falling under the influence of gravity back to Earth. Sulphur dioxide, for example, is deposited as a gas and as a salt.

The gases present in acid deposition are found to occur naturally in the environment. They are given off from a number of sources including volcanic eruptions and the rotting of vegetation. They become a problem when humans produce the gases in large amounts, and at high concentrations by the burning of fossil fuels. The distances that pollutant gases travel means that acid deposition is an international or transboundary problem. This means that acid pollutants are not necessarily deposited in the same country where they were produced.

## ACIDIC EMISSIONS

Rain water is naturally acidic as a result of carbon dioxide dissolved in water and from volcanic emissions of sulphur. However, it is the chemical conversion of sulphur and nitrogen emissions from power stations, factories, vehicles and homes, where fossil fuels are burnt, that we call acid rain. These waste gases are carried by the wind, sometimes over long distances, and can in time be converted into sulphuric and nitric acids. Natural sources of sulphur dioxide ($SO_2$) include releases from volcanoes, oceans, biological decay and forest fires. Actual amounts released from natural sources in the world are difficult to quantify; in 1983 the United Nations Environment Programme estimated a figure of between 80 million and 288 million tonnes of sulphur oxides per year. Man-made sulphur dioxide emissions result from combustion or burning of fossil fuels, due to varying amounts of sulphur being present in these fuels. Worldwide emissions of $SO_2$ are thought to be around 69 million tonnes per year.

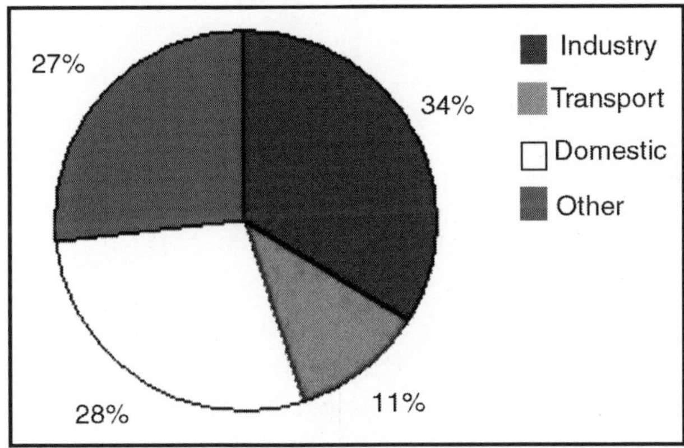


**Fig.** Sulphur Dioxide emissions 1615 Thousand Tonnes.

Levels of sulphur dioxide from combustion sources in the UK have declined in recent decades. Between 1970 and 1999, UK sulphur dioxide emissions fell by 82 per cent due to recession, restructuring of industry, substitution of fuels (for example natural gas) and air pollution control technology. Power station emissions fell by 73 per cent over the same period, but the percentage of UK emissions from power stations has actually increased to 65 per cent of the 1999 total compared to 45 per cent of the total in 1970.

Natural sources of nitrogen oxides (NOx) include volcanoes, lightening strikes and biological decay. Estimates range from between 20 million and 90 million tonnes per year NOx released from natural sources, compared to around 24 million tonnes from human sources world-wide. Nitrogen oxides are produced when fossil fuels are burned.

The major sources of NOx in the UK in 1999 were road transport (44 per cent), power stations (21 per cent) and industry (including iron and steel, and refineries) (12 per cent). Emissions of nitrogen oxides from road transport increased during the 1970s and 1980s before decreasing again during the 1990s. For example, in 1970, emissions of NOx from road transport in the UK were 0.769 million tonnes by in 1990 they had risen to over 1.31 million tonnes NOx. Since then, however, emissions from transport have been declining due to improvements in vehicle technology, such as the use of catalytic converters, and the use of cleaner fuels. In 1999 they were 0.714 million tonnes, lower than in 1970.

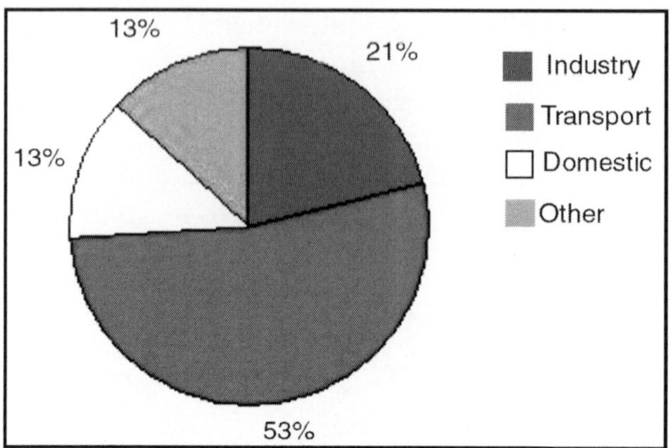


**Fig.** Nitrogen Oxides Emissions 1753 Thousand Tonnes ($No_2$ and No)

The geographical distribution of human acidic emission sources is not even. Nitrogen and sulphur emission sources are heavily concentrated in the Northern Hemisphere, particularly in Europe and North America. As a result, precipitation is generally more acidic in these countries, with an acidity in the range pH 4.1 to pH 5.1. 'Normal' or 'unpolluted' rainfall has a pH of 5.6.

## BUILDINGS

Buildings have always been subject to attack by weathering; the effects of rain, wind, sun, and frost. Acid rain can accelerate the rate of this damage. Throughout the world, emissions of sulphur dioxide and nitrogen oxides contribute to the international problem of acidification. Acid deposition affects most materials to some degree. Limestone, marble and sandstone are particularly vulnerable, whilst granitic-based rocks are more resistant to acidity. Other vulnerable materials include carbon-steel, nickel, zinc, copper, paint, some plastics, paper, leather and textiles. Stainless steel and aluminium are more resistant metals. Structural damage to underground pipes, cables and foundations submerged in acid waters can also occur, in addition to damage to buildings, bridges and vehicles upon ground.

Whilst dry deposition contributes to the corrosion of materials, in most areas with substantial rainfall it is the effect of wet deposition on building surfaces which is more damaging. Building stone can be damaged when calcium carbonate in stone dissolves in acid rain to form a crust of calcium sulphate or gypsum. The sulphated layers are more readily washed away by rainfall or removed by the action of frost and other weather conditions, resulting in more stone being exposed. This permanent alteration of stone surfaces by the action of acid deposition is known as sulphation.

Sulphur dioxide is the main pollutant in respect to corrosion but others also take their toll, including nitrogen oxides, carbon dioxide, ozone (on organic materials) and sea salt from sea spray. Research has revealed that when nitrogen dioxide is present with sulphur dioxide, increased corrosion rates occur. This is because the nitrogen dioxide oxidises the sulphur dioxide to sulphite thereby promoting further sulphur dioxide absorption. The interactions between building materials and pollutants are very complex and many variables are involved. Deposition of pollutants onto surfaces depends on atmospheric concentrations of the pollutants and the climate and microclimate around the surface. Once the pollutants are on the surface, interactions will vary depending on the amount of exposure, the reactivity of different materials and the amount of moisture present. The last factor is particularly important because the sulphur dioxide that falls as dry deposition is oxidised to sulphuric acid in the presence of moisture on the surface.

The effects of acid deposition on modern buildings are considerably less damaging than the effects on ancient monuments. Limestone and carbonate stones which are used in most heritage buildings in the UK are the most vulnerable to corrosion and need continued renovation. Cathedrals such as York Minster and Westminster Abbey have been severely eroded in recent years.

A five-year research programme in the UK has suggested that if sulphur dioxide emissions were reduced by 30 per cent, savings over 30 years could be as high as £9.5 billion. Many other countries have noticed an acceleration of damage to their cultural heritage. The Taj Mahal in India, the Colosseum in Rome and monuments in Krakow, Poland are continuing to deteriorate. In Sweden, medieval stained glass windows are thought to have been affected by acid rain.

**CARS**

Vehicle nitrogen oxides emissions make a significant contribution to acid deposition. The major gases leading to the formation of acid deposition are sulphur dioxide ($SO_2$) and nitrogen oxides (NOx). Vehicles do not produce much sulphur dioxide as petroleum contains very little sulphur. In 1999, vehicles contributed 1 per cent of the UK sulphur dioxide emissions whilst power stations being the major source of the total UK sulphur dioxide emissions contributed 65 per cent. Nitrogen oxides from vehicles, however, are a major contributor to acid deposition. In 1999, road transport accounted for 44 per cent of nitrogen oxides emissions in the UK.

A motor vehicle produces air pollutants when fuel is burnt to give mechanical power. In a totally efficient combustion process, hydrocarbons in fuel, and oxygen will react to form carbon dioxide and water. However, the combustion process is never perfect; some of the hydrocarbon fuel is only partially burnt forming carbon monoxide and water, whilst some of the hydrocarbons are not combusted at all. These can, and often are, emitted from the exhaust as unburned hydrocarbons. During the combustion process the temperature can reach 2500°C. At these temperatures nitrogen and oxygen from the air in the combustion chamber react to form nitrogen oxides.

Nitrogen oxides pollution from petrol-driven vehicles can be significantly reduced by fitting a catalytic converter to the exhaust system. This is a relatively low cost method of pollution control which has little effect on vehicle performance and fuel consumption. All new cars sold in Britain from January 1993 onwards have catalytic converters.

## CHIMNEYS AND STACKS

Although media attention has shifted towards other environmental issues such as global warming, acid deposition continues to be a problem at the beginning of the $21^{st}$ century. In fact, acid rain is not a new problem at all. It dates from the middle of the $19^{th}$ century when a Scottish chemist, Robert Angus Smith, began to study the effect of air pollution in Manchester, UK, and used the term 'acid rain' to describe his findings. What is very new is the scale of the problem. In Smith's time, acid rain fell both in towns and cities and downwind from them, but now, the pollution is spread far and wide, within and between nations. It has now become an international or transboundary problem.

In December 1952 the Great London Smog accounted for over 4,000 deaths due to high levels of smoke and sulphur dioxide pollution. To prevent such a catastrophe ever occurring again, the UK Government passed two Clean Air Acts, the first in 1956 and the second in 1968. The Clean Air Act of 1968 brought in the basic principle for the use of tall chimneys or stacks for industries burning solid, liquid or gaseous fossil fuels. At the time of this legislation it was recognised that smoke pollution could be controlled, but that sulphur dioxide removal was generally impracticable. Hence, the higher the chimney, the better the dispersal of the air pollution.

Urban air quality improved following the Clean Air Acts. Research into and monitoring of airborne pollutants reinforced the view that tall chimneys ensured adequate dispersion in the atmosphere, reducing significantly the amount of sulphur dioxide and nitrogen oxides pollution at ground level. However, concern grew regarding the eventual fate in the atmosphere of the pollution emitted. Aircraft measurements in the 1970s revealed that as much as 70 per cent of sulphur dioxide emitted in the UK was leaving the country. It became clear that this pollution was the primary cause of the acidification of thousands of freshwater lakes in Scandinavia, and the damage to hundreds of thousands of trees.

Whilst chimneys and stacks are still used today to disperse pollution from fossil fuel burning, sulphur dioxide and nitrogen oxides pollution from power stations and industry has been falling since the 1990s as a consequence of improved emission control technology in response to international agreements on transboundary pollution negotiated in the 1970s and 1980s.

## CRITICAL LOADS

Critical loads have been defined as: "the highest load that will not cause chemical changes leading to long-term harmful effects in the most sensitive ecological systems". Critical loads are the maximum amount of pollutants that ecosystems can tolerate without being damaged. The definition has been redrafted in order to fit specialist areas of interest, most particularly the acidification of freshwater and soils.

The concept behind critical loads is based upon a dose-response relationship where the threshold of harmful response (within the ecosystem) is triggered by a certain load of pollutant - the critical load. However, it is not always easy to apply without careful consideration of the nature of the affected ecosystem and the threshold effects of harmful pollutants. In order for critical loads to be used, target loads need to be set for different areas in order to try and halt the acidification processes. Target loads have been defined as "the permitted pollutant load determined by political agreement". Therefore, target loads can be either higher or lower than the critical load values.

For example, the target load may be lower so as to give a safety margin or the target load may be higher for economic reasons. The reasoning behind this is that critical loads only show where there is an acidification problem and to what degree damage is occurring. Target loads are used in order that emissions can be reduced just as to meet the targets and limit the amount of damage.

The critical loads for total acidity of sulphur and nitrogen need to be determined so that a coherent international agreement can be reached with regard to abatement policies. There are numerous methods that are available for obtaining critical loads. In order to obtain values for the critical loads, an ecosystem has to be chosen and then a suitable indicator species is selected to represent the ecosystem. A chemical limit is subsequently defined as the concentration at which the indicator species will die. In forests the indicators are trees, and in freshwaters they are fish.

Through the Convention on Long-Range Transboundary Pollution, member countries of the United Nations Economic Commission for Europe (UNECE) agreed that the critical loads approach provided an effective scientific approach for devising strategies for the abatement of acid deposition. The UK Government accepted that the critical loads approach was the best way to establish abatement strategies in relation to sulphur dioxide and nitrogen oxides emissions. It was recognised that critical loading maps were essential in providing information on the geographical distribution of pollutant-sensitive locations. The Critical Loads Advisory Group (CLAG) was set up to produce critical load maps for the United Kingdom. The National Centre for Critical

Loads Mapping was subsequently established at the Institute for Terrestrial Ecology, now the Centre for Ecology and Hydrology.

Critical load maps of soil acidity have been produced for the UK at a grid resolution of 1 km squares. Maps for the critical acidity of freshwater environments are based on a single water sample from a single site in each of the 10 km squares used, assumed to be the most sensitive surface water within the grid square. These critical loads maps, when combined with acid deposition values, produce exceedance maps which show where and by how much the critical loads are being exceeded.

Maps are available for soils in the UK relating to acidity and sulphur deposition showing areas that are sensitive to acidification. These correspond to areas where there have been reports of acidification. In the UK a national target load map for the year 2005 has been produced for soils on a 20km by 20km grid system, showing the target loads that need to be met for such areas.

In 1997, critical loads for acidification were exceeded in 71 per cent of UK ecosystems. As sulphur deposition continues to fall, this value is expected to fall to below half by 2010, when nitrogen deposition will dominate. Critical loads for eutrophication (nutrient depletion) in 1997 were exceeded in about a quarter of UK 1km by 1km squares with sensitive grasslands and a little over half with heathland. Again, this is expected to decline over the next 10 to 15 years.

## DOING OUR BIT FOR ACID RAIN

Every day we need energy to heat and light our homes, and to drive our cars. Most of the energy we use for this comes from the burning of fossil fuels which release pollutant gases, such as sulphur dioxide and nitrogen oxides, into the atmosphere.

Through international protocols and the use of emission control technology introduced since the end of the 1980s, countries have witnessed reductions in acidic emissions from power stations and road vehicles. Nevertheless, it is only by reducing our dependence on fossil fuels and energy consumption that a long-term reduction in acid deposition can be maintained.

The individual has little influence on how his/her energy is produced, for example by coal or gas fired power stations, or alternatively by wind or solar power. However, the individual does have control on how he or she uses that available energy. Through the implementation of simple measures we can all effectively bring about a reduction in energy consumption, and help to reduce the amount of emissions which may lead to acid rain. Using less energy also means savings on fuel bills.

Heating (space and water) accounts for approximately 25 per cent of UK energy use. On average 55 per cent of fuel bills are spent on space heating, but in an uninsulated house about half of this heat escapes through the walls! Water heating can account for up to 20 per cent of the average fuel bill but we are often wasteful of this resource. Energy use in these two areas can be cut whilst still providing the heating that you require. Energy-saving light bulbs are now widely available in supermarkets and electrical stores.

The initial cost of energy saving light bulbs are relatively high compared to standard light bulbs, but the lower running costs and longer lifetimes mean that the initial cost can be recouped within a couple of years. The energy use and efficiency of household appliances, such as fridges, freezers, cookers, washing machines and televisions depends on the age, model and manufacturer. In the UK 20 per cent of electricity is used by domestic appliances. Retailers in Europe are required to label all new fridges and freezers with an eco-label.

Transport is the fastest growing energy-consumption sector in the UK and the number of cars on the road is projected to increase by 17 per cent by 2010. It is therefore an area that requires great attention to reduce fuel consumption and hence pollution. Transport pollution is emitted at ground level from a mobile source, and is therefore a larger problem than other pollution sources. As an alternative to driving the car, walk, cycle or use public transport where it is suitable and safe for you to do so, particularly for short trips where using the car is not really necessary and an alternative exists. Where walking or cycling is impractical, consider taking public transport if it is available and convenient to use. A bus full of passengers is more than twice as fuel-efficient as a family car. In addition, if you and your friends drive to work consider the option of car sharing.

Everyone contributes to national and global emissions of pollutant gases, but it is not only governments that can take action to reduce the environmental damage caused. For their policies to work effectively and for their targets to be achieved the actions of the individual are required. The cumulative energy reductions by individuals would reduce the need for energy consumption, conserve stocks of raw materials such as coal, oil and gas, and bring about a reduction in pollutant gas emissions which cause acid rain.

## FOSSIL FUELS

Conventional power stations burn coal, oil or gas to produce electricity. Coal, oil and gas are called fossil fuels because they form over millions of years through the decay, burial and compaction of rotting vegetation on land (coal), and marine organisms on the sea floor (oil and gas). Burning fossil fuels in this way releases large quantities of sulphur dioxide and nitrogen oxides which can cause acid rain.

Coal is a solid fuel formed over millions of years by the decay of land vegetation. Over time, successive layers become buried, compacted and heated, a process through which the deposits are turned into coal. Coal is widely used in the generation of electricity because it is a highly concentrated energy source. However, it is not a particularly "clean" fuel, releasing more acidic pollution than either oil or gas. Coal was the first fossil fuel to be exploited on a large scale during the $19^{th}$ century with the beginning of the Industrial Revolution. Before the commercial introduction of electricity, coal was primarily used in industrial boilers to create steam energy to power machinery.

Oil is formed from the remains of marine microorganisms (microscopic animals and plants) deposited on the sea floor. As they accumulate over millions of years they gradually infiltrate the microscopic cavities of the sea floor sediment and rock where they decay. The resulting oil remains trapped in these spaces, forming oil reserves which can be extracted through large drilling platforms. The use of oil increased significantly after the Second World War. In the early 1970s, approximately 40 per cent of global fossil fuel use came from oil, but during the 1990s this figure has decreased. Improved energy efficiency has caused oil consumption to decline in many developed, industrialised countries, as well as shifts to other fuels such as natural gas and nuclear energy. Decreasing use of oil is also resulting from tougher environmental restrictions concerning its use in some regions.

Natural gas is formed in the same way as oil, from the remains of marine microorganisms. From the mid-1960s, up until the present day, there has been a dramatic increase in the amount of proven reserves of natural gas. Consequently, natural gas has become the fastest-growing energy resource. The present global use of natural gas is approximately 20 per cent of all fossil fuel use, and this figure is predicted to rise in the future. Natural gas provides an alternative to oil or coal in the provision of energy, and in terms of acidic pollution it is a cleaner fuel. Some estimates indicate the reserves of natural gas may be available for up to 400 years.

## FRESHWATER LAKES

Natural acidification of freshwater environments has been taking place since the last ice age. However, the recent rapid acidification of many of lakes throughout the world can not be attributed to natural causes, but instead to the effects of acidic pollution from the burning of fossil fuels by mankind.

The acidity of water of freshwater lakes and streams is predominantly determined by the soil and rock types of an area, since 90 per cent of the water entering these water courses has passed through the ground. Only 10 per cent of water in lakes and streams comes directly from rainfall. Consequently, areas that are most susceptible to freshwater acidification have an acidic geology such as granite and a peat-based soil. Areas that are affected by acidification include Scandinavia, Central Europe, Scotland, Canada, and the United States. Nevertheless, acid rain entering freshwater that is naturally acidic as a consequence of the underlying geology can place additional stresses on ecosystems dependent upon such freshwater environments. Lakes and streams that are generally regarded as acidified contain very nutrient-poor water.

Acid rain can enter the watercourse either directly or more usually through the catchment. If the catchment has alkaline-rich soil then the acid rain may be neutralised, and water entering the lake is of low acidity. However, if the catchment has a thin, alkaline-poor soil then acid water is passed to the lake. Acidification of a lake occurs over time. At first the natural buffering capacity of the lake neutralises the additional acidity entering the lake but at some point, the lake buffering capacity runs out and the acidity of the water increases rapidly. In time, the lake water stabilises at a certain acidity, maintaining a small number of species of plants and animals, but usually lacking many fish.

In addition to acid rain, a number of other factors can influence the acidity of freshwater, including the introduction of livestock into the catchment area and the use of nitrogen fertilizers.

The onset of acidification brings about a clearer bluer water body due to the settling out of decaying organic matter. Whilst the total amount of living matter remains largely unchanged, the diversity of different species drops considerably. Rushes thrive in acidified freshwater. White Sphagnum moss may invade lakes and form a thick green carpet over the bottom of the lake on account of the clearer waters allowing more light to reach the moss. Soft bodied animals such as leeches, snails and crayfish are early victims, often being one of the first signs of the commencement of acidification. Few insect species are very resistant to acidification and species such as mayfly disappear even under moderate acidification. However, species such as dragonfly larvae, water beetle and bloodworms can grow abnormally large in their population size when competition is removed. Salmon, trout and roach are particularly at risk from freshwater acidification, pike and eel being relatively resistant.

Acidification of freshwaters was first identified in Scandinavia during the early 1970s. Since then thousands of lakes and rivers there have become acidified. Much of this freshwater acidification has been the result of transboundary pollution blowing across the North Sea from the UK. Sweden has over 85,000 lakes that are greater than one hectare in size. Of these, 14,000 are acidified by man-made air pollution, 4,000 being severely acidified. Acid sensitive species are absent from around 40 per cent of Sweden's rivers and streams.

## IMPACTS OF ACID RAIN

Acid deposition, more commonly known as acid rain, results from man-made emissions of sulphur dioxide and nitrogen oxides through the burning of fossil fuels for energy and transportation. Acid rain has negative effects on the environment in which we live. Since acid rain is a transboundary pollution problem, acidic emissions produced by one country can be deposited in another. Sweden and Norway, for example, both receive more than 90 per cent of their sulphur pollution from abroad.

If large quantities of acid rain are deposited they may have detrimental consequences for wildlife, forests, soils, freshwater and buildings. Acid rain acidifies the soils and waters where it falls, killing off plants and animals. Surface water acidification can lead to a decline in, and loss of, fish populations and other aquatic species including frogs, snails and crayfish. Acid rain affects trees, usually by weakening them through damage to their leaves. Certain types of building stone can be dissolved in acid rain.

## INTERNATIONALAGREEMENTS

Because pollutants can be carried many hundreds of kilometres by winds, acid pollutants emitted in one country may be deposited as acid rain in other countries. Acid deposition has become an international problem. This problem is highlighted by the fact that emissions of a particular pollutant from one country does not equal the deposition of that pollutant in the same country. Some countries emit small quantities of pollutants yet deposition can be several times greater. Other countries emit more pollution than is deposited in their country because of prevailing wind directions.

In 1979, the United Nations Economic Commission for Europe (UNECE) implemented the Convention on Long-Range Transboundary Pollution. In 1985 most UNECE members adopted the Protocol on the Reduction of Sulphur Emissions, agreeing to reduce sulphur dioxide emissions by 30 per cent (from 1980 levels) by 1993. This was called the 30 per cent club. All of the countries that signed the Protocol achieved this reduction, and many of those that did not sign, have met these reductions. In June 1994, a number of European countries signed the Second Protocol for sulphur. Most of the western European countries agreed to reduce sulphur emissions by between 70 and 80 per cent by the year 2000 whilst eastern European countries generally have a lower target of between 40 and 50 per cent.

Overall, emissions of sulphur dioxide in Europe are estimated to have fallen by 25-30 per cent between 1980 and 1990, and by 40 per cent by the year 2000. Further falls in sulphur dioxide emissions are expected over the next decade.

The Sofia Protocol for reducing nitrogen oxide emissions was adopted in 1988. This required all countries that signed the Protocol to stabilise emissions of nitrogen oxide, although some countries committed themselves to 30 per cent reductions by 1998. However, many of these countries are unlikely to meet these targets, due to increases in road traffic, despite European Union legislation requiring cars built after 1993 to be fitted with a catalytic converter. In 1988 a Directive was introduced for European Community (EC) countries which required power stations to reduce emissions of sulphur dioxide and nitrogen oxides. For the UK, reductions of sulphur dioxide by 60 per cent by 2003 and nitrogen oxides by 30 per cent by 1998 have been set. The UK is well on course to exceed both targets through new gas-fired power stations replacing coal fired power stations, and flue gas desulphurisation equipment fitted to some of the existing coal-fired power stations.

The most recent UNECE Convention on Long Range Transboundary Air Pollution protocol was signed by 27 countries in December 1999. The Gothenburg Protocol, designed to Abate Acidification, Eutrophication and Ground-level Ozone aims to cut emissions of four pollutants: sulphur dioxide ($SO_2$), nitrogen oxides (NOx), volatile organic compounds (VOCs), and ammonia ($NH_3$), by setting country-by-country emission ceilings to be achieved by the year 2010.

## INDUSTRIAL EMISSION CONTROLS

Acidic emissions of sulphur dioxide and nitrogen oxides arise from many industrial sources as a result of combustion processes. In the UK power stations contributed 65 per cent of all sulphur dioxide emitted in the UK in 1999. Other industries were responsible for 22 per cent. Industries also emit nitrogen oxides which can also cause rainfall to become more acidic. While road transport is the major source of nitrogen oxides in the UK, power stations accounted for 21 per cent and other industries 13 per cent in 1999. There are many technologies which can be used in industry to reduce the emissions of pollutants to the atmosphere and these can be applied before, during or after combustion.

Examples of pre-combustion sulphur control technology (removing sulphur before burning) include coal scrubbing and oil desulphurisation. Another removal process is to change the design of the boiler and to install pressurised fluidised bed combusters (FBC) which removes sulphur from coal during the burning process. Another process which removes sulphur dioxide from coal during combustion is the Integrated Gasification Combined Cycle. Coal is gasified under pressure with a mixture of air and steam which results in the formation of gas which can then be burned to produce electricity.

One of the post-combustion sulphur controls (removing sulphur after burning) is Flue Gas Desulphurisation (FGD). In FGD processes, waste gases are scrubbed with a chemical absorbent such as limestone to remove sulphur dioxide. There are many different FGD processes, the main ones being the limestone-gypsum process and the Wellman-Lord regenerative process. The limestone-gypsum FGD involves mixing limestone and water with the flue gases to produce a slurry which absorbs the sulphur dioxide. The slurry is then oxidised to calcium sulphate (gypsum) which can then be used in the building trade.

Unlike sulphur, it is not possible to reduce the nitrogen content of the fuel before combustion by physical cleaning as it is combined within the organic matter of the fuel, and at present there are no commercially available methods to reduce organic nitrogen. Instead, nitrogen oxides can be removed during combustion. Low nitrogen oxides burners ensure that the fuel is burnt in low oxygen concentrations, such that any nitrogen oxides produced are reduced to nitrogen gas. Once initial combustion has taken place, further air is added to the combustion chamber to ensure that the fuel is completely burnt. Advanced low nitrogen oxides burners can reduce emissions by up to 30 per cent. Such burners can be installed on either new or existing combustion plants.

Emissions of nitrogen oxides, like for sulphur dioxide, can also be reduced by treating the flue gases. One method involves mixing the flue gases with ammonia, converting the nitrogen oxides to nitrogen and water. This process is suitable for fitting to existing plant and new build applications, and can achieve an emissions reduction of up to 80 to 90 per cent.

Some fuels, for example natural gas, are naturally less polluting in terms of acidic emissions, whilst traditional coal power generation is more polluting, depending on the amount of sulphur there is in the coal being burnt. To help reduce atmospheric emissions of sulphur dioxide and nitrogen oxides in the UK, many of the more recent power stations have been built to operate on gas rather than coal.

## INDUSTRY AND POWER GENERATION

Industry and power generation are main sources of sulphur dioxide emissions, the precursor for sulphuric acid in acid rain. In the UK power stations and all other types of industry account for 90 per cent of all sulphur dioxide pollution. During the Industrial Revolution industries were often located in urban areas. Following the UK Clean Air Acts in the 1950s and 1960s, and with the decline in heavy industry, few large industries and power stations are located in towns and cities today. Many large industries are now located in the more rural areas of the UK. However, most urban areas have some smaller industries and possibly a power station. The larger industrial sources, even though located out of town, may still have environmental impacts on urban areas. If a town or city lies in the prevailing wind path of these industries, it may still experience some acid deposition, causing faster erosion of certain building materials and damage to urban vegetation.

Power stations contribute significantly to the total emissions of nitrogen oxides in the UK. In 1999, 21 per cent of nitrogen oxides came from this source and a further 13 per cent from other industries, iron and steel and refineries. The major source of nitrogen oxides pollution in the UK is now road transport (44 per cent). Like sulphur dioxide, nitrogen oxides are also converted into acidic compounds when combined with water in the atmosphere, and contribute to acid rain.

## LIMING

The only sure way to prevent acidification of freshwater bodies is to reduce the emissions of acid pollutants in the first place. There is a relationship between sulphur and nitrogen emissions, acid deposition, acids in water run-off and loss of alkalinity. If acidification of freshwaters is to be prevented then the deposition of sulphur and nitrogen needs to be reduced. The technical means are available to reduce emissions, such as flue gas desulphurisation, low nitrogen oxides burners, use of low sulphur coal and oil and increasing energy efficiency.

At present the main way of reversing acidification in freshwaters is liming the water body or its surrounding catchment. The main liming method is to add the lime directly to the water body. Liming of water directly, however, causes aluminium and other metals to come out of solution and fall to the bottom of the lake, causing toxicity problems for organisms living on the lake bed. Lime can be added to the catchment, although this can have an adverse effect on wetland species of plants.

The advantages, however, are that the effects are longer lasting and metals are prevented from leaching into the lake water from the soil. The effects of liming are almost entirely favourable within the lake. The alkalinity of the lake is increased, the pH increased and heavy metal concentrations decrease back to within safe limits for fish life. The number of species of fish and plankton increase as does the total production of living matter.

Acidified lakes in Sweden have been restored in the short term by liming. Each year thousands of tonnes of limestone are sprayed on Swedish lakes and watercourses, by means of trucks, boats or helicopters. Liming on such a large scale, however, is expensive, costing million of pounds.

Liming provides only a temporary solution, hence it is far better to attack the source of the problem by reducing emissions of acidifying pollutants.

## MEASURING ACID RAIN

The pH scale is used to measure the acidity or alkalinity of an aqueous solution and is determined by the hydrogen ion content (H+). This scale was invented by a Danish scientist called Sorenson in 1909. The pH scale ranges from 0, which is strongly acid, to 14 which is strongly alkaline, the scale point 7 being neutral. Examples of solutions with differing pH values include car battery acid (pH 1), lemon juice (pH 2), beer (pH 4), natural rain (pH 5-6), milk (pH 6), washing-up liquid (pH 7), seawater (pH 8), milk of magnesia (pH 10) and ammonia (pH 12),

The pH scale is logarithmic rather than linear, and so there is a ten fold increase in acidity with each pH unit, such that rainfall with pH 5 is ten times more acidic than pH 6, rainfall with pH 4 is 100 times more acidic than pH 6 and rainfall with pH 3 is 1000 times more acidic than pH 6.

Rainfall acidity is measured in pH units.'Normal' or'unpolluted' rainfall has a pH of 5.6. This is slightly acidic due to the presence of carbon dioxide in the atmosphere which forms weak carbonic acid in water. It is not uncommon for acidified rainwater to have a pH of 4, about 30 times as acidic as normal rainwater.

## MODELLING ACID RAIN

Because we can't monitor acid rain in every place where it occurs, models are used to simulate acid deposition and the likely impacts from acidification over large areas. Models, which run on computers, can be used to estimate the dispersion of air pollutants through the atmosphere. These models are called dispersion models. Complex dispersion models used to predict acid rain simulate the emission, transport, chemical changes and deposition of pollutants. They track the pollutants from their source to the points of deposition.

Models have been used to simulate acid deposition throughout Europe. Modellers from the European Monitoring and Evaluation Programme (EMAP) produce maps of pollutant concentrations and depositions for the whole of Europe. They also produce tables which quantify the contribution of each country to the total European deposition. Monitoring networks provide measured information on acid rain to compare with model results and confirm model reliability.

## MONITORING ACID RAIN

The simplest way to monitor the acidity of acid rain (from wet deposition) is to collect rain samples and to measure the pH of the water. Since 1987, such monitoring of acid rain has taken place in and around Greater Manchester in the UK, the Greater Manchester Acid Deposition Survey (GMADS). Rain collectors have been sited in Greater Manchester following strict criteria. The collectors must be 100m away from small point sources, 100m from small mobile sources, 1km from major roads, 5km from large surface works, and 10km from large point sources.

Across the UK, the Acid Deposition Monitoring Network monitors the composition and acidity of precipitation at 32 sites to provide an accurate measurement of pollutant deposition in rain and snow to assist in implementing a critical loads approach to environmental protection. All precipitation samples are analysed for conductivity, pH, sulphate, nitrate and chloride concentration, and a variety of metal ions (ammonium, sodium, calcium, potassium, and magnesium). Combining these measurements of acid deposition with the estimated critical loads in the form of maps can highlight those areas where the tolerance to acidity is being exceeded, and where damage from acidification is most likely to be found.

## NATURAL SOURCES

Although human pollution, through the burning of fossil fuels, has contributed to acid deposition, rainwater is naturally acidic as a result of carbon dioxide in the air dissolving in the water. In addition, natural sources of sulphur and nitrogen emissions can contribute further to the acidity of rainwater.

Natural sources of sulphur dioxide include release from volcanoes, biological decay and forest fires. Actual amounts released from natural sources in the world are difficult to quantify; in 1983 the United Nations Environment Programme estimated a figure of between 80 million and 288 million tonnes of sulphur oxides per year. Natural sources of nitrogen oxides include volcanoes, oceans, biological decay and lightning strikes. Estimates range between 20 million and 90 million tonnes per year nitrogen oxides released from natural sources.

**Fig.** Lightning is a Major NaturalSource of Nitrogen Oxides Pollution.

## NITROGEN OXIDES

Nitrogen oxides is a collective term used to refer to two species of oxides of nitrogen: nitric oxide (NO) and nitrogen dioxide ($NO_2$). Nitric oxide is a colourless, flammable gas with a slight odour. Although somewhat toxic, its odour is insufficient to provide warning. Nitrogen dioxide is a reddish brown, non-flammable, gas with a detectable smell. In significant concentrations it is highly toxic, causing serious lung damage with a delayed effect. Nitrogen dioxide is a strong oxidizing agent that reacts in the air to form corrosive nitric acid, as well as toxic organic nitrates. It also plays a major role in the atmospheric reactions that produce ground-level ozone or smog.

Globally, quantities of nitrogen oxides produced naturally by bacterial and volcanic action, and lightning, outweigh man-made emissions. Man-made emissions are mainly due to fossil fuel combustion from both stationary sources, such as power generation (24 per cent), and mobile sources, such as transport (49 per cent). Other atmospheric contributions come from non-combustion processes, for example nitric acid manufacture, welding processes and the use of explosives. In the atmosphere, nitrogen oxides mix with water vapour producing nitric acid. This acidic pollution can be transported by win over many hundreds of miles, and deposited as acid rain.

## RAINFALL ACIDITY

Rainfall is naturally acidic due to the presence of carbon dioxide in the atmosphere which combines with rainwater to form weak carbonic acid. However, the combustion of fossil fuels produces waste gases such as sulphur dioxide, oxides of nitrogen and to a lesser extent, chloride. These pollutants can be converted, through a series of complex chemical reactions, into sulphuric acid, nitric acid or hydrochloric acid, increasing the acidity of the rain or other type of precipitation, such as snow and hail.

Carbon dioxide + Water' Carbonic acid (weak)
Sulphur dioxide + Water' Sulphuric acid
Nitrogen oxides + Water' Nitric acid

## SOILS

Soil is the basis of wealth upon which all land-based life depends. Acid deposition is known to wash essential nutrients from soils, and aluminium which is normally bound in the soil may be released into ground water. Soil acidification may affect the health of trees and other vegetation.

Soils containing calcium and limestone are more able to neutralise sulphuric and nitric acid depositions than a thin layer of sand or gravel with a granite base. If the soil is rich in limestone or if the underlying bedrock is either composed of limestone or marble, then the acid rain may be neutralised. This is because limestone and marble are more alkaline and produce a higher pH when dissolved in water. The higher pH of these materials dissolved in water offsets or buffers the acidity of the rainwater producing a more neutral pH.

In regions where the soil is not rich in limestone or if the bedrock is not composed of limestone or marble, then no neutralising effect takes place, and the acid rainwater accumulates in the bodies of water in the area. This applies to much of the northeastern United States where the bedrock is typically composed of granite. Granite has no neutralising effect on acid rainwater. Therefore, over time, more and more acid precipitation accumulates in lakes and ponds. Such areas or catchments are termed acid-sensitive (poorly buffered), and can suffer serious ecological damage due to acid rain.

To grow, trees and other vegetation need healthy soil to develop in. Long-term changes in the chemistry of some sensitive soils occur as a result of acid rain. As acid rain moves through the soils, it can strip away vital plant nutrients such as calcium, potassium and magnesium through chemical reactions, thus posing a potential threat to future forest productivity. Furthermore, the number of microorganisms present in the soil also decreases as the soil becomes more acidic. This further depletes the amount of nutrients available to plant life because the microorganisms play an important role in releasing nutrients from decaying organic material. Trees growing in acidified soil are more susceptible to viruses, fungi and insect pests. Other plant life may grow more slowly or die as a result of soil acidification.

Poisonous metals such as aluminium, cadmium and mercury are leached from soils through reacting with acids. This happens because these metals are bound to the soil under normal conditions, but the added dissolving action of acids causes rocks and small-bound soil particles to break down. In addition, the roots of plants trying to survive in acidic soil may be damaged directly by the acids present. Finally, if the plant life does not die from these effects, then it may be weakened enough so that it will be more susceptible to other harsh environmental influences like cold winters or high winds.

## SULPHUR DIOXIDE

Sulphur dioxide ($SO_2$) is a colourless gas, belonging to the family of gases called sulphur oxides (SOx). It reacts on the surface of a variety of airborne solid particles, is soluble in water and can be oxidised within airborne water droplets.

Natural sources of sulphur dioxide include releases from volcanoes, oceans, biological decay and forest fires. The most important man-made sources of sulphur dioxide are fossil fuel combustion, smelting, manufacture of sulphuric acid, conversion of wood pulp to paper, incineration of refuse and production of elemental sulphur. Coal burning is the single largest man-made source of sulphur dioxide accounting for about 50 per cent of annual global emissions, with oil burning accounting for a further 25 to 30 per cent.

The major health concerns associated with exposure to high concentrations of sulphur dioxide include effects on breathing, respiratory illness, alterations in pulmonary defences, and aggravation of existing cardiovascular disease. In the atmosphere, sulphur dioxide mixes with water vapour producing sulphuric acid. This acidic pollution can be transported by wind over many hundreds of miles, and deposited as acid rain.

## TRANSBOUNDARY POLLUTION

Stationary emissions sources, such as coal-fired and oil-fired power stations, and mobile sources, such as cars, ships and aircraft emit a complex mixture of pollutants, including sulphur dioxide and nitrogen oxides (the precursors to acid rain). It is now well established that this air pollution is transported over hundreds or even thousands of kilometres. Consequently, when acidic pollution is finally deposited, its environmental impacts are felt in areas far removed from their sources. Since this air pollution has no regard for national boundaries, it has been termed transboundary pollution.

Because acidic pollution is transboundary, there is no clear relationship between how much pollution a country emits and how much is deposited there. Throughout Europe, the prevailing wind direction is generally westerly or southwesterly. Consequently, much of the pollution emitted in the UK travels across the North Sea and is deposited in Scandinavia. Whilst the UK emits much more pollution than it receives through acid deposition, Norway and Sweden experience proportionally a much greater amount of acid rain compared to their lower emissions.

To control the spread of transboundary pollution the United Nations Economic Commission for Europe (UNECE) implemented the Convention on Long-Range Transboundary Pollution (1979). Since that time, emissions of sulphur dioxide across Europe have been lowered dramatically, but increases in the volume of traffic have meant that emissions of nitrogen oxides have not fallen as quickly. Consequently, whilst other environmental issues such as global warming and ozone depletion have received more attention in recent years, transboundary acid rain remains a problem today.

## TREES

Acid rain can have serious impacts on trees and forests. Acid rain does not usually kill trees directly. Instead, it is more likely to weaken them by damaging their leaves, limiting the nutrients available to them, or poisoning them with toxic substances slowly released from the soil. The main atmospheric pollutants that affect trees are nitrates and sulphates. Forest decline is often the first sign that trees are in trouble due to air pollution.

Scientists believe that acidic water dissolves the nutrients and helpful minerals in the soil and then washes them away before the trees and other plants can use them to grow. At the same time, the acid rain causes the release of toxic substances such as aluminium into the soil. These are very harmful to trees and plants, even if contact is limited. Forests in high mountain regions receive additional acid from the acidic clouds and fog that often surround them. These clouds and fog are often more acidic than rainfall. When leaves are frequently bathed in this acid fog, their protective waxy coating can wear away. The loss of the coating damages the leaves and creates brown spots. Leaves turn the energy in sunlight into food for growth. This process is called photosynthesis. When leaves are damaged, they cannot produce enough food energy for the tree to remain healthy. Once trees are weak, diseases or insects that ultimately kill them can more easily attack them. Weakened trees may also become injured more easily by cold weather.

While forestry has long been considered to be adversely affected by acid rain, recent studies show it to be part of the acidifying process. The rough canopies of mature evergreen forests are efficient scavengers of particulate and gaseous contaminants in polluted air. This results in a more acidic deposition under the forest canopies than in open land. Chemical processes at the roots of trees, evergreens in particular, further acidify the soil and soil water in forest catchments. When the forests are located on low-alkaline soils, these processes can lead to a significant acidification of the run-off water and consequent damage to associated streams and lakes.

## VEHICLE EMISSION CONTROLS

Through emissions of nitrogen oxides, cars and other road vehicles are major contributors to acidic emissions which cause acid rain. In all countries of the industrialised world, the number of vehicles on the roads has been continually increasing since the 1970s. With a large rise in traffic amounts, it becomes increasingly important to keep pollutant emissions to a minimum. There are presently a number of ways in which road traffic pollution can be reduced, including the use of emission control technology solutions. Since January 1993, all new cars sold in the European Union have been fitted with a catalytic converter. Most catalytic converters lead to a dramatic reduction in emissions of nitrogen oxides, as well as other harmful pollutants.

Exhaust Gas Recirculation involves returning exhaust air to the fuel inlet, which results in a reduction in peak engine temperatures and emissions of nitrogen oxides from petrol vehicles. Smaller, lighter cars use less fuel and hence produce less pollution. Technological development using lighter materials for construction may therefore reduce emissions.

The upon technologies all provide a reduction in emissions from vehicles. Electric transport is an alternative development that could lead to a large reduction in acidic pollution at ground level, if it became more wide spread. Electric transport produces no emissions at the point of use, although pollution is emitted during the production of electricity from power stations. The main drawback for electric vehicles is the need to recharge batteries. In addition, although they have lower fuel and maintenance costs than petrol and diesel, at present they require a higher capital investment.

The technical fixes such as those outlined upon need to be combined with management area to reduce traffic in city centres, education to encourage the public to use their cars less, and the further development of alternative fuels that are not harmful to the environment.

## WILDLIFE

Freshwater acidification due to acid rain can lead to a decline in, and loss of, fish populations. Below pH 4.5 no fish are likely to survive. Fish loss is occurring in many countries, including Scandinavia, Scotland, Wales and North America. A decrease in pH is often associated with an increase in toxic metal availability, being particularly true for aluminium and mercury.

Decreased pH and elevated aluminium have been shown to increase fish mortality, decrease fish growth, decrease egg production and embryo survival, and result in physiological impairment of adult fish. In general, embryos, fry, and juveniles are less acid-tolerant than adult fish. Aluminium can precipitate onto fish gills, inhibiting diffusion and resulting in respiratory stress.

Acid deposition is a possible cause of declines in amphibian populations. The larval stages of aquatic amphibian species are most affected by acidic water. Many frog species use temporary ponds, but these tend to be small and shallow, and are easily affected by precipitation chemistry because their only sources of water are rainfall and snowmelt.

Frogs that use large, permanent bodies of water for breeding generally lay their eggs in the summer, so they do not experience the acid pulses from snowmelt. However, the eggs and larvae of these species are even more sensitive to subtle changes in pH levels than those of species that breed in the temporary ponds. As for fish, the toxic effect of decreased pH levels on amphibians is complicated when concentrations of metals, such as aluminium, in the water increase, but as a general rule, embryos of sensitive amphibian species are killed by water with a pH of 4.5 or lower, while embryos of tolerant species can survive down to a pH of 3.7.

Fluoride and heavy metals can accumulate in acidified soil to levels that are toxic to soil invertebrates. Species sensitive to metals are replaced by ones that are more metal-tolerant. For example, soft-bodied species such as earthworms and nematodes seem to be more readily affected by elevated metal concentrations. Invertebrates play an important role in forest floor litter decomposition. As forest floor litter builds up, mineral release is delayed, and the availability of nutrients to plants is reduced. Herbivores are ultimately affected when the quantity or quality of their food supply decreases.

# 6

# Greenhouse Gas

Greenhouse gases are gases in an atmosphere that absorb and emit radiation within the thermal infrared range. This process is the fundamental cause of the greenhouse effect. Common greenhouse gases in the Earth's atmosphere include water vapour, carbon dioxide, methane, nitrous oxide, and ozone. In our solar system, the atmospheres of Venus, Mars and Titan also contain gases that cause greenhouse effects. Greenhouse gases, mainly water vapour, greatly affect the temperature of the Earth; without them this planet would likely be much colder. Although many factors such as the sun and the water cycle are responsible for the Earth's weather and energy balance, if all else was held equal and stable, the planet's average temperature would be considerably lower without greenhouse gases.

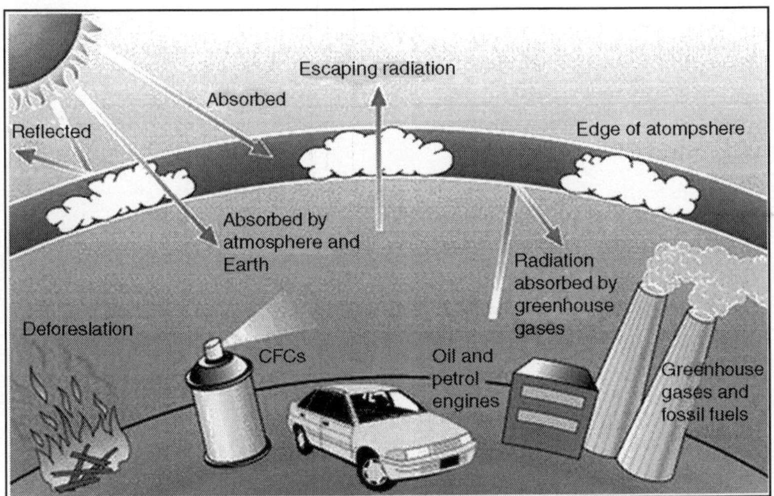


**Fig.** Simple Diagram of Greenhouse Effect.

Human activities have an impact upon the levels of greenhouse gases in the atmosphere, which has other effects upon the system, with their own possible repercussions. The 2007 assessment report compiled by the IPCC observed that "changes in atmospheric concentrations of greenhouse gases and aerosols, land cover and solar radiation alter the energy balance of the climate system", and concluded that "increases in anthropogenic greenhouse gas concentrations is very likely to have caused most of the increases in global average temperatures since the mid-20th century".

### Greenhouse Gases in Earth's Atmosphere

In order, Earth's most abundant greenhouse gases are:

- Water vapour
- Carbon dioxide
- Methane
- Nitrous oxide
- Ozone
- CFCs

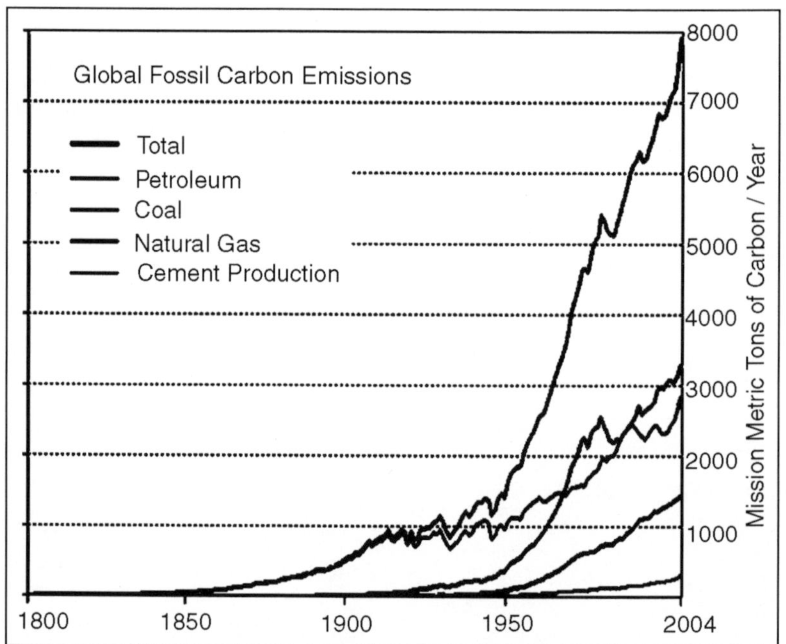


**Fig.** Modern Global Anthropogenic Carbon Emissions.

When these gases are ranked by their contribution to the greenhouse effect, the most important are:

- Water vapour, which contributes 36–72 per cent
- Carbon dioxide, which contributes 9–26 per cent
- Methane, which contributes 4–9 per cent
- Ozone, which contributes 3–7 per cent

The major non-gas contributor to the Earth's greenhouse effect, clouds, also absorb and emit infrared radiation and thus have an effect on radiative properties of the greenhouse gases.

The contribution to the greenhouse effect by a gas is affected by both the characteristics of the gas and its abundance. For example, on a molecule-for-molecule basis methane is about eight times stronger greenhouse gas than carbon dioxide, but it is present in much smaller concentrations so that its total contribution is smaller.

It is not possible to state that a certain gas causes an exact percentage of the greenhouse effect, because the influences of the various gases are not additive. The higher ends of the ranges quoted are for the gas alone; the lower ends, for the gas counting overlaps. Other greenhouse gases include sulfur hexafluoride, hydrofluoro- carbons and perfluorocarbons.

Although contributing to many other physical and chemical reactions, the major atmospheric constituents, nitrogen ($N_2$), oxygen ($O_2$), and argon (Ar), are not greenhouse gases. This is because homonuclear diatomic molecules such as $N_2$ and $O_2$ and monatomic molecules such as Ar have no net change in their dipole moment when they vibrate and hence are almost totally unaffected by infrared light. Although heteronuclear diatomics such as carbon monoxide (CO) or hydrogen chloride (HCl) absorb IR, these molecules are short-lived in the atmosphere owing to their reactivity and solubility. As a consequence they do not contribute significantly to the greenhouse effect and are not often included when discussing greenhouse gases. Late 19$^{th}$ century scientists experimentally discovered that $N_2$ and $O_2$ did not absorb infrared radiation (called, at that time, "dark radiation") and that water as a vapour and in cloud form, $CO_2$ and many other gases did absorb such radiation. It was recognized in the early 20$^{th}$ century that the greenhouse gases in the atmosphere caused the Earth's overall temperature to be higher than it would be without them.

## NATURAL AND ANTHROPOGENIC

Aside from purely human-produced synthetic halocarbons, most greenhouse gases have sources from both the ecosystem in general (natural) and from human activities specifically (anthropogenic). During the pre-industrial holocene, concentrations of existing gases were roughly constant. In the more populated industrial era, human activities have added greenhouse gases to the atmosphere, mainly through the burning of fossil fuels and clearing of forests.

| Gas | Preindustrial Level | Current Level | Increase since 1750 | Radiative forcing ($W/m^2$) |
|---|---|---|---|---|
| Carbon dioxide | 280 ppm | 387ppm | 104 ppm | 1.46 |
| Methane | 700 ppb | 1,745 ppb | 1,045 ppb | 0.48 |
| Nitrous oxide | 270 ppb | 314 ppb | 44 ppb | 0.15 |
| CFC-12 | 0 | 533 ppt | 533 ppt | 0.17 |

Ice cores provide evidence for variation in greenhouse gas concentrations over the past 800,000 years. Both $CO_2$ and $CH_4$ vary between glacial and interglacial phases, and concentrations of these gases correlate strongly with temperature.

Before the ice core record, direct data does not exist. However, various proxies and modelling suggests large variations; 500 million years ago $CO_2$ levels were likely 10 times higher than now. Indeed higher $CO_2$ concentrations are thought to have prevailed throughout most of the Phanerozoic eon, with concentrations four to six times current concentrations during the Mesozoic era, and ten to fifteen times current concentrations during the early Palaeozoic era until the middle of the Devonian period, about 400 Ma.

The spread of land plants is thought to have reduced $CO_2$ concentrations during the late Devonian, and plant activities as both sources and sinks of $CO_2$ have since been important in providing stabilising feedbacks.

Earlier still, a 200-million year period of intermittent, widespread glaciation extending close to the equator (Snowball Earth) appears to have been ended suddenly, about 550 Ma, by a colossal volcanic outgassing which raised the $CO_2$ concentration of the atmosphere abruptly to 12 per cent, about 350 times modern levels, causing extreme greenhouse conditions and carbonate deposition as limestone at the rate of about 1 mm per day.

This episode marked the close of the Precambrian eon, and was succeeded by the generally warmer conditions of the Phanerozoic, during which multicellular animal and plant life evolved. No volcanic carbon dioxide emission of comparable scale has occurred since. In the modern era, emissions to the atmosphere from volcanoes are only about 1 per cent of emissions from human sources.

## ANTHROPOGENIC GREENHOUSE GASES

Besides other changes to the environment, since about 1750 human activity has increased the concentration of carbon dioxide and other greenhouse gases. Measured atmospheric concentrations of carbon dioxide are currently 100 ppmv higher than pre-industrial levels.

Natural sources of carbon dioxide are more than 20 times greater than sources due to human activity, but over periods longer than a few years natural sources are closely balanced by natural sinks such as weathering of continental rocks and photosynthesis of carbon compounds by plants and marine plankton.

As a result of this balance, the atmospheric concentration of carbon dioxide had remained between 260 and 280 parts per million for the 10,000 years between the end of the last glacial maximum and the start of the industrial era.

It is likely anthropogenic warming, such as that due to elevated greenhouse gas levels, has had a discernible influence on many physical and biological systems. Projected changes in several climate factors, including atmospheric carbon dixoide, are projected to impact various issues such as freshwater resources, industry, food and health.

The main sources of greenhouse gases due to human activity are:

- Burning of fossil fuels and deforestation leading to higher carbon dioxide concentrations. Land use change (mainly deforestation in the tropics) account for up to one third of total anthropogenic $CO_2$ emissions.
- Livestock enteric fermentation and manure management, paddy rice farming, land use and wetland changes, pipeline losses, and covered vented landfill emissions leading to higher methane atmospheric concentrations. Many of the newer style fully vented septic systems that enhance and target the fermentation process also are sources of atmospheric methane.
- Use of chlorofluorocarbons (CFCs) in refrigeration systems, and use of CFCs and halons in fire suppression systems and manufacturing processes.
- Agricultural activities, including the use of fertilizers, that lead to higher nitrous oxide ($N_2O$) concentrations.

The seven sources of $CO_2$ from fossil fuel combustion are (with percentage contributions for 2000–2004):

- Solid fuels (*e.g.* coal): 35 per cent
- Liquid fuels (*e.g.* gasoline): 36 per cent
- Gaseous fuels (*e.g.* natural gas): 20 per cent
- Flaring gas industrially and at wells: $<1$ per cent
- Cement production: 3 per cent
- Non-fuel hydrocarbons: $<1$ per cent
- The "international bunkers" of shipping and air transport not included in national inventories: 4 per cent

The U.S. EPA ranks the major greenhouse gas contributing end-user sectors in the following order: industrial, transportation, residential, commercial and agricultural. Major sources of an individual's GHG include home heating and cooling, electricity consumption, and transportation. Corresponding conservation measures are improving home building insulation, installing geothermal heat pumps and compact fluorescent lamps, and choosing energy-efficient vehicles.

Carbon dioxide, methane, nitrous oxide and three groups of fluorinated gases (sulfur hexafluoride, HFCs, and PFCs) are the major greenhouse gases and the subject of the Kyoto Protocol, which came into force in 2005.

Although CFCs are greenhouse gases, they are regulated by the Montreal Protocol, which was motivated by CFCs' contribution to ozone depletion rather than by their contribution to global warming. Note that ozone depletion has only a minor role in greenhouse warming though the two processes often are confused in the media.

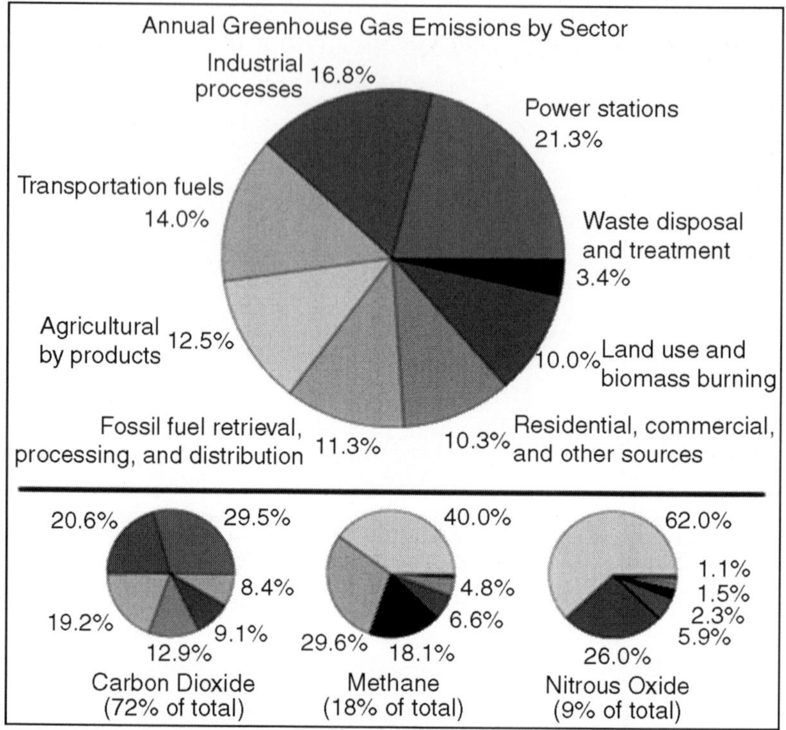


**Fig.** Global Anthropogenic Greenhouse Gas Emissions Broken Down Into **8** Different Sectors for the Year **2000**

Nitrogen trifluoride ($NF_3$) is used in the manufacture of microelectronics. It is a potent greenhouse gas, but presently its concentration is very low and it is not subject to greenhouse gas treaties. A 2008 study suggested that its concentration should be monitored, particularly as its production has increased due to use in plasma televisions.

## Role of Water Vapour

Water vapour accounts for the largest percentage of the greenhouse effect, between 36 per cent and 66 per cent for water vapour alone, and between 66 per cent and 85 per cent when factoring in clouds. Water vapour concentrations fluctuate regionally, but human activity does not significantly affect water vapour concentrations except at local scales, such as near irrigated fields.

The Clausius-Clapeyron relation establishes that air can hold more water vapour per unit volume when it warms. This and other basic principles indicate that any warming associated with the increased concentration of the other greenhouse gases also increases the concentration of water vapour as well.

In climate matters, when a warming trend results in effects that induce further warming, the process is referred to as a "positive feedback"; when the effects induce cooling, the process is referred to as a "negative feedback".

Because water vapour is the primary greenhouse gas and because warm air can hold more water vapour than cooler air, the primary positive feedback involves water vapour.

This positive feedback does not result in runaway global warming because it is offset by negative feedback, which stabilizes average global temperatures. One primary negative feedback is the effect of temperature on emission of infrared radiation: as the temperature of a body increases, the emitted radiation increases with the fourth power of its absolute temperature.

Other important considerations involve water vapour being the only greenhouse gas whose concentration is highly variable in space and time in the atmosphere and the only one that also exists in both liquid and solid phases, frequently changing to and from each of the three phases or existing in mixes.

Such considerations include clouds themselves, air and water vapour density interactions when they are the same or different temperatures, the absorption and release of kinetic energy as water evaporates and condenses to and from vapour, and behaviours related to vapour partial pressure.

For example, the release of latent heat by rain in the ITCZ drives atmospheric circulation, clouds vary atmospheric albedo levels, and the oceans provide evaporative cooling that modulates the greenhouse effect down from estimated 67°C surface temperature.

## GREENHOUSE GAS EMISSIONS

Measurements from Antarctic ice cores show that before industrial emissions started, atmospheric $CO_2$ levels were about 280 parts per million by volume (ppmv), and it appears that concentrations stayed between 260 and 280 during the preceding ten thousand years.

One study using evidence from stomata of fossilized leaves suggests greater variability, with carbon dioxide levels above 300 ppm during the period seven to ten thousand years ago, though others have argued that these findings more likely reflect calibration or contamination problems rather than actual $CO_2$ variability.

Because of the way air is trapped in ice (pores in the ice close off slowly to form bubbles deep within the firn) and the time period represented in each ice sample analyzed, these figures represent averages of atmospheric concentrations of up to a few centuries rather than annual or decadal levels.

Since the beginning of the Industrial Revolution, the concentrations of most of the greenhouse gases have increased. For example, the concentration of carbon dixode has increased by about 36 per cent to 380 ppmv, or 100 ppmv over modern pre-industrial levels.

The first 50 ppmv increase took place in about 200 years, from the start of the Industrial Revolution to around 1973; however the next 50 ppmv increase took place in about 33 years, from 1973 to 2006.

Recent data also shows the concentration is increasing at a higher rate. In the 1960s, the average annual increase was only 37 per cent of what it was in 2000 through 2007.

**Recent Rates of Change and Emission**

The sharp acceleration in $CO_2$ emissions since 2000 to more than a 3 per cent increase per year (more than 2 ppm per year) from 1.1 per cent per year during the 1990s is attributable to the lapse of formerly declining trends in carbon intensity of both developing and developed nations.

Although over 3/4 of cumulative anthropogenic $CO_2$ is still attributable to the developed world, China was responsible for most of global growth in emissions during this period. Localised plummeting emissions associated with the collapse of the Soviet Union have been followed by slow emissions growth in this region due to more efficient energy use, made necessary by the increasing proportion of it that is exported. In comparison, methane has not increased appreciably, and $N_2O$ by 0.25 per cent $y^{-}1$.

The direct emissions from industry have declined due to a constant improvement in energy efficiency, but also to a high penetration of electricity. If one includes indirect emissions, related to the production of electricity, emissions from industry in Europe are roughly stabilized since 1994.

***Asia***

Atmospheric levels of $CO_2$ continue to rise, partly a sign of the industrial rise of Asian economies led by China. Over the 2000-2010 interval China is expected to increase its carbon dioxide emissions by 600 Mt, largely because of the rapid construction of old-fashioned power plants in poorer internal provinces.

***United Kingdom***

The UK set itself a target of reducing carbon dioxide emissions by 20 per cent from 1990 levels by 2010, but according to its own figures it will fall short of this target by almost 4 per cent.

***United States***

The United States emitted 16.3 per cent more GHG in 2005 than it did in 1990. According to a preliminary estimate by the Netherlands Environmental Assessment Agency, the largest national producer of $CO_2$ emissions since 2006 has been China with an estimated annual production of about 6200 megatonnes. China is followed by the United States with about 5,800 megatonnes. However the per capita emission figures of China are still about one quarter of those of the US population. Relative to 2005, China's fossil $CO_2$ emissions increased in 2006 by 8.7 per cent, while in the USA, comparable $CO_2$ emissions decreased in 2006 by 1.4 per cent. The agency notes that its estimates do not include some $CO_2$ sources of uncertain magnitude.

These figures rely on national $CO_2$ data that do not include aviation. Although these tonnages are small compared to the $CO_2$ in the Earth's atmosphere, they are significantly larger than pre-industrial levels.

***Relative CO2 Emission from Various Fuels***

Pounds of Carbon dioxide emitted per million British thermal units of energy for various fuels:

| Fuel name | $CO_2$ emitted (lbs/106 Btu) |
|---|---|
| Natural gas | 117 |
| Liquefied petroleum gas | 139 |
| Propane | 139 |
| Aviation gasoline | 153 |
| Automobile gasoline | 156 |
| Kerosene | 159 |
| Fuel oil | 161 |
| Tires/tire derived fuel | 189 |
| Wood and wood waste | 195 |
| Coal (bituminous) | 205 |
| Coal (subbituminous) | 213 |
| Coal (lignite) | 215 |
| Petroleum coke | 225 |
| Coal (anthracite) | 227 |

**Removal from the Atmosphere and Global Warming Potential**

Aside from water vapour, which has a residence time of about nine days, major greenhouse gases are well-mixed, and take many years to leave the atmosphere. Although it is not easy to know with precision how long it takes greenhouse gases to leave the atmosphere, there are estimates for the principal greenhouse gases.

Greenhouse gases can be removed from the atmosphere by various processes:

- As a consequence of a physical change (condensation and precipitation remove water vapour from the atmosphere).
- As a consequence of chemical reactions within the atmosphere. This is the case for methane. It is oxidized by reaction with naturally occurring hydroxyl radical, OH· and degraded to $CO_2$ and water vapour at the end of a chain of reactions (the contribution of the $CO_2$ from the oxidation of methane is not included in the methane Global warming potential). This also includes solution and solid phase chemistry occurring in atmospheric aerosols.

- As a consequence of a physical interchange at the interface between the atmosphere and the other compartments of the planet. An example is the mixing of atmospheric gases into the oceans at the boundary layer.
- As a consequence of a chemical change at the interface between the atmosphere and the other compartments of the planet. This is the case for $CO_2$, which is reduced by photosynthesis of plants, and which, after dissolving in the oceans, reacts to form carbonic acid and bicarbonate and carbonate ions.
- As a consequence of a photochemical change. Halocarbons are dissociated by UV light releasing Cl· and F· as free radicals in the stratosphere with harmful effects on ozone (halocarbons are generally too stable to disappear by chemical reaction in the atmosphere).

## ATMOSPHERIC LIFETIME

Jacob (1999) defines the lifetime ô of an atmospheric species X in a one-box model as the average time that a molecule of X remains in the box. Mathematically ô can be defined as the ratio of the mass m (in kg) of X in the box to its removal rate, which is the sum of the flow of X out of the box ($F_{out}$), chemical loss of X (L), and deposition of X (D) (all in kg/sec):

The atmospheric lifetime of a species therefore measures the time required to restore equilibrium following an increase in its concentration in the atmosphere. Individual atoms or molecules may be lost or deposited to sinks such as the soil, the oceans and other waters, or vegetation and other biological systems, reducing the excess to background concentrations. The average time taken to achieve this is the mean lifetime.

The atmospheric lifetime of $CO_2$ is often incorrectly stated to be only a few years because that is the average time for any $CO_2$ molecule to stay in the atmosphere before being removed by mixing into the ocean, photosynthesis, or other processes. However, this ignores the balancing fluxes of $CO_2$ into the atmosphere from the other reservoirs. It is the net concentration changes of the various greenhouse gases by all sources and sinks that determines atmospheric lifetime, not just the removal processes.

## GLOBAL WARMING POTENTIAL

The global warming potential (GWP) depends on both the efficiency of the molecule as a greenhouse gas and its atmospheric lifetime. GWP is measured relative to the same mass of $CO_2$ and evaluated for a specific timescale.

Thus, if a molecule has a high GWP on a short time scale (say 20 years) but has only a short lifetime, it will have a large GWP on a 20 year scale but a small one on a 100 year scale. Conversely, if a molecule has a longer atmospheric lifetime than $CO_2$ its GWP will increase with time.

Examples of the atmospheric lifetime and GWP for several greenhouse gases include:

- Carbon dioxide has a variable atmospheric lifetime, and cannot be specified precisely. Recent work indicates that recovery from a large input of atmospheric $CO_2$ from burning fossil fuels will result in an effective lifetime of tens of thousands of years. Carbon dioxide is defined to have a GWP of 1 over all time periods.
- Methane has an atmospheric lifetime of 12 ± 3 years and a GWP of 72 over 20 years, 25 over 100 years and 7.6 over 500 years. The decrease in GWP at longer times is because methane is degraded to water and $CO_2$ through chemical reactions in the atmosphere.
- Nitrous oxide has an atmospheric lifetime of 114 years and a GWP of 289 over 20 years, 298 over 100 years and 153 over 500 years.
- CFC-12 has an atmospheric lifetime of 100 years and a GWP of 11000 over 20 years, 10900 over 100 years and 5200 over 500 years.
- HCFC-22 has an atmospheric lifetime of 12 years and a GWP of 5160 over 20 years, 1810 over 100 years and 549 over 500 years.
- Tetrafluoromethane has an atmospheric lifetime of 50,000 years and a GWP of 5210 over 20 years, 7390 over 100 years and 11200 over 500 years.
- Sulphur hexafluoride has an atmospheric lifetime of 3,200 years and a GWP of 16300 over 20 years, 22800 over 100 years and 32600 over 500 years.
- Nitrogen trifluoride has an atmospheric lifetime of 740 years and a GWP of 12300 over 20 years, 17200 over 100 years and 20700 over 500 years.

The use of CFC-12 (except some essential uses) has been phased out due to its ozone depleting properties. The phasing-out of less active HCFC-compounds will be completed in 2030.

## AIRBORNE FRACTION

Airborne fraction (AF) is the proportion of a emission (*e.g.* $CO_2$) remaining in the atmosphere after a specified time. Canadell (2007) define the annual AF as the ratio of the atmospheric $CO_2$ increase in a given year to that year's total emissions, and calculate that of the average 9.1 PgC $y^{-}1$ of total anthropogenic emissions from 2000 to 2006, the AF was 0.45. For $CO_2$ the AF over the last 50 years (1956-2006) has been increasing at 0.25±0.21 per cent/year.

## RELATED EFFECTS

Carbon monoxide has an indirect radiative effect by elevating concentrations of methane and tropospheric ozone through scavenging of atmospheric constituents (*e.g.*, the hydroxyl radical, OH) that would otherwise destroy them. Carbon monoxide is created when carbon-containing fuels are burned incompletely.

Through natural processes in the atmosphere, it is eventually oxidized to carbon dioxide. Carbon monoxide has an atmospheric lifetime of only a few months and as a consequence is spatially more variable than longer-lived gases.

Another potentially important indirect effect comes from methane, which in addition to its direct radiative impact also contributes to ozone formation. Shindell et al. (2005) argue that the contribution to climate change from methane is at least double previous estimates as a result of this effect.

## THE EARTH'S GREENHOUSE EFFECT

The realisation that Earth's climate might be sensitive to the atmospheric concentrations of gases that create a greenhouse effect is more than a century old. Fleming and Weart provided an overview of the emerging science. In terms of the energy balance of the climate system, Edme Mariotte noted in 1681 that although the Sun's light and heat easily pass through glass and other transparent materials, heat from other sources does not. The ability to generate an artificial warming of the Earth's surface was demonstrated in simple greenhouse experiments such as Horace Benedict de Saussure's experiments in the 1760s using a 'heliothermometre' to provide an early analogy to the greenhouse effect.

It was a conceptual leap to recognise that the air itself could also trap thermal radiation. In 1824, Joseph Fourier, citing Saussure, argued 'the temperature [of the Earth] can be augmented by the interposition of the atmosphere, because heat in the state of light finds less resistance in penetrating the air, than in repassing into the air when converted into non-luminous heat'. In 1836, Pouillit followed up on Fourier's ideas and argued 'the atmospheric stratum...exercises a greater absorption upon the terrestrial than on the solar rays'. There was still no understanding of exactly what substance in the atmosphere was responsible for this absorption. In 1859, John Tyndall identified through laboratory experiments the absorption of thermal radiation by complex molecules. He noted that changes in the amount of any of the radiatively active constituents of the atmosphere such as water or $CO_2$ could have produced 'all the mutations of climate which the researches of geologists reveal'. In 1895, Svante Arrhenius followed with a climate prediction based on greenhouse gases, suggesting that a 40 per cent increase or decrease in the atmospheric abundance of the trace gas $CO_2$ might trigger the glacial advances and retreats. One hundred years later, it would be found that $CO_2$ did indeed vary by this amount between glacial and interglacial periods. However, it now appears that the initial climatic change preceded the change in $CO_2$ but was enhanced by it.

G. S. Callendar solved a set of equations linking greenhouse gases and climate change. He found that a doubling of atmospheric $CO_2$ concentration resulted in an increase in the mean global temperature of 2°C, with considerably more warming at the poles, and linked increasing fossil fuel combustion with a rise in $CO_2$ and its greenhouse effects: 'As man is now changing the composition of the atmosphere at a rate which must be very exceptional on the geological time scale, it is natural to seek for the probable effects of such a change. From the best laboratory observations it appears that the principal result of increasing atmospheric carbon dioxide... would be a gradual increase in the mean temperature of the colder regions of the Earth.' In 1947, Ahlmann reported a 1.3°C warming in the North Atlantic sector of the Arctic since the 19th century and mistakenly believed this climate variation could be explained entirely by greenhouse gas warming.

Similar model predictions were echoed by Plass in 1956: 'If at the end of this century, measurements show that the carbon dioxide content of the atmosphere has risen appreciably and at the same time the temperature has continued to rise throughout the world, it will be firmly established that carbon dioxide is an important factor in causing climatic change'. In trying to understand the carbon cycle, and specifically how fossil fuel emissions would change atmospheric $CO_2$, the interdisciplinary field of carbon cycle science began. One of the first problems addressed was the atmosphere-ocean exchange of $CO_2$. Revelle and Suess explained why part of the emitted $CO_2$ was observed to accumulate in the atmosphere rather than being completely absorbed by the oceans. While $CO_2$ can be mixed rapidly into the upper layers of the ocean, the time to mix with the deep ocean is many centuries. By the time of the TAR, the interaction of climate change with the oceanic circulation and biogeochemistry was projected to reduce the fraction of anthropogenic $CO_2$ emissions taken up by the oceans in the future, leaving a greater fraction in the atmosphere.

In the 1950s, the greenhouse gases of concern remained $CO_2$ and $H_2O$, the same two identified by Tyndall a century earlier. It was not until the 1970s that other greenhouse gases–$CH_4$, $N_2O$ and CFCs–were widely recognised. By the 1970s, the importance of aerosol-cloud effects in reflecting sunlight was known and atmospheric aerosols were being proposed as climate-forcing constituents. Charlson and others built a consensus that sulphate aerosols were, by themselves, cooling the Earth's surface by directly reflecting sunlight. Moreover, the increases in sulphate aerosols were anthropogenic and linked with the main source of $CO_2$, burning of fossil fuels. Thus, the current picture of the atmospheric constituents driving climate change contains a much more diverse mix of greenhouse agents.

## PAST CLIMATE OBSERVATIONS, ASTRONOMICAL THEORY AND ABRUPT CLIMATE CHANGES

Throughout the 19th and 20th centuries, a wide range of geomorphology and palaeontology studies has provided new insight into the Earth's past climates, covering periods of hundreds of millions of years. The Palaeozoic Era, beginning 600 Ma, displayed evidence of both warmer and colder climatic conditions than the present; the Tertiary Period was generally warmer; and the Quaternary Period showed oscillations between glacial and interglacial conditions. Louis Agassiz developed the hypothesis that Europe had experienced past glacial ages, and there has since been a growing awareness that long-term climate observations can advance the understanding of the physical mechanisms affecting climate change. The scientific study of one such mechanism–modifications in the geographical and temporal patterns of solar energy reaching the Earth's surface due to changes in the Earth's orbital parametres–has a long history. The pioneering contributions of Milankovitch to this astronomical theory of climate change are widely known, and the historical review of Imbrie and Imbrie calls attention to much earlier contributions, such as those of James Croll, originating in 1864.

The pace of palaeoclimatic research has accelerated over recent decades. Quantitative and well-dated records of climate fluctuations over the last 100 kyr have brought a more comprehensive view of how climate changes occur, as well as the means to test elements of the astronomical theory. By the 1950s, studies of deep-sea cores suggested that the ocean temperatures may have been different during glacial times. Ewing and Donn proposed that changes in ocean circulation actually could initiate an ice age. In the 1960s, the works of Emiliani and Shackleton showed the potential of isotopic measurements in deepsea sediments to help explain Quaternary changes. In the 1970s, it became possible to analyse a deep-sea core time series of more than 700 kyr, thereby using the last reversal of the Earth's magnetic field to establish a dated chronology. This deep-sea observational record clearly showed the same periodicities found in the astronomical forcing, immediately providing strong support to Milankovitch's theory.

Ice cores provide key information about past climates, including surface temperatures and atmospheric chemical composition. The bubbles sealed in the ice are the only available samples of these past atmospheres. The first deep ice cores from Vostok in Antarctica provided additional evidence of the role of astronomical forcing. They also revealed a highly correlated evolution of temperature changes and atmospheric composition, which was subsequently confirmed over the past 400 kyr and now extends to almost 1 Myr. This discovery drove research to understand the causal links between greenhouse gases and climate change.

The same data that confirmed the astronomical theory also revealed its limits: a linear response of the climate system to astronomical forcing could not explain entirely the observed fluctuations of rapid ice-age terminations preceded by longer cycles of glaciations. The importance of other sources of climate variability was heightened by the discovery of abrupt climate changes.

In this context, 'abrupt' designates regional events of large amplitude, typically a few degrees celsius, which occurred within several decades–much shorter than the thousand-year time scales that characterise changes in astronomical forcing. Abrupt temperature changes were first revealed by the analysis of deep ice cores from Greenland.

Oeschger et al. recognised that the abrupt changes during the termination of the last ice age correlated with cooling in Gerzensee and suggested that regime shifts in the Atlantic Ocean circulation were causing these widespread changes. The synthesis of palaeoclimatic observations by Broecker and Denton invigourated the community over the next decade. By the end of the 1990s, it became clear that the abrupt climate changes during the last ice age, particularly in the North Atlantic regions as found in the Greenland ice cores, were numerous, indeed abrupt and of large amplitude. They are now referred to as Dansgaard-Oeschger events. A similar variability is seen in the North Atlantic Ocean, with north-south oscillations of the polar front and associated changes in ocean temperature and salinity. With no obvious external forcing, these changes are thought to be manifestations of the internal variability of the climate system.

The importance of internal variability and processes was reinforced in the early 1990s with analysis of records with high temporal resolution. New ice cores new ocean cores from regions with high sedimentation rates, as well as lacustrine sediments and cave stalagmites produced additional evidence for unforced climate changes, and revealed a large number of abrupt changes in many regions throughout the last glacial cycle. Long sediment cores from the deep ocean were used to reconstruct the thermohaline circulation connecting deep and surface waters and to demonstrate the participation of the ocean in these abrupt climate changes during glacial periods.

By the end of the 1990s, palaeoclimate proxies for a range of climate observations had expanded greatly. The analysis of deep corals provided indicators for nutrient content and mass exchange from the surface to deep water, showing abrupt variations characterised by synchronous changes in surface and deep-water properties. Precise measurements of the $CH_4$ abundances in polar ice cores showed that they changed in concert with the Dansgaard-Oeschger events and thus allowed for synchronisation of the dating across ice cores. The characteristics of the antarctic temperature variations and their relation to the Dansgaard-Oeschger events in Greenland were consistent with the simple concept of a bipolar seesaw caused by changes in the thermohaline circulation of the Atlantic Ocean. This work underlined the role of the ocean in transmitting the signals of abrupt climate change.

Abrupt changes are often regional, for example, severe droughts lasting for many years have changed civilizations, and have occurred during the last 10 kyr of stable warm climate. This result has altered the notion of a stable climate during warm epochs, as previously suggested by the polar ice cores. The emerging picture of an unstable oceanatmosphere system has opened the debate of whether human interference through greenhouse gases and aerosols could trigger such events. Palaeoclimate reconstructions cited in the FAR were based on various data, including pollen records, insect and animal remains, oxygen isotopes and other geological data from lake varves, loess, ocean sediments, ice cores and glacier termini.

These records provided estimates of climate variability on time scales up to millions of years. A climate proxy is a local quantitative record that is interpreted as a climate variable using a transfer function that is based on physical principles and recently observed correlations between the two records. The combination of instrumental and proxy data began in the 1960s with the investigation of the influence of climate on the proxy data, including tree rings, corals and ice cores. Phenological and historical data are also a valuable source of climatic reconstruction for the period before instrumental records became available. Such documentary data also need calibration against instrumental data to extend and reconstruct the instrumental record. With the development of multi-proxy reconstructions, the climate data were extended not only from local to global, but also from instrumental data to patterns of climate variability. Most of these reconstructions were at single sites and only loose efforts had been made to consolidate records.

Mann et al. made a notable advance in the use of proxy data by ensuring that the dating of different records lined up. Thus, the true spatial patterns of temperature variability and change could be derived, and estimates of NH average surface temperatures were obtained. The Working Group I WGI FAR noted that past climates could provide analogues. Fifteen years of research since that assessment has identified a range of variations and instabilities in the climate system that occurred during the last 2 Myr of glacial-interglacial cycles and in the super-warm period of 50 Ma. These past climates do not appear to be analogues of the immediate future, yet they do reveal a wide range of climate processes that need to be understood when projecting 21stcentury climate change.

## SOLAR VARIABILITY AND THE TOTAL SOLAR IRRADIANCE

Measurement of the absolute value of total solar irradiance is difficult from the Earth's surface because of the need to correct for the influence of the atmosphere. Langley attempted to minimise the atmospheric effects by taking measurements from high on Mt. Whitney in California, and to estimate the correction for atmospheric effects by taking measurements at several times of day, for example, with the solar radiation having passed through different atmospheric pathlengths. Between 1902 and 1957, Charles Abbot and a number of other scientists around the globe made thousands of measurements of TSI from mountain sites. Values ranged from 1,322 to 1,465 W m-2, which encompasses the current estimate of 1,365 W m-2. Foukal et al. deduced from Abbot's daily observations that higher values of TSI were associated with more solar faculae.

In 1978, the Nimbus-7 satellite was launched with a cavity radiometre and provided evidence of variations in TSI. Additional observations were made with an active cavity radiometre on the Solar Maximum Mission, launched in 1980. Both of these missions showed that the passage of sunspots and faculae across the Sun's disk influenced TSI. At the maximum of the 11-year solar activity cycle, the TSI is larger by about 0.1 per cent than at the minimum. The observation that TSI is highest when sunspots are at their maximum is the opposite of Langley's hypothesis. As early as 1910, Abbot believed that he had detected a downward trend in TSI that coincided with a general cooling of climate. The solar cycle variation in irradiance corresponds to an 11-year cycle in radiative forcing which varies by about 0.2 W m-2.

There is increasingly reliable evidence of its influence on atmospheric temperatures and circulations, particularly in the higher atmosphere. Calculations with three-dimensional models suggest that the changes in solar radiation could cause surface temperature changes of the order of a few tenths of a degree celsius. For the time before satellite measurements became available, the solar radiation variations can be inferred from cosmogenic isotopes and from the sunspot number. Naked-eye observations of sunspots date back to ancient times, but it was only after the invention of the telescope in 1607 that it became possible to routinely monitor the number, size and position of these 'stains' on the surface of the Sun. Throughout the 17th and 18th centuries, numerous observers noted the variable concentrations and ephemeral nature of sunspots, but very few sightings were reported between 1672 and 1699.

This period of low solar activity, now known as the Maunder Minimum, occurred during the climate period now commonly referred to as the Little Ice Age. There is no exact agreement as to which dates mark the beginning and end of the Little Ice Age, but from about 1350 to about 1850 is one reasonable estimate. During the latter part of the 18th century, Wilhelm Herschel noted the presence not only of sunspots but of bright patches, now referred to as faculae, and of granulations on the solar surface. He believed that when these indicators of activity were more numerous, solar emissions of light and heat were greater and could affect the weather on Earth. Heinrich Schwabe published his discovery of a '10-year cycle' in sunspot numbers. Samuel Langley compared the brightness of sunspots with that of the surrounding photosphere. He concluded that they would block the emission of radiation and estimated that at sunspot cycle maximum the Sun would be about 0.1 per cent less bright than at the minimum of the cycle, and that the Earth would be 0.1°C to 0.3°C cooler.

These satellite data have been used in combination with the historically recorded sunspot number, records of cosmogenic isotopes, and the characteristics of other Sun-like stars to estimate the solar radiation over the last 1,000 years. These data sets indicated quasi-periodic changes in solar radiation of 0.24 to 0.30 per cent on the centennial time scale. These values have recently been re-assessed. The TAR states that the changes in solar irradiance are not the major cause of the temperature changes in the second half of the 20th century unless those changes can induce unknown large feedbacks in the climate system. The effects of galactic cosmic rays on the atmosphere and those due to shifts in the solar spectrum towards the ultraviolet range, at times of high solar activity, are largely unknown. The latter may produce changes in tropospheric circulation via changes in static stability resulting from the interaction of the increased UV radiation with stratospheric ozone. More research to investigate the effects of solar behaviour on climate is needed before the magnitude of solar effects on climate can be stated with certainty.

## GREEN ECOSYSTEM

An ecosystem is a group of living and non-living components interacting together on a given physical landscape. The size of an ecosystem is arbitrary and could be as small as a few square centimetres if you are looking at a soil microbial ecosystem; as large as thousands of square kilometres if you are looking a biome like the Great Plains ecosystem; or a few hectares if you are looking at a single forest stand ecosystem.

### FORESTED ECOSYSTEM

An ecosystem model is an accurate but simplified representation of an ecosystem that can be very useful in thinking about or simulating the actions of a real ecosystem. Because any ecosystem has many different but interrelated components, the best way to understand the system is to break it down into its component parts. To get an introduction to a very simplified forest model, which gives participants and introduction to how a hardwood forest ecosystem works before and after exotic earthworms invade.

- *Step One*: The first step in building a graphical model of a hardwood forest ecosystem is to identify its major components. The components of any ecosystem are those physical things that contain energy and nutrients.

A forested ecosystem, by definition contains trees, so that is our first component. In addition to the various species and layers of trees in a forest, there are other distinct ecosystem components. The *understory* contains most of the visible plant life found between the sapling layer and forest floor. The forest floor is where one would find most of the plant roots, bulbs, fungi, seeds, years of accumulated leaves and twigs. The soil is the "dirt" under the forest floor and is composed largely of minerals of various grain size (very small grain size = clay very large grain size = sand) and organic material that has been mixed in with the mineral component. In addition, there are numerous animals that live in the forest, and of course we cannot forget people. We will add those two components to our forest ecosystem model later.

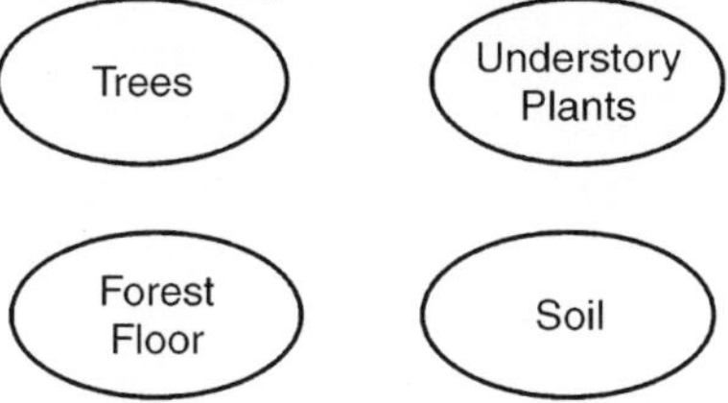


**Fig.** Components of a Graphical Forest Ecosystem

- *Step Two*: Once you have identified the components of your ecosystem model, you need to define the *processes* that connect the components. This is graphically done by using arrows to indicate the flow of nutrients or energy among the different ecosystem components. The components of our ecosystem model are now connected by processes that result in the movement of energy or nutrients among the components.

One thing to notice in our ecosystem model is that there is no process connecting the *trees* component with the understory plants component.

This is because there are no substantial processes that result in the flow of nutrients directly from a tree to an understory plant or visa-versa. There would be important relationships between the trees and understory plants.

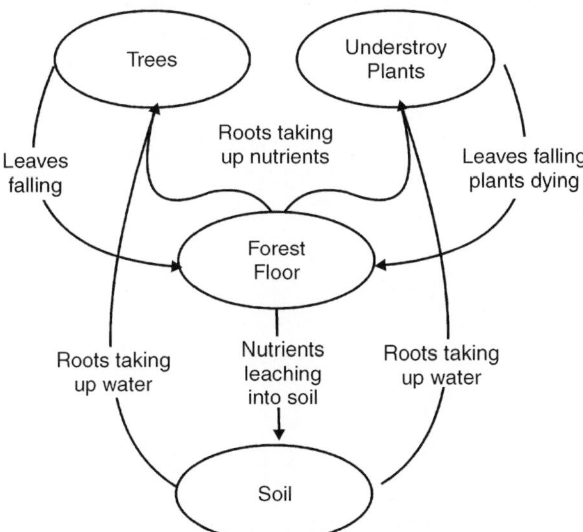


**Fig.** Movement of Energy or Nutrients Among the Components

Trees provide shade to the understory plants. But remember, in an ecosystem model, only processes that result in flow of energy or nutrients are represented. Now let's add the animals and the people components to our ecosystem. energy and nutrients flow from the trees and understory plants to the animals when they eat the leaves, twigs and buds of trees or graze on understory plants; and when the animal excrete waste products or die, energy and nutrients are returned to the forest floor component. Since *people* are really just a special kind of animal, energy and nutrients flow from the trees to people when they eat something from a tree, like maple syrup

We have added two components to our ecosystem model, along with some processes connecting them to other components.

- *Step Three*: Determine the major inputs and outputs of your ecosystem. As you are building your ecosystem model, one thing to think about is whether your ecosystems could be opened or closed. A closed ecosystem is one that has no inputs of energy or nutrients from outside the ecosystem and no outputs of energy or nutrients leaving the system. The earth is an example of a closed ecosystem with respect to nutrients and an open ecosystem with respect to energy All the nutrients that have ever been on earth are here and simply continue to cycle, there are no additions or losses. However, the earth is constantly getting inputs of energy from the sun and simultaneously radiating energy back. The earth doesn't heat up too much or cool down too much because the earth's energy balance is in a relatively stable equilibrium, meaning that the amount of energy being input and output are about equal.

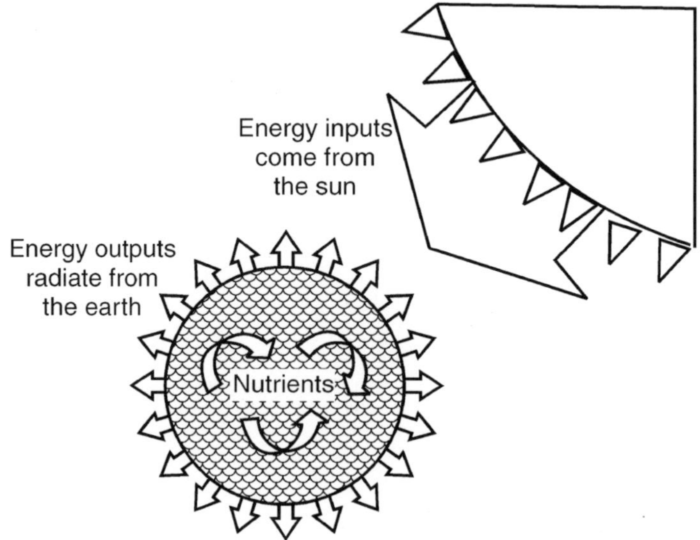


**Fig.** The Earth Ecosystem

The earth ecosystem and has no inputs or outputs of nutrients which are constantly recycled within the global ecosystem, while the earth has both inputs and outputs of energy that are in a relatively stable equilibrium.

Just as the Earth ecosystem is closed with respect to nutrients, unmanaged earthworm-free hardwood forest ecosystems are often very nearly closed nutrient ecosystems that there are very few inputs or outputs of nutrients.

Rather the nutrients are constantly recycled among the various ecosystem components. In contrast, most agricultural ecosystems require nutrient inputs from outside to function properly some typical inputs and outputs of nutrients and energy for forested ecosystems include evapotranspiration, nutrient leaching, sunlight and rain.

- *Step Four*: Once you have identified the components, processes and major inputs and outputs in your ecosystem model, then you can begin to add the actual values to these parts of your ecosystem by measuring them. The amount of litter that falls to the forest floor each year (a process), what the biomass of trees is in a given forest (a component), how much light reaches the forest over a growing season (an input), or how much nitrogen leaches from the forest (an output). Needless to say, some of these things are easier to measure than others and for most of these things it would be very hard to directly measure the value for a whole forest.

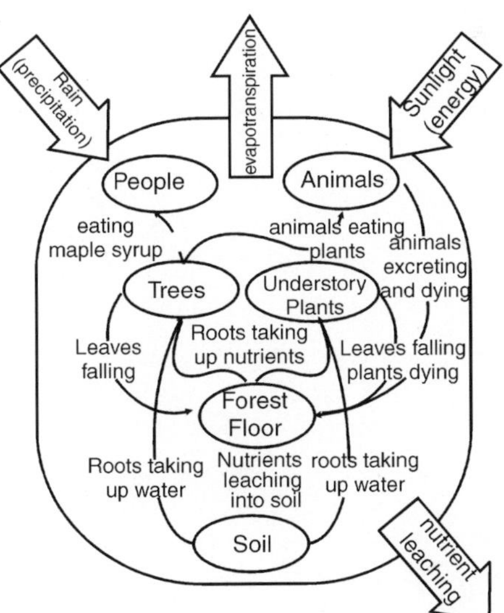


**Fig.** Evapotranspiration

It would be hard to catch every single leaf that fell from the trees in a given year and weigh them all! So, researchers estimate these values taking samples of the given measurement they want to know. In the case of leaf litter, you can put out trays in the forest and after all the leaves have fallen for the year, dry and weight the leaf litre in your trays. They you can use that value to calculate an estimate of the total leaf litter for your forest

*Step Five*: Use your ecosystem model to think about how changes can cascade through an ecosystem or to ask specific questions that can be answered with further research. When the major components, processes and inputs and outputs of an ecosystem are understood, then you can use the model how changing one part of the ecosystem affects other parts. If you harvest trees from your forest, that will decrease the amount of leaf litter reaching the forest floor each year which may lead to decreases in available nutrients for understory plants. This is the type of thing forest ecology researchers often study.

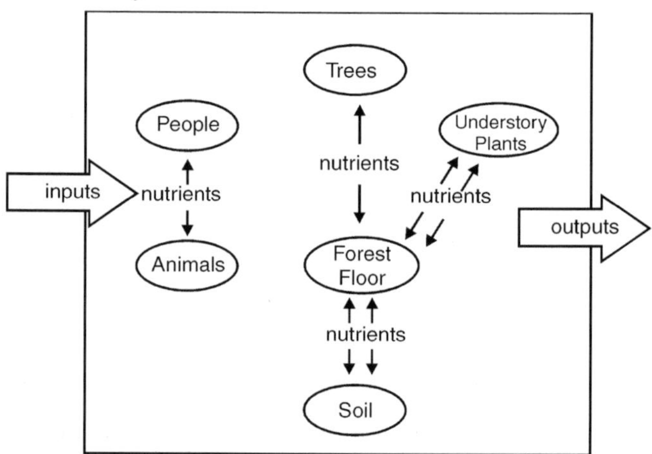


**Fig.** Hard wood forest Ecosystem Model.

An ecosystem that is in equilibrium doesn't gain or lose nutrients. Researchers may monitor soil nutrient levels for many years after trees have been harvested how the real forest behaves compared to what they thought might happen based on their forest model, their understanding of how the forest works. If the results in the real forest are very different than those predicted by their model, then they know that they don't have full understanding of how their forest works and they may go back to try to improve their model.

In general, when building a model, you start simple and add more detail as needed. To get an introduction to a very simplified forest model which gives participants and introduction to how a hardwood forest ecosystem works before and after exotic earthworms invade.

## Energy Flow Through the Ecosystem

Both energy and inorganic nutrients flow through the ecosystem. We need to define some terminology first. Energy "flows" through the ecosystem in the form of carbon-carbon bonds. When respiration occurs, the carbon-carbon bonds are broken and the carbon is combined with oxygen to form carbon dioxide. This process releases the energy, which is either used by the organism or the energy may be lost as heat. The dark arrows represent the movement of this energy.

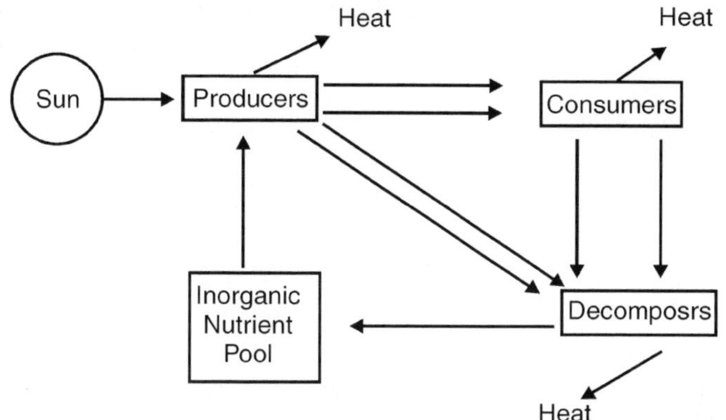


**Fig.** Energy and Inorganic Nutrients flow through the Ecosystem

The other component are the inorganic nutrients. They are inorganic because they do not contain carbon-carbon bonds. These inorganic nutrients include the phosphorous in your teeth, bones, and cellular membranes; the nitrogen in your amino acids; and the iron in your blood.

The movement of the inorganic nutrients is represented by the open arrows. Note that the autotrophs obtain these inorganic nutrients from the inorganic nutrient pool, which is usually the soil or water surrounding the plants or algae. These inorganic nutrients are passed from organism to organism as one organism is consumed by another. Ultimately, all organisms die and become detritus, food for the decomposers. At this stage, the last of the energy is extracted and the inorganic nutrients are returned to the soil or water to be taken up again. The inorganic nutrients are recycled, the energy is not.

Many of us, when we hear the word "nutrient" immediately think of calories and the carbon-carbon bonds that hold the caloric energy. When writing about energy flow and inorganic nutrient flow in an ecosystem, you must be clear as to what you are referring. Unmodified by "inorganic" or "organic", the word "nutrient" can leave your reader unsure of what you mean. This is one case in which the scientific meaning of a word is very dependent on its context. Another example would be the word "respiration", which to the layperson usually refers to "breathing", but which means "the extraction of energy from carbon-carbon bonds at the cellular level" to most scientists (except those scientists studying breathing, who use respiration in the lay sense).

To summarize: In the flow of energy and inorganic nutrients through the ecosystem, a few generalizations can be made:

- The ultimate source of energy (for most ecosystems) is the sun
- The ultimate fate of energy in ecosystems is for it to be lost as heat.

- Energy and nutrients are passed from organism to organism through the food chain as one organism eats another.
- Decomposers remove the last energy from the remains of organisms.
- Inorganic nutrients are cycled, energy is not.

## GREENHOUSE GAS CONCENTRATIONS

The atmospheric concentration of a greenhouse gas is a measure of the abundance of that gas in air, usually defined in terms of the proportion of the total volume that it accounts for. Greenhouse gases are trace gases in the atmosphere and are usually measured in parts per million by volume (ppmv), parts per billion by volume (ppbv) or parts per trillion (million million) by volume (pptv).

Despite their very low concentrations, they have the ability to regulate the global climate, by trapping heat that would otherwise escape to space. The natural greenhouse effect keeps the Earth's surface 33°C warmer than it should be. The enhancement of the greenhouse effect by man-made greenhouse gas emissions, including carbon dioxide, methane, nitrous oxide and CFCs, may have caused a global warming during the 20$^{th}$ century of 0.6°C, with another 3°C projected to occur in the 21$^{st}$ century.

The pre-industrial concentration of carbon dioxide in the atmosphere was about 280 ppmv. This concentration had remained fairly constant since the end of the last Ice Age about 14.000 years ago, when it increased from 190 ppmv. Coinciding with this prehistoric rise in atmospheric carbon dioxide was a global rise in average surface temperature of 5°C. The concern today is that the increase in atmospheric carbon dioxide since the late 18$^{th}$ century as a result of man-made carbon dioxide emissions is causing a further global warming. Atmospheric concentration of carbon dioxide is now about 370 ppmv, and is increasing 1.2 ppmv each year.

This concentration is higher than at any time during the last 160,000 years. This level of increase over the last 200 years has contributed roughly 60 per cent to the enhancement of the greenhouse effect. To prevent further increases in carbon dioxide concentrations would require a massive 60 per cent reduction of global carbon dioxide emissions.

The present global atmospheric concentration of methane is 1.75 ppmv, more than double the pre-industrial value of about 0.8 ppmv. Since the late 18th century methane has contributed about 20 per cent to the enhancement of the greenhouse effect. In order to stabilise concentrations of methane at present day levels, an immediate reduction in global emissions by 15 to 20 per cent would be needed.

The global atmospheric concentration of nitrous oxide - 314 ppbv - is now about 16 per cent greater than pre-industrial concentrations. Nitrous oxide has been responsible for about 4 to 6 per cent of the enhancement of the greenhouse effect. In order to stabilise concentrations of nitrous oxide at present day levels, an immediate reduction in global emissions by 70 to 80 per cent would be needed.

The concentrations of the CFCs in the atmosphere are very small, measured in parts per trillion (million million). However, they have contributed about 12 per cent to the enhancement of the natural greenhouse effect, because they are very good at trapping heat. Molecule for molecule some CFCs are thousands of times stronger than carbon dioxide as greenhouse gases. Concentrations of some of the CFCs are now beginning to decline since their use was phased out in the aftermath of the Montreal Protocol (1987) to prevent ozone depletion.

Concentrations of their replacements, the HCFCs however, are increasing. To date, HCFCs have contributed about 2 per cent to the enhancement of the greenhouse effect.

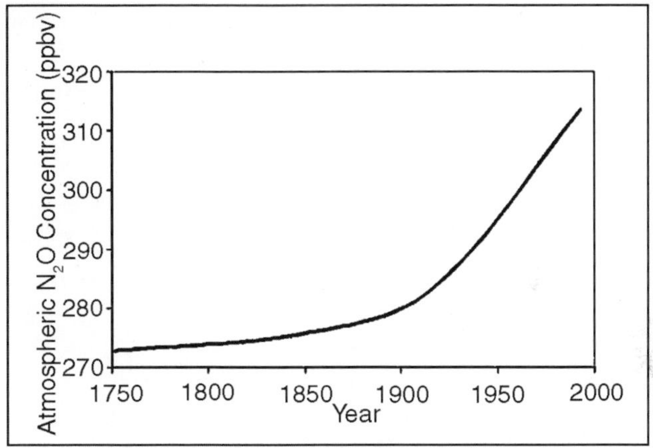


**Fig**. Increase in Atmospheric Nitrous oxide Concentration Since the 18th Century.

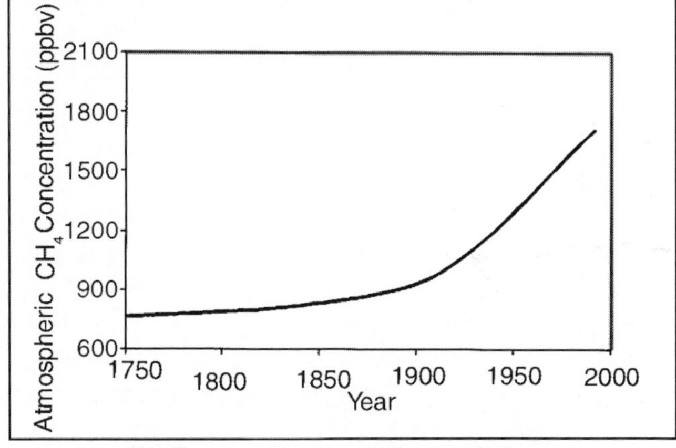


**Fig**. Increase in Atmospheric Methane Concentration Since the 18th Century.

## RADIATIVE FORCING OF GREENHOUSE GASES

Accompanying the increasing concentrations of greenhouse gases in the atmosphere has been an increase in greenhouse radiative forcing (enhanced greenhouse effect) due to the enhanced absorption of terrestrial infrared radiation. This radiative forcing can be quantified from the increases in greenhouse gases since the beginning of the Industrial Revolution.

### FACTORS AFFECTING GREENHOUSE RADIATIVE FORCING

A number of basic factors affect the behaviour of different greenhouse gases as forcing agents within the climate system. First, the absorption strength and wavelength of the absorption in the thermal infrared are of fundamental importance in dictating whether a molecule can be an important greenhouse forcing agent; this effect is modified by the overlap between the absorption bands and those of other gases present in the atmosphere.

For example, the natural quantities of $CO_2$ are so large (compared to other trace gases) that the atmosphere is very opaque over short distance at the centre of its 15μm absorption band. The addition of a small amount of gas capable of absorbing at this wavelength has negligible effect on the net radiative flux at the tropopause. Other greenhouse gases with absorption bands in the more transparent regions of the infrared spectrum, in particular between 10 and 12μm, will have a far greater radiative forcing effect. Second, the atmospheric residence time (lifetime) of a greenhouse gas can greatly influence its potential as a radiative forcing agent. Gases that remain in the atmosphere for considerable periods of time, before being removed via their sinks, will have a greater forcing potential over longer time horizons.

Third, the existing quantities of a greenhouse gas in the atmosphere can dictate the effect that additional molecules of that gas can have. For gases such as halocarbons, where the naturally occurring concentrations are zero or very small, their forcing is close to linear for present-day concentrations.

Gases such as methane and nitrous oxide are present in such quantities that significant absorption is already occurring, and it is found that their forcing is approximately proportional to the square root of their concentration.

For carbon dioxide, parts of the spectrum are already so opaque that additional molecules are almost ineffective; the forcing is found to be only logarithmic in concentration. As well as the direct effects on radiative forcing, many greenhouse gases also have indirect radiative effects on the climate through their interactions with atmospheric chemical processes.

For example, the oxidation of methane in the atmosphere leads to additional production of $CO_2$. Certain halocarbons such as the CFCs significantly affect the distribution of ozone, another greenhouse gas, in the atmosphere.

The hydroxyl (OH) radical, itself not a greenhouse gas, is extremely important in the troposphere as a chemical scavenger. Reactions with OH largely control the atmospheric lifetime, and, therefore, the concentrations of a number of greenhouse gases, in particular methane and many of the halocarbons.

## GREENHOUSE WARMING POTENTIALS

Although there are a number of ways of measuring and contrasting the radiative forcing potential of different greenhouse gases, the Global Warming Potential (GWP) is perhaps the most useful, particularly as a policy instrument.

GWPs take account of the various factors influencing the radiative forcing potential of greenhouse gases. Such measures combine the calculations of the absorption strength of a molecule with assessments of its atmospheric lifetime; it can also include the indirect greenhouse effects due to chemical changes in the atmosphere caused by the gas. A number of GWPs are listed in Table.

**Table. Global Warming Potentials of the Major Greenhouse Gases**

| Trace Gas | Global Warming Potential (relative to $CO_2$) | | |
|---|---|---|---|
| | Integration | Time Horizon, | Years |
| | 20 | 100 | 500 |
| $CO_2$ | 1 | 1 | 1 |
| $CH_4$ (incl. indirect) | 62 | 24.5 | 7.5 |
| $N_2O$ | 290 | 320 | 180 |
| CFC-12 | 7900 | 8500 | 4200 |
| HCFC-22 | 4300 | 1700 | 520 |

## ΔF-ΔC RELATIONSHIPS

The change in net radiative flux ($Wm^{-2}$) at the tropopause, ΔF, associated with a particular greenhouse gas, is usually expressed as some function of the change in atmospheric concentration of that gas. Direct-effect ΔF-ΔC relationships are calculated using detailed radiative-convective models.

The form of the ΔF-ΔC relationship depends primarily on the existing gas concentration, as explained in. For low/moderate/high concentrations, the form is well approximated by a linear/square-root/logarithmic dependence of ΔF on concentration.

*For example, the ΔF-ΔC relationship for $CO_2$ is given by*:

$$\Delta F = 6.3 \ln (C/C_0)$$

where $C_0$ is the initial $CO_2$ concentration, C is the final concentration and ln is the natural logarithm. This relationship is valid for concentrations up to 1000ppmv. Alternatively, the ΔF-ΔC relationship for CFC-12 is given by:

$$\Delta F = 0.22 (X-X_0)$$

where $X_0$ is the initial CFC-12 concentration and X is the final concentration. The relationship holds for X less than 2ppbv (2000pptv). Nevertheless, it should be appreciated that such relationships are empirical in nature and are therefore subject to uncertainties.

- First, there are uncertainties in the basic spectroscopic data for many gases.
- Second, uncertainties arise through details in the radiative-convective modelling.
- Third, the assumptions used to model $\Delta F$-$\Delta C$ relationships are subject to uncertainties, for example the assumed vertical profile of concentration, temperature and moisture changes, and the indirect effects on radiative forcing due to chemical interactions.

## GREENHOUSE RADIATIVE FORCING

From the modelled $\Delta F$-$\Delta C$ relationships, the increase in radiative forcing due to the enhanced greenhouse gas concentrations can be calculated. Instrumental records exist for the most recent decades, whilst proxy data are used to calculate greenhouse gas concentrations earlier in the study period.

Table gives the concentration changes for five of the commonly known greenhouse gases, whilst Table details their contribution to radiative forcing for a number of time intervals. Here, 1765 has been regarded as the onset of the Industrial Revolution.

**Table. Changes in Atmospheric Concentration of the Major Greenhouse Gases Since 1750**

| Year | $CO_2$(ppmv) | $CH_4$(ppbv) | $N_2O$ (ppbv) | CFC11 (pptv) | CFC12 (pptv) |
|---|---|---|---|---|---|
| 1765 | 279 | 790 | 275 | 0 | 0 |
| 1900 | 296 | 974 | 292 | 0 | 0 |
| 1960 | 316 | 1272 | 297 | 18 | 30 |
| 1970 | 325 | 1421 | 299 | 70 | 121 |
| 1980 | 337 | 1569 | 303 | 158 | 273 |
| 1992 | 355 | 1714 | 311 | 270 | 504 |

**Table. Change in Radiative Forcing ($Wm^{-2}$) due to Concentration Changes in Greenhouse Gases**

| Time period | $CO_2$ | $CH_4$ | $N_2O$ | CFC11 | CFC12 | Total† |
|---|---|---|---|---|---|---|
| 1765-1900 | 0.37 | 0.1 | 0.027 | 0.0 | 0.0 | 0.53 |
| 1765-1960 | 0.79 | 0.24 | 0.045 | 0.004 | 0.008 | 1.17 |
| 1765-1970 | 0.96 | 0.30 | 0.054 | 0.014 | 0.034 | 1.48 |
| 1765-1980 | 1.20 | 0.36 | 0.068 | 0.035 | 0.076 | 1.91 |
| 1765-1990 | 1.50 | 0.42 | 0.10 | 0.062 | 0.14 | 2.45 |

† Direct radiative forcing greenhouse gases (*i.e.* excludes radiative effects of ozone loss).

Changes in $CO_2$ concentration over the last two centuries have contributed most to the greenhouse radiative forcing. Over the period 1765 to 1990, $CO_2$ forcing has accounted for 61 per cent of the total enhanced greenhouse forcing. Nevertheless, other greenhouse gases, in particular the halocarbons, are now accounting for an increasing proportion of the total greenhouse forcing, due to their relatively larger GWPs. The total increase in radiative forcing since 1765 is shown in Figure.

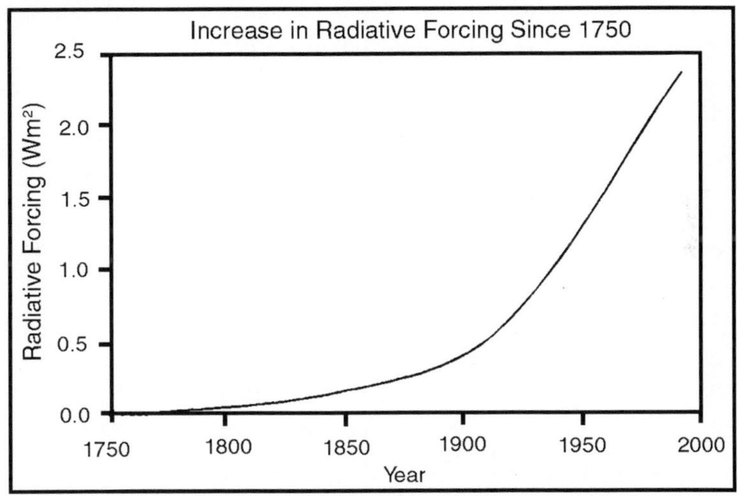


**Fig.** Increase in Radiative Forcing since 1750

The total increase in direct radiative greenhouse forcing is approximately 2.5Wm-2. This value should be compared to the solar constant, 1368Wm-2, the amount of total solar radiation intercepted by the Earth. The average amount of radiation arriving at the top of the troposphere is about 270Wm-2. This figure is lower than the solar constant, for it takes into account the latitudinal and temporal variations in insulation.

## RADIATIVE FORCING OF OZONE

The $\Delta F$-$\Delta C$ relationship for atmospheric ozone is more complex than for other trace gases because of its marked vertical variations in absorption and concentration. Changes in ozone can cause greenhouse forcing by influencing both solar and infrared radiation.

The net change in radiative forcing is strongly dependent on the vertical distribution of ozone concentration changes, and is particularly sensitive to variations around the tropopause.

Decreases in stratospheric ozone, principally over the Antarctic at altitudes between about 14 and 24km, have been occurring since the 1970s due to the anthropogenic release of CFCs and halons.

These changes in ozone substantially perturb both solar and long-wave radiation.

While the solar effects due to ozone loss are determined by the total column ozone amounts, the long-wave effects are determined both by the amount and its vertical location (Lacis *et al.*, 1990). In general, stratospheric ozone loss will tend to increase the solar forcing, resulting in surface-troposphere warming, whilst decreasing greenhouse forcing, with consequent surface-troposphere cooling.

Most models unambiguously demonstrate that changes in greenhouse forcing are dominant compute a global mean forcing of -0.2 ?0.1$Wm^{-2}$ between 1970 and 1990. Such a value represents a significant offset to the positive greenhouse forcing from changes in halocarbons over the same period, estimated at 0.22$Wm^{-2}$.

The build up of tropospheric ozone due to chemical reactions involving precursors produced in various industrial processes has potentially important consequences for radiative forcing.

Hauglustine *et al.* (1994) used a 2-D radiative-convective model to estimate that changes in tropospheric ozone since pre-industrial times have contributed a global mean forcing of 0.55$Wm^{-2}$.

This agrees well with other studies, both modelling and observational. Nevertheless, the increases in tropospheric ozone will be highly regional and so will the positive greenhouse forcing associated with them.

## ATMOSPHERIC AEROSOLS

Atmospheric aerosol particles are conventionally defined as those particles suspended in air having diametres in the region of 0.001 to 10 μm. They are formed by the reaction of gases in the atmosphere, or by the dispersal of material at the surface. Although making up only 1 part in $10^9$ of the mass of the atmosphere, they have the potential to significantly influence the short-wave radiative transfer. The recent addition of anthropogenic aerosols to the atmosphere has introduced a negative change in radiative forcing which partially offsets the positive greenhouse forcing.

## SOURCES AND SINKS OF AEROSOLS

Atmospheric aerosol particles may be emitted as particles or formed in the atmosphere from gaseous precursors (secondary sources).

Summarises the estimated recent annual emissions into the troposphere or stratosphere from the major sources of atmospheric aerosol, both natural and anthropogenic, primary and secondary. These include sulphates from the oxidation of sulphur-containing gases, nitrates from gaseous nitrogen species, organic materials from biomass combustion and oxidation of VOCs+, soot from combustion, and mineral dust from aeolian (wind-blown) processes. It should be noted that each aerosol flux estimate is a best guess, and uncertainty ranges of ±100 per cent are not untypical.

Removal of aerosol mass is mainly achieved by transfer to the Earth's surface or by volatilisation. Such transfer is brought about by precipitation (wet deposition) and by direct uptake at the surface (dry deposition). The efficiency of both these deposition processes, and hence the time spent in the atmosphere by an aerosol particle, is a complex function of the aerosol's physical and chemical characteristics (*e.g.* particle size), and the time and location of its release.

For fine sulphate aerosols (0.01 to 0.1μm) released into or formed near the Earth's surface, an average lifetime is typically of the order of several days. This time scale is dominated mainly by the frequency of recurrence or precipitation.

Conversely, particles transported into or formed in the upper troposphere are likely to remain there for weeks or months because of the less efficient precipitation scavenging.

Stratospheric aerosols, formed as a consequence of large volcanic eruptions can remain there for up to one or two years.

**Table. Recent Global Annual Emissions Estimates from Major Aerosols, Mt**

| Source | Flux Estimate | Particle Size† |
|---|---|---|
| Natural | | |
| Primary | | |
| Mineral aerosol | 1500 | mainly coarse |
| Sea salt | 1300 | coarse |
| Volcanic dust | 33 | coarse |
| Organic aerosols | 50 | coarse |
| Secondary | | |
| Sulphates from biogenic gases | 90 | fine |
| Sulphates from volcanic $SO_2$ | 12 | fine |
| Organic aerosols from VOCs | 55 | fine |
| Nitrates from NOx | 22 | mainly coarse |
| Total | 3062 | |
| Anthropogenic | | |
| Primary | | |
| Industrial dust | 100 | coarse and fine |
| Soot | 10 | mainly fine |
| Biomass burning | 80 | fine |
| Secondary | | |
| Sulphates from $SO_2$ | 140 | fine |
| Organic aerosols from VOCs | 10 | fine |
| Nitrates from NOx | 40 | mainly coarse |
| Total | 380 | |

† fine: <1μm diametre; coarse: >1μm diametre

Owing to the short lifetime of aerosol particles in the atmosphere, particularly in the troposphere, and the non-uniform distribution of sources, their geographical distribution is highly non-uniform. As a consequence, the relative importance of the numerous sources shown in Table varies considerably over the globe. In certain areas of the Northern Hemisphere, for example Europe and North America, industrial sources are relatively much more important.

# 7

# Atmosphere

## ATMOSPHERE

Some of the atmosphere' s profoundest mysteries are far above any possible ascent of man-carrying balloons or propellerdriven airplanes—beyond reach, even, of the twenty-five-mile sounding balloon. In the upper stratosphere there are levels warmed by the action of sun rays on ozone.

Beyond the stratosphere, in the ionosphere, there are levels where solar radiation separates innumerable free ions and electrons out of the extremely rarefied air. Some traces of our earth's atmosphere, becoming ever more and more tenuous with increasing distance, probably extend out to cosmic depths of two thousand miles.

Our present knowledge of the upper stratosphere and the ionosphere beyond is mostly indirect painstakingly gleaned, that is, from such intangible phenomena as sound reflection, light reflection, radio reflection, semi-cosmic clouds, meteors, polar lights, and so on.

Or, to put it plainly, our present picture of the upper atmosphere is largely limned with guesswork; and subject to constant correction as Doctor X announces a new and inexplicable observation, as Doctor Y propounds a new and scarcely intelligible theory, or as Doctor Z bobs up to confound Doctor Y. This state of affairs is likely to continue until the sounding rocket at last carries recording instruments, or possibly human observers, above the lower stratosphere into black depths beyond the blue.

Nevertheless, innumerable passive observations, as well as certain active experiments with sound waves and radio waves, have already revealed many interesting facts about the upper atmosphere.

## STRUCTURE OF THE ATMOSPHERE

The Atmosphere is divided into layers according to major changes in temperature. Gravity pushes the layers of air down on the earth's surface. This push is called air pressure. 99 per cent of the total mass of the atmosphere is below 32 kilometres.

## TROPOSPHERE

Troposphere – 0 to 12 km – Contains 75 per cent of the gases in the atmosphere. This is where you live and where weather occurs. As height increases, temperature decreases. The temperature drops about 6.5 degrees Celsius for every kilometre above the earth's surface.

### Tropopause

Tropopause - located at the top of the troposhere. The temperature remains fairly constant here. This layer separates the troposphere from the stratosphere. We find the jet stream here. These are very strong winds that blow eastward.

### Stratosphere

Stratosphere – 12 to 50 km - in the lower part of the stratosphere. The temperature remains fairly constant (-60 degrees Celsius). This layer contains the ozone layer. Ozone acts as a shield for in the earth's surface. It absorbs ultraviolet radiation from the sun. This causes a temperature increase in the upper part of the layer.

### Mesophere

Mesophere – 50 to 80 km - in the lower part of the stratosphere. The temperature drops in this layer to about -100 degrees Celsius. This is the coldest region of the atmosphere. This layer protects the earth from meteoroids. They burn up in this area.

### Thermosphere

Thermosphere – 80 km and up - The air is very thin. Thermosphere means "heat sphere". The temperature is very high in this layer because ultraviolet radiation is turned into heat. Temperatures often reach 2000 degrees Celsius or more. This layer contains:

### Ionosphere

Ionosphere - This is the lower part of the thermosphere. It extends from about 80 to 550 km. Gas particles absorb ultraviolet and X-ray radiation from the sun. The particles of gas become electrically charged (ions). Radio waves are bounced off the ions and reflect waves back to earth. This generally helps radio communication. However, solar flares can increase the number of ions and can interfere with the transmission of some radio waves.

### Exosphere

Exosphere - the upper part of the thermosphere. It extends from about 550 km for thousands of kilometres. Air is very thin here. This is the area where satellites orbit the earth.

## MAGNETOSPHERE

Magnetosphere - the area around the earth that extends beyond the atmosphere. The earth's magnetic field operates here. It begins at about 1000 km. It is made up of positively charged protons and negatively charged electrons.

This traps the particles that are given off by the sun. They are concentrated into belts or layers called the Van Allen radiation belts. The Van Allen belts trap deadly radiation. When large amounts are given off during a solar flare, the particles collide with each other causing the aurora borealis or the northern lights.

## LAYERS OF THE ATMOSPHERE

In a way, Earth's atmosphere is like an ocean of air that surrounds the land. It extends roughly 500 kilometres above the surface of the Earth and is split into layers. The layers are defined by sharp changes in temperature when moving vertically through the atmosphere. They are from the bottom of the atmosphere up to space: troposphere, stratosphere, mesosphere, thermosphere, and exosphere.

## GASES IN THE ATMOSPHERE

Earth's atmosphere is made of a mixture of gases. Surprisingly, oxygen is not the most abundant gas in the atmosphere. That honour belongs to nitrogen at roughly 78 per cent of all the gases. Oxygen comes in second at around 21 per cent. The last 1 per cent or so of the atmosphere is made up of such small amounts of other gases that it doesn't make sense to list each gas separately, so they are lumped together and called trace gases. Water vapour is another gas in Earth's atmosphere and it ranges from 0 per cent–4 per cent of the gases in the atmosphere, depending on temperature.

The other gases in the atmosphere adjust to fit the water vapour. Ozone is another important gas in the atmosphere. The stratosphere has the highest amount of naturally occurring ozone. Ozone in the troposphere is considered a pollutant. Ozone is made of three atoms of oxygen bonded together to make one molecule of ozone.

Ozone is destroyed by chemicals caused by chloroflurocarbons (CFCs). Researchers think that chlorine from these types of chemicals breaks the bonds between the oxygen atoms. This produces one molecule of regular oxygen, and one molecule of chlorine and oxygen bonded together. Along comes a free floating oxygen that bumps out the chlorine and bonds with the other oxygen. Now there is another regular molecule of oxygen and a free floating atom of chlorine.

This chlorine is now ready to go and destroy another molecule of ozone. This is bad for the ozone layer. Ozone absorbs ultraviolet radiation from the sun. Without it, skin cancer rates would increase.

## PRESSURE IN THE ATMOSPHERE

In general, as one moves away from the surface of the Earth, the air pressure lessens. As one moves higher into the atmosphere, the amount of air pressing down on an object is less than it would be if it were at the surface. Kind of like a football tackle. The player at the bottom of the pile feels the most pressure because of all the other players piled up on top.

The player on the top of the pile feels the least pressure since there aren't any players higher up. Changes in pressure can occur because of changes in temperature. This also results in changes in density of the gas. Warm air is less dense than cold air, therefore it rises and creates areas of low pressure. Cold air is more dense than warm air, therefore it sinks and creates areas of high pressure. This is a very important concept to understand in weather.

## ENERGY TRANSFER IN THE ATMOSPHERE

The energy that heats Earth's atmosphere comes from the sun in the form of radiation. Radiation is the transfer of energy in the form of electromagnetic waves. These waves travel from the sun, through space, down through Earth's atmosphere, where they hit the surface of Earth. The surfaces on Earth absorb the radiation and become warm. These warm surfaces are in direct contact with the air above the ground. The air above the ground becomes warm because the ground conducts the heat from the ground to the air. Conduction is the transfer of energy due to direct contact. This warm air is less dense than the cooler air surrounding it, and therefore will rise higher in the troposphere. This is an example of convection. Convection is the transfer of energy due to the flow of a heated material. Using radiation, conduction and convection, the lower portion of the troposphere is heated. The temperature difference between the warm and cold air creates a difference in density; warm air is less dense than cold air. As the warm air rises, it creates a low pressure area on the surface of Earth. As cold air sinks, it creates a high pressure area on the surface of Earth. When convection occurs, winds, storms, and clouds form.

| Type of Energy Transfer | How Energy is Transferred |
|---|---|
| Radiation | Electromagnetic waves |
| Conduction | Direct Contact |
| Convection | Differences in Density |

## WIND-FORMATION

Wind is caused by the uneven heating and cooling of the Earth. As warm air rises, it creates a low pressure area on the surface of Earth. As cold air sinks, it creates a high pressure area on the surface of Earth. As the warm air rises it pushes cooler air in to take the warm air's place. This circular movement of air is called a convection current. The horizontal movement of air on the surface of the Earth is known as wind.

## Wind-local Winds

Local winds are directly related to convection currents. These local winds blow around in a rather small scale (on the order of a town or so). In order to have wind, differences in pressure caused by differences in density must be present.

These differences are caused by the uneven heating and cooling rates of land and water. Land both heats and cools more quickly than water does. That means that in the day, the land will heat up rather quickly, but a water source such as a lake will take considerably more time to heat up.

On the other hand, once water is heated, it will take a relatively long time for it to cool down. You might have noticed this in the early summer (like in June) when you want to go swimming in a lake. Walking across the parking lot or sand, your bare feet feel strong heat, but the water itself will feel pretty cold. If you wait until evening, the water will feel warmer than it did during the day because it has now had all day to heat up.

The sand or parking lot will feel cold since most of the heat has dissipated. This uneven heating means that air over land during the day will be warmer than the air over water during the day. Warm air over the land will be rising, where it will cool and sink back to the Earth over the water. Air from over the water will rush in to take the place of the rising warm air over the land.

This breeze is called a sea breeze (or lake breeze) since it is coming from over the water. At night, the air over the water is warmer than the air over the land (since it has had all day to heat up, and it takes water longer to cool down than land does) so the air over the water rises, where it cools and sinks back down over the land. This air rushes in to take the place of the air rising over the water. This breeze is called a land breeze since it is coming from the land.

## Wind-Global Winds

Global winds are also caused by convection. These are belts of wind that circle the globe. A couple of things to think about when talking about global wind belts.

- Air would flow from the pole to the equator in a straight line, if it could (because cold air is more dense than warm air so it sinks at the poles, and rushes along the surface of the Earth towards the equator).
- Air doesn't flow in that straight path because of Earth's rotation. Earth's rotation causes the Coriolis Effect. The Coriolis Effect is the deflection of wind to the right (northern hemisphere) or left (southern hemisphere).
- If the Earth had no land masses, the winds would spiral around the globe in both hemispheres. Since the Earth has land masses, however, this swirling pattern is disrupted.

- This disruption causes the winds to be organized into belts or zones of wind. Moving either north or south of the equator they are: trade winds, prevailing westerlies and polar easterlies. These winds are named for the direction they come from. Right around the equator (from about 10° N latitude to about 10° S latitude) are the doldrums, or windless zone. Being a sailor stuck in the doldrums would be a very bad idea!

**Colour of the Sky**

It is easy to see that the sky is blue. The light from the Sun looks white. But it is really made up of all the colours of the rainbow. A prism is a specially shaped crystal. When white light shines through a prism, the light is separated into all its colours. Like energy passing through the ocean, light energy travels in waves, too. Some light travels in short, "choppy" waves. Other light travels in long, lazy waves. Blue light waves are shorter than red light waves. All light travels in a straight line unless something gets in the way to—

- Reflect it (like a mirror)
- Bend it (like a prism)or
- Scatter it (like molecules of the gases in the atmosphere)

Sunlight reaches Earth's atmosphere and is scattered in all directions by all the gases and particles in the air. Blue light is scattered in all directions by the tiny molecules of air in Earth's atmosphere. Blue is scattered more than other colours because it travels as shorter, smaller waves. This is why we see a blue sky most of the time. Closer to the horizon, the sky fades to a lighter blue or white. The sunlight reaching us from low in the sky has passed through even more air than the sunlight reaching us from overhead.

As the sunlight has passed through all this air, the air molecules have scattered and rescattered the blue light many times in many directions. Also, the surface of Earth has reflected and scattered the light. All this scattering mixes the colours together again white and less blue.

**Tropospheric Ozone, the Polluter**

Ozone occurs naturally at ground-level in low concentrations. The two major sources of natural ground-level ozone are hydrocarbons, which are released by plants and soil, and small amounts of stratospheric ozone, which occasionally migrate down to the earth's surface. Neither of these sources contributes enough ozone to be considered a threat to the health of humans or the environment. But the ozone that is a byproduct of certain human activities does become a problem at ground level and this is what we think of as 'bad' ozone. With increasing populations, more automobiles, and more industry, there's more ozone in the lower atmosphere. Since 1900 the amount of ozone near the earth's surface has more than doubled. Unlike most other air pollutants, ozone is not directly emitted from any one source. Tropospheric ozone is formed by the interaction of sunlight, particularly ultraviolet light, with hydrocarbons and nitrogen oxides, which are emitted by automobiles, gasoline vapours, fossil fuel power plants, refineries, and certain other industries.

High ozone levels usually occur during the warm, sunny summer months (from May through September). Typically, ozone levels reach their peak in mid to late afternoon, after the sun has had time to react fully with the exhaust fumes from the morning rush hours. A hot, sunny, still day is the perfect environment for ozone pollution production. In early evening, the sunlight's intensity decreases and the photochemical production process that forms ground level ozone begins to subside.

**Negative Impacts of Tropospheric Ozone**

While stratospheric ozone shields us from ultraviolet radiation, in the troposphere this irritating, reactive molecule damages forests and crops; destroys nylon, rubber, and other materials; and injures or destroys living tissue. It is a particular threat to people who exercise outdoors or who already have respiratory problems. Ozone affects plants in several ways. High concentrations of ozone cause plants to close their stomata. These are the cells on the underside of the plant that allow carbon dioxide and water to diffuse into the plant tissue. This slows down photosynthesis and plant growth.

Ozone may also enter the plants through the stomata and directly damage internal cells. Rubber, textile dyes, fibres, and certain paints may be weakened or damaged by exposure to ozone. Some elastic materials can become brittle and crack, while paints and fabric dyes may fade more quickly. When ozone pollution reaches high levels, pollution alerts are issued urging people with respiratory problems to take extra precautions or to remain indoors. Smog can damage respiratory tissues through inhalation.

Ozone has been linked to tissue decay, the promotion of scar tissue formation, and cell damage by oxidation. It can impair an athlete's performance, create more frequent attacks for individuals with asthma, cause eye irritation, chest pain, coughing, nausea, headaches and chest congestion and discomfort. It can worsen heart disease, bronchitis, and emphysema. So why can't we take all of this "bad" ozone and blast it up into the stratosphere.

The answer lies in the vast quantities needed and ozone's instability in the dynamic atmosphere. Ozone molecules don't last very long, with or without human intervention. The vehicle necessary to transport such enormous amounts of ozone into the stratosphere does not exist, and, if it did, it would require so much fuel that the resulting pollution might undo any positive effect.

Rather than seek such grandiose solutions, we need to decrease the production of those chemicals that break down ozone in the stratosphere and help create ozone in the troposphere. The dual ozone problems—pollution or smog in the troposphere and depletion of the ozone layer in the stratosphere—are indeed very different. But the problems have common ties in that they both are related to air pollutants that come from industry, transportation, and other human activities.

**Concluding Thoughts**

Citizens live in areas that are impacted by tropospheric ozone pollution. They are familiar with "smog-alerts," local government pleas to reduce vehicle traffic, and news reports about cities that have failed to meet EPA standards for ozone pollution levels. As you work through these activities with your class, you can easily connect the instructional activities to your students' surroundings by collecting and discussing news reports about smog issues in your own or nearby metropolitan areas. You may further wish to engage students in a discussion (or perhaps even a research project) about the impact of private automobiles on ozone levels and the prospects for alternative-fuel vehicles to reduce vehicle emissions and tropospheric ozone.

**Stratospheric Ozone, the Protector**

The debate over the existence of an ozone problem breeds media coverage. However, the real story is not whether stratospheric ozone levels are decreasing, but what those decreases may mean for life on earth. As the percentage of ozone in the atmosphere decreases, the amount of UV-B radiation reaching the surface increases. It's the UV-B radiation, not the ozone itself that concerns scientists, because the invisible wavelengths are linked to skin cancers and other biological damage. Measuring UV-B is tricky. Levels are affected by time of day, day of the year, latitude, weather conditions, and the amount of ozone aloft. UV is the part of the electromagnetic spectrum made up of wavelengths between 280 and 400 nanometers (billionths of a meter). Most of this is UV-A light, only mildly associated with sunburn and DNA damage and relatively benign to most plant life.

But the ill effects increase more than a thousandfold in the shorter wavelengths referred to as UV-B. Below 300 nanometers, the rays are sparse but very damaging; near 315 nanometers they're more numerous but much less destructive. Close to 310 nanometers lies the middle ground, where the number and impact of rays combine to cause the greatest harm to humans and plants. Engineers face enormous challenges when designing instruments that can measure individual wavelengths, yet such precision is necessary to determine the amount of dangerous light entering the atmosphere.

**The Story of the Ozone Hole**

Although often referred to as the ozone 'hole', it is really not a hole but rather a thinning of the ozone layer in the stratosphere. As suggested use the term 'hole' in reference to the seasonal thinning of the ozone layer. The appearance of a hole in the earth's ozone layer over Antarctica, first detected in 1976, was so unexpected that scientists didn't pay attention to what their instruments were telling them; they thought their instruments were malfunctioning.

When that explanation proved to be erroneous, they decided they were simply recording natural variations in the amount of ozone. It wasn't until 1985 that scientists were certain they were seeing a major problem. Why did it take scientists so long to solve this mystery? To begin with, observations that challenge preconceived ideas don't always get taken seriously, even in science. Two decades ago scientists did not suspect the importance of the chemical processes that rapidly destroy ozone in the Antarctic stratosphere.

When they saw dramatic fluctuations in ozone levels, they assumed their instruments were in error, or that whatever was happening was due to natural processes like sunspot activity or volcanic eruptions. They didn't realise that chlorine was the main culprit and that most of the chlorine in the stratosphere comes from human activity. The largest source is a class of chemical compounds known as chlorofluorocarbons (CFCs).

Because of their chemical stability, low toxicity, and valuable physical properties, these chemicals, versatile and stable in the lower atmosphere, at least, have been extensively used since the 1960s as refrigerants, industrial cleaning solvents, propellants in aerosol spray cans, and to make Styrofoam.

### Global Warming

Global Warming is caused by many things. The causes are split up into two groups, man-made or anthropogenic causes, and natural causes.

### Natural Causes

Natural causes are causes created by nature. One natural cause is a release of methane gas from arctic tundra and wetlands. Methane is a greenhouse gas. A greenhouse gas is a gas that traps heat in the earth's atmosphere. Another natural cause is that the earth goes through a cycle of climate change. This climate change usually lasts about 40,000 years.

### Man-made Causes

Man-made causes probably do the most damage. There are many man-made causes. Pollution is one of the biggest man-made problems. Pollution comes in many shapes and sizes. Burning fossil fuels is one thing that causes pollution. Fossil fuels are fuels made of organic matter such as coal, or oil. When fossil fuels are burned they give off a green house gas called $CO_2$. Also mining coal and oil allows methane to escape. How does it escape? Methane is naturally in the ground.

When coal or oil is mined you have to dig up the earth a little. When you dig up the fossil fuels you dig up the methane as well. Another major man-made cause of Global Warming is population. More people means more food, and more methods of transportation, right? That means more methane because there will be more burning of fossil fuels, and more agriculture. Now your probably thinking, "Wait a minute, you said agriculture is going to be damaged by Global Warming, but now you're saying agriculture is going to help cause Global Warming?" Well, have you ever been in a barn filled with animals and you smell something terrible? You're smelling methane. Another source of methane is manure.

Because more food is needed we have to raise food. Animals like cows are a source of food which means more manure and methane. Another problem with the increasing population is transportation. More people means more cars, and more cars means more pollution. Also, many people have more than one car. Since $CO_2$ contributes to global warming, the increase in population makes the problem worse because we breathe out $CO_2$.

Also, the trees that convert our $CO_2$ to oxygen are being demolished because we're using the land that we cut the trees down from as property for our homes and buildings. We are not replacing the trees (an important part of our eco system), so we are constantly taking advantage of our natural resources and giving nothing back in return.

## EFFECT FOR GLOBAL WARMING

### The Greenhouse Effect

When sunlight reaches Earth's surface some is absorbed and warms the earth and most of the rest is radiated back to the atmosphere at a longer wavelength than the sun light. Some of these longer wavelengths are absorbed by greenhouse gases in the atmosphere before they are lost to space.

The absorption of this longwave radiant energy warms the atmosphere. These greenhouse gases act like a mirror and reflect back to the Earth some of the heat energy which would otherwise be lost to space. The reflecting back of heat energy by the atmosphere is called the "greenhouse effect".

The major natural greenhouse gases are water vapour, which causes about 36-70 per cent of the greenhouse effect on Earth (not including clouds); carbon dioxide $CO_2$, which causes 9-26 per cent; methane, which causes 4-9 per cent, and ozone, which causes 3-7 per cent.

It is not possible to state that a certain gas causes a certain percentage of the greenhouse effect, because the influences of the various gases are not additive. Other greenhouse gases include, but are not limited to, nitrous oxide, sulfur hexafluoride, hydrofluoro-carbons, perfluorocarbons and chlorofluorocarbons.

**Global Warming Causes by Greenhouse Effect**

Greenhouse gases in the atmosphere act like a mirror and reflect back to the Earth a part of the heat radiation, which would otherwise be lost to space. The higher the concentration of green house gases like carbon dioxide in the atmosphere, the more heat energy is being reflected back to the Earth. The emission of carbon dioxide into the environment mainly from burning of fossil fuels (oil, gas, petrol, kerosene, etc.)

**Effects of global warming**

*There are two major effects of global warming*:

- Increase of temperature on the earth by about 3° to 5° C (34° to 41° Fahrenheit) by the year 2100.
- Rise of sea levels by at least 25 meters (82 feet) by the year 2100.

The effects of global warming: Increasing global temperatures are causing a broad range of changes. Sea levels are rising due to thermal expansion of the ocean, in addition to melting of land ice. Amounts and patterns of precipitation are changing. The total annual power of hurricanes has already increased markedly since 1975 because their average intensity and average duration have increased.

Changes in temperature and precipitation patterns increase the frequency, duration, and intensity of other extreme weather events, such as floods, droughts, heat waves, and tornadoes. Other effects of global warming include higher or lower agricultural yields, further glacial retreat, reduced summer stream flows, species extinctions. As a further effect of global warming, diseases like malaria are returning into areas where they have been extinguished earlier.

Although global warming is affecting the number and magnitude of these events, it is difficult to connect specific events to global warming. Although most studies focus on the period up to 2100, warming is expected to continue past then because carbon dioxide (chemical symbol $CO_2$) has an estimated atmospheric lifetime of 50 to 200 years. For a summary of the predictions for the future increase in temperature up to 2100.

**Atmospheric Composition**

Nitrogen and oxygen are the main components of the atmosphere by volume. Together these two gases make up approximately 99 per cent of the dry atmosphere. Both of these gases have very important associations with life. Nitrogen is removed from the atmosphere and deposited at the Earth's surface mainly by specialized nitrogen fixing bacteria, and by way of lightning through precipitation. The addition of this nitrogen to the Earth's surface soils and various water bodies supplies much needed nutrition for plant growth. Nitrogen returns to the atmosphere primarily through biomass combustion and denitrification.

Oxygen is exchanged between the atmosphere and life through the processes of photosynthesis and respiration. Photosynthesis produces oxygen when carbon dioxide and water are chemically converted into glucose with the help of sunlight. Respiration is a the opposite process of photosynthesis. In respiration, oxygen is combined with glucose to chemically release energy for metabolism. The products of this reaction are water and carbon dioxide. Water vapour varies in concentration in the atmosphere both spatially and temporally. The highest concentrations of water vapour are found near the equator over the oceans and tropical rain forests. Cold polar areas and subtropical continental deserts are locations where the volume of water vapour can approach zero per cent.

*Water vapour has several very important functional roles on our planet*:

- It redistributes heat energy on the Earth through latent heat energy exchange.
- The condensation of water vapour creates precipitaion that falls to the Earth's surface providing needed fresh water for plants and animals.
- It helps warm the Earth's atmosphere through the greenhouse effect.

The fifth most abundant gas in the atmosphere is carbon dioxide. The volume of this gas has increased by over 35 per cent in the last three hundred years. This increase is primarily due to human induced burning from fossil fuels, deforestation, and other forms of land-use change. Carbon dioxide is an important greenhouse gas. The human-caused increase in its concentration in the atmosphere has strengthened the greenhouse effect and has definitely contributed to global warming over the last 100 years.

Carbon dioxide is also naturally exchanged between the atmosphere and life through the processes of photosynthesis and respiration. Methane is a very strong greenhouse gas.Methane concentrations in the atmosphere have increased by more than 150 per cent. The primary sources for the additional methane added to the atmosphere (in order of importance) are: rice cultivation; domestic grazing animals; termites; landfills; coal mining; and, oil and gas extraction. Anaerobic conditions associated with rice paddy flooding results in the formation of methane gas. However, an accurate estimate of how much methane is being produced from rice paddies has been difficult to ascertain. More than 60 per cent of all rice paddies are found in India and China where scientific data concerning emission rates are unavailable. Nevertheless, scientists believe that the contribution of rice paddies is large because this form of crop production has more than doubled since 1950.

Grazing animals release methane to the environment as a result of herbaceous digestion. Some researchers believe the addition of methane from this source has more than quadrupled over the last century. Termites also release methane through similar processes. Land-use change in the tropics, due to deforestation, ranching, and farming, may be causing termite numbers to expand. If this assumption is correct, the contribution from these insects may be important. Methane is also released from landfills, coal mines, and gas and oil drilling. Landfills produce methane as organic wastes decompose over time. Coal, oil, and natural gas deposits release methane to the atmosphere when these deposits are excavated or drilled.

The average concentration of the greenhouse gas nitrous oxide is now increasing at a rate of 0.2 to 0.3 per cent per year. Its part in the enhancement of the greenhouse effect is minor relative to the other greenhouse gases. However, it does have an important role in the artificial fertilization of ecosystems. In extreme cases, this fertilization can lead to the death of forests, eutrophication of aquatic habitats, and species exclusion.

Sources for the increase of nitrous oxide in the atmosphere include: land-use conversion; fossil fuel combustion; biomass burning; and soil fertilization. Most of the nitrous oxide added to the atmosphere each year comes from deforestation and the conversion of forest, savanna and grassland ecosystems into agricultural fields and rangeland. Both of these processes reduce the amount of nitrogen stored in living vegetation and soil through the decomposition of organic matter. Nitrous oxide is also released into the atmosphere when fossil fuels and biomass are burned. However, the combined contribution to the increase of this gas in the atmosphere is thought to be minor.

The use of nitrate and ammonium fertilizers to enhance plant growth is another source of nitrous oxide. How much is released from this process has been difficult to quantify. Estimates suggest that the contribution from this source represents from 50 per cent to 0.2 per cent of nitrous oxide added to the atmosphere annually. Ozone's role in the enhancement of the greenhouse effect has been difficult to determine. Accurate measurements of past long-term levels of this gas in the atmosphere are currently unavailable. Moreover, concentrations of ozone gas are found in two different regions of the Earth's atmosphere. The majority of the ozone found in the atmosphere is concentrated in the stratosphere at an altitude of 15 to 55 kilometres above the Earth's surface. This stratospheric ozone provides an important service to life on the Earth as it absorbs harmful ultraviolet radiation.

In recent years, levels of stratospheric ozone have been decreasing due to the buildup of human created chlorofluorocarbons in the atmosphere. Since the late 1970s, scientists have noticed the development of severe holes in the ozone layer over Antarctica. Satellite measurements have indicated that the zone from 65° North to 65° South latitude has had a 3 per cent decrease in stratospheric ozone since 1978. Ozone is also highly concentrated at the Earth's surface in and around cities. Most of this ozone is created as a by product of human created photochemical smog. This buildup of ozone is toxic to organisms living at the Earth's surface.

### Atmospheric Effects on Incoming Solar Radiation

Three atmospheric processes modify the solar radiation passing through our atmosphere destined to the Earth's surface. These processes act on the radiation when it interacts with gases and suspended particles found in the atmosphere. The process of scattering occurs when small particles and gas molecules diffuse part of the incoming solar radiation in random directions without any alteration to the wavelength of the electromagnetic energy.

Scattering does, however, reduce the amount of incoming radiation reaching the Earth's surface. A significant proportion of scattered shortwave solar radiation is redirected back to space. The amount of scattering that takes place is dependent on two factors: wavelength of the incoming radiation and the size of the scattering particle or gas molecule.

In the Earth's atmosphere, the presence of a large number of particles with a size of about 0.5 microns results in shorter wavelengths being preferentially scattered. This factor also causes our sky to look blue because this colour corresponds to those wavelengths that are best diffused. If scattering did not occur in our atmosphere the daylight sky would be black.

### THE GREENHOUSE EFFECT

The greenhouse effect is a naturally occurring process that aids in heating the Earth's surface and atmosphere. It results from the fact that certain atmospheric gases, such as carbon dioxide, water vapour, and methane, are able to change the energy balance of the planet by absorbing longwave radiation emitted from the Earth's surface. Without the greenhouse effect life on this planet would probably not exist as the average temperature of the Earth would be a chilly -18° Celsius, rather than the present 15° Celsius. As energy from the Sun passes through the atmosphere a number of things take place.

A portion of the energy (26 per cent globally) is reflected or scattered back to space by clouds and other atmospheric particles. About 19 per cent of the energy available is absorbed by clouds, gases (like ozone), and particles in the atmosphere. Of the remaining 55 per cent of the solar energy passing through the Earth's atmosphere, 4 per cent is reflected from the surface back to space. On average, about 51 per cent of the Sun's radiation reaches the surface. This energy is then used in a number of processes, including the heating of the ground surface; the melting of ice and snow and the evapouration of water; and plant photosynthesis.

## STRATOSPHERE

The colour of the sky, of course, comes from the blue light which the air molecules scatter, and at sunset the light slanting through miles and miles of dense air is predominantly of penetrating red. The duration of twilight is a key, imperfect but not entirely valueless, to the density of the upper air; for the twilight afterglow, so infused with simple and reposeful beauty, is nothing but the shining of air particles high overhead in the rays of the sunken sun. Knowing the time after sunset through which twilight persists, perhaps an hour or two, and the consequent dip of the sun below the horizon, perhaps eighteen degrees or so, scientists have calculated that air sufficiently dense for this reflection extends to a height of about fifty miles.

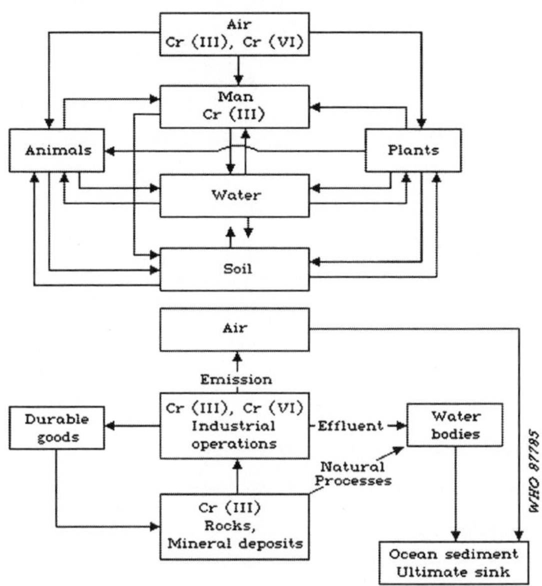


**Fig.** General Structure of the Air Ocean

This fifty-mile altitude is perhaps the upper limit of air as we can perceive it and imagine it the end of the stratosphere and the beginning of the ionosphere. Also to be seen at dawn and dusk under favourable conditions is the great domed shadow of the earth itself on the upper atmosphere, rising or sinking in that quarter of the sky opposite from the sun a few minutes after sunset or before sunrise.

It is seen best in crystal-clear weather from high altitudes. The curving shadow-are the sun lights the upper atmosphere with pale rose; below it, there is only dull-gray darkness.* Another sunset phenomenon that tells somewhat of the nature of the upper air is the rose-tinted glow known as the 'purple light,' seen after sunset on very clear evenings above the point where the sun disappeared. It is caused by the refraction and reflection of the longer-wave light through the upper atmospheric layers around the bulge of the earth. An observer out in space (say on the moon during a lunar eclipse) would see it as a continuous ruddy aureole around the whole globe.

Though the stratosphere and the black depths above it are cloudless in the ordinary sense, clouds of a very diaphanous kind are sometimes observed floating at these upper levels. At heights of fifteen to twenty miles patches of parti-coloured filmy iridescence, called nacreous or'mother-of-pearl' clouds, sometimes appear in the crystal-clear skies attending an invasion of cold, dry polar air after the passage of a cyclonic rain area.

These are distinct from, and much higher than, the cirriform clouds that sometimes show iridescence around their edges; in a true nacreous cloud, floating ten miles above the troposphere, the play of colour extends clear into the cloud centre. Nacreous clouds are perhaps composed of microscopic ice crystals or minute, super-cooled water droplets; and they show the existence of mid-stratospheric winds that are mostly from westerly and northwesterly in north temperate and sub-arctic latitudes, ranging in violence from flat calm to more than a hundred and fifty miles an hour. Considerably higher, marking the beginning of the ionosphere, are noctilucent clouds, occasionally seen after sunset in arctic and sub-arctic latitudes at heights around fifty to fiftyfive miles -- heights determined, of course, by triangulation from two or more ground stations situated along known base lines. Noctilucent clouds are excessively thin and wispy in appearance, and their colour shades from golden near the horizon to white and bluish-white above.

Unlike any of the lower clouds, they probably come from outer space and probably consist of excessively fine cosmic dust. Perhaps unable to penetrate the denser atmosphere below fifty miles, they always seem to hang just above this level. Appearing as faintly luminous clouds in the night sky, they are, nevertheless, actually lit by sun rays, and perhaps energized by solar ultra-violet, beyond the earth's shadow. Noctilucent clouds give evidence of superhurricane winds at the base of the ionosphere -- drifts from easterly and northerly that sometimes blow at two hundred or three hundred or even six hundred miles an hour!

In 1879, Josef Stefan discovered in old Vienna an important law of physics that might seem at first sight far removed from any usefulness to aerologists. Stefan's law is, in brief, that both the sort and intensity of light rays (both visible and invisible) emitted by any hot body are related to its temperature. The temperature of the sun is known; therefore, physicists can say how far its radiations extend into the ultra-violet. But it was found that, as we measure them on earth, solar ultra-violet rays stop short of this theoretical spectrum (which extends, with decreasing intensity, even down to X-rays). Hence something in the atmosphere must be absorbing them. That something has turned out to be ozone—the rare triatomic form of familiar oxygen.

Some slight traces of ozone occur near the earth's surface -popularly associated with pine woods and sea breezes. Measurable amounts are formed by electric sparks, and particularly by lightning flashes. But most of the atmosphere's ozone (still only a minute percentage of the other aerial gases) occurs at heights around fifteen to twenty miles—the zone of the nacreous clouds. At heights of twenty-five to thirty miles the ozone layer is more tenuous, but more important because it receives the full force of solar ultra-violet. It is a fortunate thing for all God's creatures that this great natural screen of ozone hangs up there thirty miles aloft, incredibly thin and diffused, transparent and invisible, and until recently unknown. For without it, the far ultra-violet and near X-rays shot forth from the sun would destroy all life on the sunlit earth within a few seconds. But the high-altitude ozone screen has an important effect on the upper air itself. In the words of Haurwitz, 'ozone as a powerful absorber of solar radiation determines to a large extent the temperature of the upper atmosphere.' We have already seen that temperature may increase somewhat with elevation in the lower stratosphere. In the upper stratosphere, where ozone enters powerfully into the picture, temperatures may be much higher—say up to summer heat (around +85°F) or even near boiling (around +160°F). These apparent high temperatures have been discovered by nothing more tangible than sound waves.

## REFLECTION OF SOUND

It is no new fact that powerful noises, such as heavy gunfire, are sometimes distinctly heard at distant places, though entirely unheard at nearer points. Back in 1666, Londoners heard the British Fleet cannonading the Dutch line-of-battle in the Channel, while the sound apparently skipped nearer places along the English coast. This circumstance puzzled Samuel Pepys and other philosophers of that day; and was not explained, in fact, until 1903, by a German named von dem Borne.

Since von dem Borne's time it has often been noticed that the sound of a loud explosion, though heard at great distances, passes completely over intervening points. This peculiar circumstance is best accounted for by supposing that there is in the upper atmosphere some sort of layer which acts in effect as a great sound reflector, and directs outgoing sound waves downward again towards the earth. In order to fit in with all the other known facts about the upper air, this reflecting 'roof' must take the form of a very warm and very diffuse upper-air layer—it is, in fact, the upper part of the ozone layer. As the velocity of sound is greater in warmer air, the outgoing sound waves travel faster in their upper portions (which encounter high temperatures first), and are hence gradually bent back to earth. The net result is the same as if they had been reflected back from the warm upper layer.

Many experiments to check the existence and height of this sound-reflecting layer were performed in Germany and England soon after the First World War Whipple, for example, set up some super-sensitive microphones while British artillery was firing on an island in the lower Thames. He heard the gunfire at Birmingham, two hundred and thirty miles away; the sounds had skipped intervening places entirely and seemed to come from the upper air.

Studies of the skip-distance patterns of sound from dynamite explosions, extending out to a distance of six hundred miles, point to the same general conclusion. The only way they can well be accounted for is by assuming high atmospheric temperatures at twenty to thirty miles altitude. The actual reflection summit, at about twenty-five miles, corresponds closely with what is known and assumed about the ozone layer.

The actual pattern of audibility of some great explosion, such as an ammunition dump or a shipload of TNT, is somewhat as follows: The blast is of course heard, and almost felt, out to a distance of perhaps twenty-five or fifty miles. From there to around a hundred miles is a 'skip zone,' or ring of comparative silence. From perhaps a hundred to a hundred and fifty miles extends the first zone of reflected audibility; then a second skip zone; then a second audibility zone from perhaps two hundred and fifty to three hundred miles; also, if the explosion is powerful enough, third and perhaps fourth or fifth skip and audibility zones. The sound of this battle practice was plainly audible at Santa Barbara, one hundred and twenty miles from the ships. It came in as a dull, jarring, almost soundless rumble that rattled house walls and windows like a mild earthquake. These upper-air sound reflections are subject to curious vagaries. There are differences between eastward and westward transmission, varying with seasonal effects on the upper-wind distribution. The reflecting layer is warmer and contains more ozone in winter; it is colder and less effective in summer. The curved reflecting layer sometimes serves to intensify the downcoming sound in a peculiar way. In one case, firing with sixteeninch guns actually broke windows in a city nearly two hundred miles away! The noisiest sound ever heard by civilized men blasted forth when a volcanic island called Krakatoa, between Sumatra and Java, blew up in 1883. The actual blasted-out material, pumice and rock, smashed and covered whole forests on islands miles away. The volcano eruption shot to a height of seventeen miles, and its black pall of concentrated dust extended out one hundred and fifty miles. An appalling ocean tidal wave rolled out, a hundred feet high, for thousands of miles, swamping hundreds of ships and drowning thirty-six thousand people.

The *actual sound* of this prodigious blast was faintly heard at Rodriguez Island, nearly three thousand miles away, and powerfully audible at many places more than a thousand miles distant! The atmospheric wave-disturbance, in sub-audible form, traveled several times around the earth. Also over the whole earth, carried by tropospheric and stratospheric winds during the next few months, spread thin brownish clouds of Krakatoa's dust.

## LIGHT

Definite and useful indications about the upper atmosphere are given by meteors, those swift points of light, erroneously called 'shooting stars,' that streak across the starry sky every few minutes on a clear, moonless night. As these bits of cosmic débris bore into the atmosphere at speeds of ten to forty miles per second, it is no wonder that they are intensely heated by friction and soon burst into flame. If they are observed or photographed simultaneously against a fixed-star background from two or more known locations, their altitude at successive instants can be determined, and it is found that most of them begin to flame about one hundred miles or more above the ground and burn completely out before reaching the thirty-mile level. The larger and slower ones, that is, may continue down to thirty miles, but most of the smaller and faster ones burn out around fifty miles. This knowledge, of course, gives a rough idea of atmospheric density at these heights.

Meteors are seen at their best and brightest in the small hours of the morning; for at that time one is on the 'forward' side of the earth as it swings at a comfortable eighteen miles a second along its orbit, and the fiery collisions of the meteors with our outer atmosphere are more or less head-on. Perhaps ten or twenty million meteors enter the earth's atmosphere every day. When we meet a definite swarm of them, this number is vastly increased; during the Leonid swarm of 1833 they appeared as a night-long shower of celestial snow.

Most visible meteors are not much larger than pinheads (ranging, perhaps, from birdshot to buckshot in size) and are completely burned up in the higher atmosphere. But occasionally a larger meteorite, or 'fireball,' weighing several pounds or even several tons and burning brightly enough momentarily to rival the sun, crashes through to the earth's surface. Larger meteors leave trails in the upper air that gleam phosphorescently at night, and resemble lines of white smoke by day. These trails have been seen to drift rapidly before dissolving, and their motion confirms the idea that enormously high winds can occur at heights above thirty miles. eteors also confirm another idea about the far upper atmosphere; for the calculated heights at which they should blaze agree with observed heights only if one assumes high temperatures at altitudes around thirty to fifty miles. This result, of course, strengthens the testimony of sound waves and ozone.

Despite their tenuous mystery, introducing new problems, the northern polar lights and their southern counterpart also offer some evidence about the upper air. In northern countries the active aurora is a sight never to be forgotten; widespreading curtains of flame appear and vanish like the draperies of an enormous stage, while arcs and bands of red and green merge into higher rays of violet and blue marching like searchlight beams across the depths of space. Carl Störmer has observed and photographed the aurora borealis many times in Norway. He finds that it occurs in two general varieties: the lower and more common types, mostly apple-green arcs and bands, flame in the earth's shadow zone at heights around sixty or seventy to one hundred miles. The higher types, mostly blue, violet and gray-violet rays, flicker at around two hundred miles, or even occasionally as high as four hundred miles, perhaps far up where the sun is shining across the pole above the blackness of arctic night. These heights correspond roughly, with the observed heights of the ionized layers that influence radio transmission.

The spectrum of the aurora shows that the rarefied gases of these far heights are mainly oxygen and nitrogen, rather than the helium and hydrogen formerly thought to predominate in the upper air. There is, in particular, one prominent green line that results when rarefied monatomic oxygen is electrically excited. Moreover, it has been found that this diffuse green glow is always present in the night sky, over all the earth, even above the equator.

It appears to come from atomic oxygen superheated by solar ultra-violet at heights around one hundred to two hundred miles. So the natural human conception of is somewhat of an illusion. Actually about one third of the moonless night sky's luminosity comes from the stars; more than one half comes from this highaltitude oxygen glow.

The primary cause of the aurora is probably 'searchlight' beams of electrons and ions shot out of the sun—beams that rotate with the sun's rotation, somewhat after the fashion of light beams from a rotating beacon, and sweep past the earth. Sometimes, when a brilliant aurora extends visibly to temperate latitudes, this solar 'beam-excitation' can almost be seen. Thus at midnight the northern aurora may appear from the south as successive outward pulses of light, originating at the left (west) of the sky and flashing progressively towards the right (east) of the sky. These pulses were very apparent in an aurora I witnessed some years ago in New York State. It began with great hanging curtains, but changed in an hour or so to a broad, white band along the northern horizon, surmounted by radial beams of white and pale green. From observed angles and the known distance of the magnetic pole, this aurora appeared to be about a thousand miles away from me, about a thousand miles wide, and several hundred miles high. The actual speed of the light pulses in the heavens appeared to be around one hundred or two hundred miles a second; and at the earth's distance a rotating solar beam should, in fact, sweep past at about two hundred and fifty miles a second.

## IONOSPHERE

Within recent years the aurora-haunted depths of the ionosphere have been probed daily and almost hourly by means of upward-sounding radio signals. Radio waves of course speed outward in all directions from the transmitter, along the ground and through the air, at the velocity of light. The ground part of the wave, however, usually dies out within a few score miles, leaving only the 'sky' wave for long-distance transmission. And the sky wave, to reach any distant place on earth, must somehow continue to bend around the bulge of our great globe. This bending of the outward-speeding sky wave is accomplished by what amounts to reflection in the ionized layers—layers where solar ultraviolet has disassociated free electrons and ions, increasing the electrical conductivity—of the earth's upper atmosphere.

Actually, just as we saw in the case of the warm ozone layer and sound, an ionized layer does not reflect radio waves; it refracts them by virtue of the fact that the outspeeding electromagnetic vibrations tend to move faster in the more intense ionization of greater heights within the 'reflecting' layer, and hence bend gradually downward again towards the ground. Nevertheless, this refraction process has the same net effect as reflection, so that one may think of the ionized layers as radioreflecting shells or 'roofs' in the upper atmosphere.

Within the last few years it has become possible to measure the height and other characteristics of the ionized layers almost at will. By means of a special transmitter, a radio 'tick,' or signal lasting less than one five-thousandth of a second, is shot upward. In something like one thousandth of a second it plunges a hundred-odd miles into space and returns to a special receiver on the ground, accurately recording the height from which it was in effect reflected—the height of an ionized layer. By this method the main reflecting layers have been probed and checked time and time again. These changes explain why a powerful broadcasting station is heard only a hundred miles or so by day, yet reaches out a thousand miles or more by night. When solar disturbances are active, some of the layer heights vary in erratic and spectacular fashion, flopping up and down by several miles in a few seconds and playing havoc with long-distance reception.

**OUTLINE**

Before leaving the upper atmosphere it might be well to gather up some of the essential facts—the aerological elements, as a scientist would say—into one unified picture. First of all there is the matter of pressure in the air ocean above us. Pressure decreases very uniformly with increasing height. This rule never varies; pressure cannot possibly increase with height. A long and complicated formula has been deduced, called the hypsometric formula, which shows the exact variation of pressure with altitude. The pressure at a given altitude is not always the same; it is influenced somewhat by the air temperature, and also very slightly by the amount of moisture in the air.

The complete hypsometric formula must therefore include temperature and moisture terms. In general, however, pressure decreases in the lower levels, by about one inch of mercury per thousand feet of altitude (or, in the metric system, by about one hundred millibars per kilometre of altitude). At upper levels, of course, the pressure decrease with altitude is much less. At sea level the average pressure is about one thousand millibars, or about thirty inches of mercury. At about three and one half miles altitude, pressure has fallen to five hundred millibars or fifteen inches, and half of the mass of the air ocean is therefore below one. This accounts for the extraordinarily clear, transparent air characteristic of high mountain summits and high altitudes in general; it also accounts for the difficulty in breathing that most humans begin to feel at ten or fifteen thousand feet altitude. At ten miles altitude the pressure fraction has fallen to a mere tenth; at twenty miles, it is something like a hundredth; at thirty miles, something like a thousandth.

Temperature, unlike pressure, varies with height in very erratic fashion. Even in the troposphere, temperature may decrease with increasing altitude, it may remain constant, or it may even increase as one goes higher. On the average, though temperature does decrease with height in the troposphere, by the amount of about 3°F per 1000 feet or about 6°C per kilometre. But above the troposphere, throughout the lower stratosphere, temperature is fairly constant regardless of altitude. This isothermal layer extends upward to about fifteen or twenty miles in winter; and perhaps extends higher, to about twenty-five or thirty miles, in summer.

Above this isothermal layer, temperature increases until, at heights around thirty miles (maximum absorption of ultra-violet by ozone), the almost vacuous air is probably as warm as the tropics at sea level (say around + 85°F or + 30°C), or perhaps up to near-boiling heat (say around + 160°F or + 70°C). Above this warm layer there is perhaps a cooler one. Then, at still greater heights above sixty miles (maximum absorption of ultra-violet by atomic oxygen), temperature may increase to several hundred degrees—the heat of a roasting oven—at least in the daytime. But at night these far heights are probably much colder—say below zero F. And above this upper warm zone, at altitudes of two or three hundred miles, for all we know, the temperature may decrease again to something near absolute zero.

The composition of the upper air has long been a bone of contention. Most meteorological texts printed prior to 1930 or so show the atmosphere above sixty miles (one hundred kilometres) as consisting largely of hydrogen and helium. More recent theoretical views of the matter, such as those of Chapman and Milne, Gutenberg and Haurwitz, have disproved this idea completely. In the light of present knowledge it appears that the composition of the atmosphere up to at least thirty miles, and probably up to sixty miles—so far as the proportions of various gases are concerned—must be very much like that of air in the lower altitudes.

At lower levels, in any case, the air consists on the average of about 78 per cent of nitrogen, 21 per cent of oxygen, nearly 1 per cent of argon, 0.03 per cent of carbon dioxide, 0.01 per cent of hydrogen, and even more minute fractions of neon and helium. At lower levels there is also, of course, a variable amount of water vapour, ranging up to about four per cent of the air in volume—a factor of great practical importance in tropospheric weather which will be dealt with at some length. In general this water vapour decreases with height, being almost absent from the stratosphere; but even up into the ionosphere, there are probably minute traces of it.

In the upper stratosphere the proportion of helium may increase somewhat. In the ionosphere the oxygen percentage may decrease, most of what oxygen there is takes the monatomic form, and the percentage of nitrogen probably increases. Helium and hydrogen are probably not present in these black depths above the blue, warmed to roasting heat by the super-glaring sun—for if present, they would soon diffuse away into cosmic space.

## CHANGES OF ATMOSPHERIC PRESSURE

Although atmospheric pressure, or weight of the air per unit horizontal area directly above the place where the pressure is measured, is not a weather element of any particular importance so far as our feelings are concerned, such as temperature and humidity are, nevertheless it is the element that more than any other locates the weather of today and tells us what tomorrow's will be, and therefore is of great importance. Latterly this importance has appreciably decreased, however, because other methods have come into prominence for finding the positions, magnitudes, speeds and directions of travel of the contrasting masses of air whose conflicts and whose passage keep the weather in ceaseless ebb and flow from fair to foul and foul to fair. Still these other methods have not yet been sufficiently developed to justify ignoring the distribution of atmospheric pressure, and how it is increasing here and decreasing yonder, as aids to a foreknowledge of the weather of the morrow. Neither is it at all likely that so great a development and use of these other means, while entirely possible, will be effected at an early date, if indeed they ever are so developed. Therefore a knowledge of the distribution of atmospheric pressure, the near equivalent of the distribution of the air itself, and of its changes, still are, and for a long while must remain, matters both of great practical importance and of keen theoretical interest.

### Barometre Corrections

Since the pressure of the air at any particular time and place is, as stated, the weight of all the atmosphere above a horizontal unit area then and there it follows that a knowledge of the values of this pressure over a given region, however extensive, tells us, when the proper corrections have been made, just what the distribution of the air itself is over that same region. But what are these correc- tions? That depends on the kind of instrument used to measure the pressure and on what we want to find, pressure or quantity of air, for we must not forget that the same quantity or mass of air, as also of other things, varies in weight, or pressure, with position not with every change of position but with most changes. But we can think easier and straighter if we consider one thing at a time, and know exactly what that one thing is. Let us first, then, compare the *masses* of air over different places. We can do this by means of a mercurial barometre which is only a device for balancing the pressure exerted by a column of mercury at its bottom, tending to make it flow out, against the pressure there of the atmosphere that keeps it back in. Clearly the pressure exerted by the mercury at the bottom of the column, when corrected for the capillary action of the tube, is proportional to the product of its density by the height of the column by the force of gravity, or pull oil a unit mass by the earth, at that place.

Similarly the pressure there of the atmosphere is proportional to the weight, or mass times the force of gravity, of all the air, if still, above that level (the level of the bottom of the mercury column) per horizontal square inch, or other specified area. When the air is in motion, however, as it always is, a further correction of the reading of the barometre is indicated, since the weight of a given mass varies with the direction and speed of its motion. But since the value of this correction (never known accurately) at most is negligibly small it will not be discussed here just mentioned for the sake of completeness.

Similarly the effect of the variation of gravity with height also is negligible. In this case then we may consider sauce for the goose as sauce for the gander too, in the sense that gravity applies in equal (almost) measure and exactly alike for both the mercury and the air. Hence the mass of the air above any given horizontal area, such as a square foot, say, at the level of the mercury in the barometre basin, is proportional to the height of the mercury column, or reading, as we say, of the barometre, corrected for whatever slight difference in that height there may be due to the pull, or capillary action, between the mercury and the wall (glass) of the barometre tube, and corrected also for the density of the mercury corrected to a common or standard density, commonly that which it has at the melting point of ice, 32° Fahrenheit, or 0° Centigrade.

In this way the readings of all the mercurial barometres are made strictly intercomparable, and the ratios of their several readings to each other exactly the same (to within an unknown but always negligible amount) as that of the masses of air above them, so that a geographic map on which these values, if simultaneous, are written, or otherwise represented, is also a map of the distribution of the atmosphere over that same region at the time the readings were made. To obtain the instantaneous distribution of the air the barometre readings necessarily must all be taken at that particular time, whatever it is, 8 A.M. 75th meridian time, or what not.

If, however, aneroid barometres are used instead of mercurial for obtaining the mass distribution of the atmosphere, then their readings, which indicate pressures directly, must be corrected for the local values of gravity, because the atmospheric pressure, being the weight per unit horizontal area of a vertical column of air, varies directly with the force of gravity which, in turn, increases with latitude. This quantity, or mass, distribution of the atmosphere is of considerable importance, especially through its influence on the rate of transmission, in or out, of radiation. A thin atmosphere and a thick one are to the earth much like a sheet and a blanket, respectively, to one sleeping in the open.

Although this mass distribution of the atmosphere is very important as a factor that helps to make the weather, especially as to temperature, what it is, still very little use ever is made of it. The distribution of pressure, or push, in the air is on the contrary frequently, carefully and extensively plotted because it is the unequal distribution of push on the atmosphere (or pressure, for they are equal to each other) that makes the winds to blow and thereby, in large measure, but not wholly, change the weather of one place by bringing to it the air and the weather of another. Clearly, then, a knowledge of this distribution must be, and is, a great help to the fore- caster when judging of the weather of the morrow and the day after. That is why the barometre is so extensively used in making meteorological observations and used almost exclusively for measuring the atmospheric pressure.

Now the pressure exerted by a column of mercury, the column in this case that is balanced by the pressure of the air, is proportional to the height of the column (corrected for capillary drag), to the density of the mercury (which in turn varies with the temperature) and to the local force of gravity. Hence, to compare the *pressures* at different places, one with another, it is necessary to correct the readings of the barometres to what they would be if all the instruments had the same temperature, say the temperature of melting ice, and were actuated by the same force of gravity its value at sea level at latitude 45°N or some other agreed-upon value.

The readings of aneroid barometres require no corrections for this purpose since, as previously stated, these instruments measure pressure directly. To sum up: If we are measuring the mass distribution of the air the readings of the mercurial barometre must not be altered for gravity differences, while those of the aneroid must be so changed.

On the other hand, when it is the distribution of pressure we are finding, the element we nearly always seek, it is the readings of the mercurial barometre that must be corrected to what they would be under some common force of gravity and not those of the aneroid barometre—they are true pressure values, if the instruments are correct, as they stand, and therefore must not be changed.

The next and final step is to reduce all these pressure values, or corrected readings of the barometre, to what they would be if taken at some common level. This is done because the travel or flow of the air, due to difference of pressure, is, of course, along the route of least resistance, which route, as a rule, is over a nearly level surface, that is, a surface everywhere normal to the direction of the force of gravity, such as the mean surface of the ocean. Actually the direction of least resistance to a sample of air through the surrounding atmosphere, the direction indeed of no resistance at all, save alone that incident to viscosity, is along what the physicist calls an isentropic surface. However, this exact and refined treatment of the subject will not here be followed to its conclusion, though to do so affords both surprises and mental satisfaction.

But let us have as little boggy ground as possible under our mental feet. Why is the level surface usually the one of least resistance to a sample of moving air? Because ordinarily the change of temperature of the atmosphere with change of height is such that work would have to be done on the sample to carry it to a greater height or push it to a lower level, but no work (save that due to viscosity which is the same in all directions) to move it horizontally.

Normally this distribution of temperature is such that a sample of air taken upward would cool, incident to its expansion under a decreased pressure, to a lower temperature than that of the then adjacent air. Being at the same pressure as this adjacent air but cooler, it also would be denser, so that an actual lift would be required to carry it above its initial level. On the other hand, if taken downwards the compression due to the increased mass of air above it would warm it to a higher temperature than that of the other air at its new position.

But being now warmer than the adjacent air, and having the same pressure, it also must be lighter. Hence, normally, only an actual downward push can force a sample of air at any height to a lower plane. In short, air normally can be taken out of its own level only at the cost of work, work against gravity if pushed up, and work against the buoyance of the denser air beneath if forced down.

Clearly then no work against gravity, nor any against buoyancy, is required when, and only when, air is moved over a level surface, or, to be exact, over an isentropic surface—a surface which, as already explained, usually is very nearly level. Obviously any direction over this surface is a direction of least resistance.

Therefore if we would know the magnitude of the cause of the winds, and from that magnitude infer their strength, we must find the values of the atmospheric pressure at the *same* level at various places over the region in question.

If we had a great number of these pressure readings somewhat evenly distributed over an area of considerable extent, say the size of the United States, or larger, we then could draw on a map of that region a line connecting all the places that, at a particular instant, had the same pressure. This line, according to present practice, we would call all "isobar," or line of equal pressure.

By drawing a number of such isobars, corresponding to small but constant differences in pressure, we construct a map or diagram from which the horizontal pressure gradient, or horizontal push per mile, say, can be found. Note that no two isobars (lines of equal pressure) can cross each other. Such crossing would signify two different pressures at the same time and place, a manifest impossibility. The direction of the flow of the air under these unequal pressures alone would be horizontal, as previously explained, and along the course of the shortest distance between the lines of equal pressure, that is, normal to these lines. A sample of air would not flow off slopingly to this normal to the isobars, though that too would be going from a greater to a lesser pressure, because every component of pressure directed to the right along or parallel to an isobar is exactly balanced by a component of pressure in the opposite direction. The only unbalance is strictly normal to the direction of the isobars, and therefore this is the direction of flow, or would be, if there were nothing to disturb or prevent such flow.

As a matter of fact this flow is modified by the rotation of the earth, and so profoundly modified as to change it from the direction of the push to the right angles thereto, except in the lower air where the drag of the surface makes the change in direction less than a right angle. But this fascinating paradox must be passed by. Its explanation would be out of place here; besides only the language of mathematics properly fits it anyway.

Since the movement of the air under differences of pressure is horizontal, or very nearly so, as just explained, it is clear that our isobars must be drawn for a common level if we would know from them what the values of the pressure gradients are that are making the winds of the moment to blow, fixing the intensity of a passing storm and largely determining whither and how fast it shall go. But what should be this common level, and how can we find the value of the pressure there?

These questions are old; their answers would be new. For the mariner they are simple enough: sea level is the place, and read your barometre where you are (on that level) is the way. But for the landlubber it is a different and awkward matter, since in general no two of his observing stations are at the same level. For the purpose of weather forecasting one can chart on his map the departures of the barometre readings of the several stations from their respective normal values for the season, or day, in question, and then connect by a smoothly curved line the places of equal, like departures.

Such a map can be, and has been, used with fair success, but it never became a general favourite, though it is the only pressure map that deals with known accurate magnitudes. One thing that renders such a map less helpful to the forecaster than at first one might expect it to be is the fact that the normal, or average, pressure for the given month, say, is one thing for weather of one kind and another for weather of a different kind. In short, the general average or normal for all days is not applicable alike to all sorts of weather conditions—the normal itself, or the observed departure from it, needs to be corrected for each special reading, but by how much we do not know. The pressure-departure map therefore is not entirely satisfactory and can not be made so.

Another promising procedure, and one that has been seriously attempted, is to use the pressures that exist at some definite height above sea level, half a mile, one mile, or what not. Wherever this level is above the surface, surely at that level there is air. Surely, too, instruments can be carried to that level and the actual pressure there measured with considerable accuracy. However, this direct manner of obtaining the desired pressure values is now too expensive to be practicable. Therefore the method used has been to compute the pressure at a given height from the pressure and temperature at the surface with a general formula and then correct the value found according to the circumstances, especially as to season and direction of the wind, as empirically deduced from data obtained under various conditions by kite, balloon and airplane. But at best more or less doubt as to accuracy always attaches to the values thus obtained. Hence weather maps that ostensibly show the distribution of atmospheric pressure over a horizontal surface at some definite height above sea level have not yet come into general use.

There remains, in this connection, one other obviously possible procedure, the one that has been, and now is, all but universally and solely used. This is, to reduce each pressure reading made on land to what it presumably would be at sea level immediately beneath the point of observation if all below that point, and round about, were not rock and soil, but air in that state and condition appropriate to the location, season and kind and distribution of the weather. The calculation, only approximate of course, is based on the known height of the station above sea level, and the observed temperature and pressure of the air, and then empirically more or less corrected. The chief source of error is the local surface temperature. Sometimes this is abnormally high and sometimes abnormally low. In the first case the reduction to sea level is made on the assumption that the hypothetical air beneath the place of observation is correspondingly warm and light, and in the second that it is correspondingly cold and heavy. The one yields a pressure too small and the other a pressure too great. Occasionally this error is so big that even the sense of the pressure distribution is reversed—actual cyclones and anticyclones, as indicated by the direction of the winds, changing by reduction to anticyclones and cyclones, respectively, on the map. Pity then the forecaster. But don't pity him overly much. He doesn't deserve it. He just winks the other eye, having in mind any poor devil trying to learn the game from such charts, mentally alters the map in accordance with his background of experience and all data available, some of which are never charted, and goes ahead serenely, if not always confidently. At his own game, and with loaded dice, of course he beats the innocent novice—until he, too, gets wise.

The one thing we definitely know about all reductions of the barometre to sea level, or any other level, is this: They invariably are in error. Generally, however, they are usable, and we do not know of any practical way to improve them. We know too that a "level" surface is not the correct surface on which to draw isobars; but then the surface that is the proper one, the isentropic surface, does not "stay put," as we say, but perpetually so warps and waves as to make pressure reduction to it quite out of the question. As a rule, however, this correct surface is, as previously stated, nearly level. Hence, and because reduction of pressure to a common height usually is only slightly erroneous, it follows that the sea-level map meets the needs of the weather forecaster fairly well, and better indeed than any other that as yet it has been practical to construct.

In this connection let us confess a little inconsistency and confusion of practice in our sea-level reductions. We reduce the pressure, or barometre, readings to sea level, but we leave the temperatures unchanged. Hence our weather maps show sea-level pressures and "mile-high" temperatures (themselves often exaggerated by surface conditions) side by side. This for the forecaster. When the climatologist sums the weather data all up he commonly reduces both temperature and pressure to sea level. Consistency, thou mayest be a jewel, but art not always prized.

## UPPER AND LOWER ATMOSPHERE

The composition of the lower atmosphere up to at least 6 or 7 miles in temperate regions, and 8 or 9 within the Tropics, is well known, except in respect to condensation nuclei and certain impurities. Throughout this region too the percentages of the several gases, except water vapour, are practically constant owing to their continual mixing incident to convection and turbulence. Somewhere in the upper air, however, the percentages of the lighter gases must increase and those of the heavier decrease under the action of gravity since at these levels vertical convection is very feeble.

But as the upper air grows thinner the more and more rapidly does our knowledge of it become less. We know that its outermost portion is the very extensive region of the aurora, hundreds of miles thick; that near the base of the auroral region there is enough ionization to make wireless communication around the world entirely practicable; and that far below this Kennelly-Heaviside layer, in turn, and yet well within the upper air occurs most of the ozone, the triatomic oxygen that indirectly is so vital to all terrestrial life. We seldom give any thought to this upper air, but it is so important that we really must know more about it.

## MASS OF THE ATMOSPHERE

From the known percentages of the several constituents of dry air, given above, their molecular weights, and various other pertinent facts such as the amount of water vapour present, height of the barometre, volume of land above sea level, and distribution of temperature with height, it is easy to compute the approximate mass of the atmosphere as a whole and of each of its several gases. The results are given in the following table, in which the factor 108 means: Add eight ciphers, or multiply by 100,000,000.

**Table. Mass of the Atmosphere and of its Constituents in Tons**

| *Substance* | *Volume per cent dry air, at surface* | *Total mass* |
|---|---|---|
| Total atmosphere | | 56,328,000 × 108 tons |
| Dry air | 100.00 | 56,181,850 " " " |
| Nitrogen | 78.03 | 42,684,725 " " " |
| Oxygen | 20.99 | 12,782,647 " " " |
| Argon | 0.9323 | 682,125 " " " |
| Water vapor | | 146,150 " " " |
| Carbon dioxide | 0.03 | 23,874 " " " |
| Hydrogen | 0.01 | 1,423 " " " |
| Neon | 0.0018 | 759 " " " |
| Krypton | 0.0001 | 141 " " " |
| Helium | 0.0005 | 88 " " " |
| Ozone | 0.00006 | 33 " " " |
| Xenon | 0.000009 | 19 " " " |

Since the values in this table were determined more or less independently it could not be expected that the percentages found of the constituents would add up to exactly 100, nor that the sum of the computed masses of the several parts would precisely equal the mass of the whole. These deviations, however, are very small—probably within the present limits of experimental errors.

The numbers here given that express the masses of the atmosphere and its several constituents are useful as quantitative values and for exact comparisons, but so great, even though in terms of tons, that we can form no distinct conceptions of them—they are just awfully big! A clearer idea may be gotten from the fact that the total mass of the atmosphere is the equivalent, roughly, of that of a block of granite a thousand miles long, a thousand miles broad and half mile thick; while the least abundant of the constituent, xenon, if loaded on cars, 19 tons to the car, would freight a train reaching 40 times around the earth along a great circle, and which, traveling 20 miles all hour, would be 6 years in passing any point on the road.

## TROPOSPHERE AND STRATOSPHERE

At about the close of the last century, soundings of the atmosphere were begun with the aid of small balloons carrying light devices that automatically registered the temperature and pressure throughout both the ascent and the descent—a kind of exploration that soon was taken up at various places. From these records, in turn, the heights corresponding to given points on the traces were readily computed. In this way we gradually have come to know the average temperature and pressure of the air at every height from the surface of the earth up to at least 12 to 15 miles, under different weather conditions, for all the seasons, and in many parts of the world. Pretty soon a means of registering the humidity was further added to the apparatus carried, so that we now have a fair knowledge also of the average vertical distribution of water vapour that corresponds to each particular type of weather. Obviously, too, these sounding balloons, as they are called, afford some knowledge of the direction and velocity of the wind at various levels for the particular time and place at which a flight is observed.

From the data thus obtained several interesting generalizations soon became evident. The most conspicuous of all, and for a long while the most doubted, because it was not understood, was the fact that at 6 or 7 miles above sea level, in middle latitudes, and generally 8 to 10 in tropical regions, the temperature no longer decreases with increase of height, but remains substantially constant. How far this equal temperature, or isothermal condition, extends is, of course, unknown, but it does go at least to the greatest altitudes yet attained, that is, 15 to 20 miles. It may extend to the limit of the atmosphere of appreciable density, or it may not. We have no direct and positive evidence of either alternative. This much we know.

From the surface up to a considerable distance above sea level, generally 6 to 10 miles, depending mainly on the latitude, the temperature decreases at the average rate of about 1°Fahrenheit in 300 feet; then almost abruptly, as a rule, practically ceases to change with further ascent. Clearly, therefore, this unsuspected and striking phenomenon divides the atmosphere into two great parts; a lower portion in which temperature rapidly decreases with increase of height, and an upper in which the temperature is nearly independent of height. In the lower, convection, or ascent and descent of the air, can occur under certain conditions, because although ascending air cools by expansion its ascent brings it into air that also is colder.

Sometimes the ascending air, especially when saturated with water vapour, cools less rapidly with increase of height than does the air through which it is passing, in which case convection is certain and often vigourous. Similarly, a local mass of air, particularly when cloud-laden, may warm less rapidly on sinking, as such masses do just after sun down, than the air it is falling through, whereupon convection again is inevitable. In short, this lower, cooling-with-altitude portion of the atmosphere is a region of convections and overturnings—not at all times, but under humidity conditions that will be explained later. It therefore has been called the troposphere, the sphere, or spherical shell, in which turning, or convection, occurs.

The troposphere, where the temperature is constant with height, marked or vigourous convection does not and can not occur. It can not occur here because rising air rapidly cools with ascent, owing to expansion incident to decrease of pressure on it by the weight of the air passed through. Such cooling would keep the rising air all the time colder and therefore heavier, volume for volume, than the surrounding air and thereby quickly reduce its motion first to zero and then reverse it. Formerly this was called the isothermal region, and to some extent it is still so called, owing to the fact that vertically, so far as explored, the temperature is substantially constant.

Since convection here is impossible masses of air forced into this region would spread out in horizontal strata, and indeed there commonly are evidences of the existence of just such strata, though we seldom if ever are sure of their origin. This filler structure suggested the other, and now the all but exclusive, name, "stratosphere," of this region. Thus the troposphere and stratosphere are quite distinct from each other. Clouds and every sort of precipitation, rain, snow, graupel, sleet and hail, and every kind of storm, are forever agitating the former, but never for a moment disturb the serenity or overcome the stability of the latter. They are sharply separated, the one from the other, along an approximately horizontal but invisible surface called the tropopause, and so called because that is where convection ceases, as above explained.

Perhaps it will be interesting to recall here that sometimes we use the expression, "the sphere," to mean the earth as a whole; that in our first approach to particulars we divide the sphere into the lithosphere, or solid portion of the earth, the hydrosphere, or water portion, and the atmosphere; and that the atmosphere in turn consists of two great concentric shells, the troposphere and the stratosphere. The lithosphere and the hydrosphere also have interesting structures about which many a fascinating tale has been told, but this is not the place to repeat them; ours is the story of the atmosphere and of it alone. When we examine the troposphere closely we find that it too has structure.

Some of these parts are fleeting, but others are at least semi-permanent. One of the more nearly constant is the trade wind, or, better, the trade winds, as there are several winds properly so designated. Advantage has been taken of them, of course, in shipping, especially by sailing vessels, but this is not the source of their name. It does not derive from any idea of commerce, but from the fact that their characteristic is persistent blowing along a particular trade, that is, tread, track or way. The trades are east winds (winds from easterly points) over the tropical oceans, or, more exactly, between the latitudes 30°N and 30°S. There really are five such winds that are well defined, one over the north Atlantic Ocean between latitude 30°N, roughly, and the Atlantic doldrums, or region of calms near the equator, and another over the south Atlantic between latitude 30°S, also roughly, and the same doldrums. Two other trade winds are similarly located over the Pacific Ocean, and there is one over the Indian Ocean, south of the equator, between Australia and Madagascar.

These wind currents, the trades, are very shallow near their poleward boundaries or edges, but grow gradually deeper and deeper to a maximum of at least 4 or 5 miles as their equatorial borders along the doldrums are approached. They are well defined and important structures of the troposphere.

**ANTITRADES**

Immediately over the trade winds are the antitrades, or winds that blow away from the equator and at the same time turn more and more nearly eastward and come down lower and lower with increase of latitude until they merge with, and become a part of, the prevailing westerlies-the winds beyond latitude 30°N or 30°S, as the case may be, that from the surface up to at least well into the stratosphere usually are from westerly points. The antitrade (there is one over each trade wind) represents a part of the return to higher latitudes of its accompanying trade, which through most of its route approaches closer and closer to the equator.

The rest of the trade, joined by a greater or less amount of tropical air, turns around the western end of a high pressure ridge, or ocean area of prevailing light and variable winds and clear skies. The frequent and deep southerly winds onto the United States from the Gulf of Mexico and the Atlantic Ocean are of one of those great joint trade and tropical currents.

Here then are two major features or elements in the structure of the troposphere, the trade and the antitrade, the east wind and the west wind, that all who pilot the argosies of the skies must know, and know how to use to proper advantage.

And the troposphere has many other parts. Two of the strangest and most important of these, strange because though apparently impossible companions they yet are always together, are the cold wind and the warm wind that peacefully flow beside each other—that, in technical terms, are each in dynamical equilibrium with the other. We know that a column of warm air will not stand up in equilibrium with an adjacent column of cold air, and we know that we know it by the way cold out-doors air pushes up and away the warm air in a heated chimney or smoke stack.

Hence we are unwilling at first to believe that a cold wind can blow alongside of a warm wind and not drive it away that the twain can meet on equal terms and each hold its own. But they can and do, because an object, such as a mass of air, whenever moving over the earth tends always, because of the earth's rotation, to turn to one side or the other (the right, going with the wind, in the northern hemisphere, the left in the southern) of the course it is on. And the magnitude of this tendency (the force the moving object would exert on a frictionless vertical surface that would hold it to a fixed geographic direction) is proportional both to its mass and to its velocity. In the case, therefore, of adjacent stationary columns of warm and cold air only gravity is operative and the heavier (because colder) column underruns and buoys up the lighter. This is why air rushes up heated chimneys and smoke stacks why they draw, and the higher the better.

In the case of the winds, however, the situation is quite different, for in proportion to their mass and their velocity they now exert (or would, against a suitable restraining wall) a horizontal or deflective force that also must be considered. If, then, the winds were in the right direction, and had the proper positions with reference to each other, the colder from the east, say, and the warmer just south thereof and from the west, or, more generally, if they were passing each other counter clockwise (in the northern hemisphere, clockwise in the southern) they might have such velocities that the resulting deflective force would just balance the difference in pressure due to difference in density.

Ordinarily this condition of equilibrium implies a very gently sloping interface between the warm current and the cold, with the former overrunning the latter. Owing, however, to surface friction and various obstacles, such as islands, and mountains, to the free flow of the winds, and to a change, for one reason or another, in the relative amounts of air on the two sides, perfect equilibrium never is long maintained. It does not break down abruptly and completely, but gives way quite slowly, and in so doing often leads to the development of the general cyclonic storm of the middle and higher latitudes.

## COLD FRONT

When the breakdown between the warm and the cold currents is well developed, with the mass of cold and, of course, polar air pushing its way equatorwards while the warmer air flows beside it poleward, we speak of the interface between them, or, rather, of the intersection of that interface with the surface of the earth, as a cold front. We also call this locus a windshift line because here the direction of the wind changes, as one system, the warm winds, departs and another, the cold, comes on.

It likewise is called a squall line because all along it the winds are turbulent and squally and, in summer, often accompanied by thunder storms. This cold front, then, is the boundary, along the surface, between distinct wind systems between a system of relatively cold and dry winds under skies and from higher latitudes, and a system of comparatively warm and humid winds under clouded skies and from lower latitudes two more of the great parts in the structure of the troposphere.

### Warm Front

The warmer of the two passing winds just mentioned soon reaches, in the course of its poleward travel, the colder air of higher latitudes. But here, as elsewhere, two masses of air, having different temperatures can not be in equilibrium with each other at rest and standing side by side. They must be, and are, in motion, and in such manner as to retain for a time, and to a greater or less extent, their individuality. Here again the colder air is wedged under the warm which flows not along this wedge, but up its slope slantingly as a rule, but up, nevertheless. The line along which this part of the interface between the cold and the warm winds cuts the surface is called the warm front.

Thus the cold and the warm sectors of the traveling cyclone maintain their independence as polar winds and tropical airs, respectively, until gradually separated from their source, after which they soon are brought into like conditions each with the air of its new location and merged with it. In this way polar air becomes tropical air, and tropical winds polar winds, back and forth ceaselessly and indefinitely; but always as entities in their travels as elements in the structure of the stratosphere just as rivers and lakes are entities in the ceaseless round of water from oceans to continents and continents to oceans.

**Inversions**

Normally the temperature of the lower atmosphere, or troposphere, decreases at every level with increase of height, but there are exceptions to this rule; and, besides, the rate of increase always is more or less irregular. A very common exception occurs next to the surface. Indeed, during still clear nights this exception is itself the rule, and practically without exception. At such times, owing to the rapid net loss of heat by the surface through radiation, the adjacent air, by contact with the cold surface, becomes cooled to a distinctly lower temperature than the air at a slightly higher level.

The height to which this cooling extends depends upon the amount of air movement and consequent turbulence or mixing. If this is slight the cooling is restricted to near the surface, but is all the more pronounced. If the movement of the air is appreciable the cooling is less pronounced but extends higher. Finally if there is considerable wind the mixing so distributes the loss of heat as to be small at any level, even at the surface. When, however, the cooling is marked and restricted to a shallow layer the temperature increases through this layer with increase of height instead of decreasing, as is the rule, from the ground up. This is an inversion of the usual temperature gradient, or a temperature inversion, or, for short, just an "inversion."

For a while the air next above a surface inversion layer can slowly blow along over it without rapidly mixing with the colder air or carrying it away thus providing another case of structure in the troposphere. Appreciable winds, however, do wear away such a layer rather rapidly. They also commonly tend to produce, and frequently do produce, a temperature inversion some distance, perhaps a thousand feet or more, above the ground. The surface inversion, discussed above, occurs, as explained, on still clear nights and is due to loss of heat by radiation, and therefore might be called a radiation inversion. The inversion now under consideration may occcur any time the wind blows, because it is due to the turbulence or vigourous mixing of the lower air incident to surface friction and the interference to free flow by trees, house and other irregularities.

Turbulence produces a temperature inversion the turbulence inversion in this way: Before the wind sets in, the temperature of the lower thousand feet or more of the air may and often does, decrease slowly with increase of height, perhaps only one degree Fahrenheit in 400 to 500 feet. It even may increase with height, especially of early mornings, through the lowest levels, as just explained.

In either case a complete mixing of the air from the surface up to the level of 1000 feet, says, brings all the air, not to a common temperature, as the stirring of water smooths out any thermal inequalities it may have had, but to such temperature that wherever, within that layer, a portion of its air may be taken, up, down or sidewise, it will, on arrival, have precisely the same temperature as the then surrounding air at the same level. That is, it will come to such temperature that nowhere will a moving portion of its either give heat to, or take heat from, the air with which it at any time is in contact. Obviously, because if the agitated air is not yet in a state of temperature equilibrium with a moving portion of itself, all that need be done to make it so is to stir it up further and mix it more thoroughly.

Now an isolated mass of air that neither gains heat from nor loses heat to the atmosphere through which it passes, obviously must cool with ascent, owing to its loss of heat incident to its expansion against the decreasing pressure. Where the air is unsaturated, so that no fog or cloud is formed, and thoroughly stirred up, as above explained, the rate of this decrease, and therefore the rate of temperature decrease in and of the layer itself is, approximately, 1°F per 190 feet, allowing for the average amount of humidity.

This, as explained, is a much faster rate of decrease of temperature than usually exists in the lower air. Hence the mixing of the lower air by turbulence so redistributes its heat as to make the lower portion warmer than it was, or otherwise would be, and the upper portion colder. Immediately above the topmost portion of the turbulent layer the air is undisturbed by vertical convection and therefore distinctly warmer than the upper portion of the agitated stratum. The turbulence inversion (inversion caused by turbulence) is a sharp partition between two portions of the troposphere, one the lower, full of irregularities, the other smooth-flowing.

Owing to its comparatively low temperature the top of the turbulence layer often is covered with a broad but shallow cloud of the stratus type which, because located at an inversion level, is warmest over its upper surface. But whether clouded or clear this inversion level is difficult of passage by air from either side. If air should pass this level going up it would at once be surrounded by other air much lighter, because warmer, than itself and therefore it would drop back. Similarly, air passing it from above would be promptly pushed up again by the denser, because colder, atmosphere it was replacing. In short, this inversion stratum, though very thin and commonly invisible, is an impassable ceiling to rising air from below and an impenetrable floor to falling air from above. Gradually, however, through heat conduction, thermal convection, and in other ways, the inversion is smoothed away, and interchange across this level thus made possible wherever the air has within and of itself an adequate supply of water or, really, steam power wherever its water vapour is sufficient to give abundant condensation, as in a cumulus cloud, and thereby a quantity of heat sufficient to keep it all the way to great altitudes continuously warmer and lighter than the surrounding medium.

It is interesting that as the temperature changes almost abruptly, perhaps several degrees, with change of height at the level of this turbulence inversion, so also must the density of the air change abruptly. And, furthermore, if the pressure gradient or push that causes the winds, is practically the same, as it seems to be, on either side of, and close to, this level, then there also must be here a nearly sudden jump in the wind velocity, from slower in the under, denser air to faster in the lighter air immediately above. However, this change of velocity, seldom more than one per cent of the whole, always is too small to be of any practical importance.

## OVERFLOW STRATA

As explained above, any layer of air that is thoroughly mixed up has a certain rate of decrease of temperature with increase of height, and such that the temperature of an isolated mass of like air rising or falling through it will change at the same rate. In such a layer vertical convection is as easy as horizontal gliding over a smooth surface.

Left to itself, though, and if unclouded, its vertical temperature gradient or lapse rate (lapse, for short) gradually changes, owing largely to gain and loss of heat by radiation, until it becomes decidedly less, and the layer thereby impenetrable to dry or unsaturated air. Saturated air, however, may be buoyed up to considerable heights, as already explained, because the heat of condensation, or heat rendered sensible as a result of condensation, so reduces the lapse rate that the ascending air is warmer and therefore lighter than the air surrounding it. The extent to which this rising air is warmer and lighter than the air through which it is passing depends, of course, on the amount of condensation. This in turn depends on the amount of water vapour present, and that depends on the temperature. Hence, in general, and starting from the same level, saturated warm air is pushed up to greater heights in the process of convection than is saturated cold air.

In any case, though, the amount of condensation per given increase of height, and therefore the quantity of heat available for further convection, becomes less and less with gain of altitude, and finally, at one level or another, insufficient to induce further ascent. At this level then, whatever it is, the rising air spreads out in a sheet or stratum that differs in humidity, temperature and lapse rate from the atmosphere of every other level, both higher and lower. In this way, that is, from convections, great and small, including the over- and underrunning associated with general or cyclonic storms, the troposphere is largely built up of overflow strata. They are not, of course, the same from day to day nor from place to place, but everywhere they are always more or less distinct and numerous. They often mark the levels of cloud layers of the sheet or stratus forms, of alto-cumuli and of the windrow or billow clouds, due to the waves caused by the flow of one stratum over another, much as water waves are induced by wind. Occasionally, two adjacent strata differ from each other so radically that a balloon can float a long distance with the bag in the one and the basket in the other.

Of course the identity of each particular stratum ultimately is lost through mixing with others above and below it, whether caused by the vigourous stirring incident to a general storm, or by virtue of the perpetual diffusion and prevalent turbulence over every interface. But so long as it does exist it may be pushed up bodily to greater heights by underrunning air, or depressed to lower levels by an overflow current, with, in either case, a change in the temperature gradient or lapse rate (except in the very unusual case when it initially is that of completely stirred-up air) and a corresponding change in its stability and resistance to penetration by convection in either direction, upward or downward. If the layer is pushed down, without lateral contraction or expansion, by an overflow of air above it the pressure on it will be increased by the same amount throughout, but it will be compressed most, and thereby heated most, on the upper side where the initial pressure is least, and compressed and heated least on the under side where the initial pressure is greatest.

This changes the temperature gradient in the stratum so depressed, except rarely, as above explained, and in such manner as to render the layer increas ingly difficult of penetration a firmer floor and a more rigid ceiling. On the other hand, if the stratum is lifted to a higher level it is cooled most on top and least at the bottom and its effectiveness as a barrier to convection corresponding decreased.

It should be noted that in atmospheric convection, and the consequent production of air strata, water vapour plays a most important role. With increase of humidity, under constant temperature and pressure, the density of the air steadily decreases, just as it would with increase of temperature at constant pressure. This fact probably accounts for many small waterspouts starting at the surface starting there because the lower air becomes relatively light through high humidity, analogous to the starting of dust whirls over a dry region due to decrease of density incident to increase of temperature. There is, however, a fundamental difference between the surface waterspout and the dust whirl.

The latter consists of dry air made light by increase of temperature, and can ascend (be pushed up) until it has lost a certain amount of its original heat and no further. Not so with saturated or highly humid air. It, too, like the dry air, ascends because it is lighter than the adjacent air around about, but it maintains this relative lightness a much longer time through the latent heat rendered sensible by progressive condensation, and thereby reaches far higher levels. In respect to convection dry air and humid air are like unto two men in business, one with a working capital but no reserve assets; the other having, in addition to his ordinary current needs a much greater fund that may be drawn upon whenever required. The one, like dry air, may start well, but his power to expand soon is exhausted. The other, by drawing on his reserve, can take advantage of every opportunity and thus rise to a far higher level of success. Just as it takes money to rise high in the business world, so too it requires water vapour to make any considerable ascent in the atmosphere.

*The konisphere (dust sphere) and its layers*. Not all we breathe is air. With every breath we inhale a million microsticks and -stones and a host of other things that are no part of a pure atmosphere. "Where do they come from?" The heavens above and the earth beneath. Every wind that sweeps a desert catches up tons, and sometimes millions of tons, of pulverized rock to spread far and wide. Fragments of vegetable fibre litter the soil the world over and are wafted hither and yon as even the gentlest breeze may blow.

Pollen of conifers, ragweeds, and a thousand other trees and plants we must take into our lungs from spring to fall every day we breathe the open air. And our bronchial tubes need chimney sweeps (luckily provided by Nature) to get rid of their coatings of soot from kitchens, factories and forest fires. Even the ocean, through its evaporated spray, makes a salt mine of the air that we breathe. Then, too, lightning sprays nitrogen acids into the atmosphere, while soft coal and volcanic vents similarly add the sulphur acids but all are too dilute really to bother us. Spores and microbes of many kinds we just have to inhale, for they are everywhere.

And as if all this were not enough the earth, every now and then, explodes at some great volcano and hurls tons upon tons of rock powder into the air where it drifts far away for weeks, months or years, according to its degree of fineness and initial height attained. Finally, in addition to all this dust of its own the world stirs up, the atmosphere to its outermost limits is filled with the ashes, so to speak, of daily millions of incinerated meteors, or shooting stars.

That is how the earth got its konisphere (dust shell). If it had no atmosphere it would have no konisphere, but having an atmosphere it must also have a coexistent and coextensive konisphere.But this konisphere is not uniform; it has distinct layers that, like other phenomena, show structure in the atmosphere.

*The more pronounced of these layers are*:

- The turbulence layer, that is, the layer of air next to the earth that, owing to surface friction, any appreciable wind fills with turbulence. Incident to this churning up of the air there also is a stirring up of the dust. At such times this is thedustiest portion of all the atmosphere, and it carries the largest particles. Its depth is, of course, that of the turbulence, and therefore may be anything from two or three hundred feet up to two or three thousand; while the amount of dust, as determined by the strength of the wind and condition of the surface, can vary from practically nothing at all, as over snow fields, to that of the terror of the desert—the blinding and stifling sand storm. Its upper boundary is rather sharply marked and often distinctly visible from any higher level. This layer includes also the city pall, that 4 tons a day, per square mile (average for Chicago, and there are worse places) smudge of soot and dirt that shuts out so much of the health-giving radiation of the sun.

- The convection layer, or stratum of diurnal convection, marked by the dust carried up from near the surface by warm ascending currents. It therefore is deepest and dustiest during summer droughts, and over arid regions. Its upper surface, perhaps two miles high, frequently is seen by the aviator or balloonist almost as distinctly as the surface of an ocean, and even to resemble that surface through the emergence above it here and there of cumulus clouds that look like so many islands. This is the next dustiest of the shells of the konisphere, but even so its burden seldom is heavy enough to bother in any way those who move about in its densest portion—at the surface of the earth.
- The tropic layer, or layer coincident with the troposphere, and therefore 6 to 7 miles deep in middle latitudes and two or three miles deeper, on the average, in tropical regions. Its top is the limit of even occasional convection, and the dust of its upper levels relatively both sparse and fine. Its upper surface rarely has been observed, since aeroplanes and balloons seldom pass that level. We know where that surface is, however, because every ascending current of air necessarily carries with it some of the dust of the lower levels, and therefore dust of terrestrial origin must extend to the upper limit of convection, that is, to the top of the troposphere, and no farther. We also have observational evidence of this upper surface through the effect of the dust particles on sunlight.
- The stratic layer, or the region of all the atmosphere of appreciable density beyond the troposphere. The dust of this region is of two parts; one roughly constant in amount, the other extremely variable. The first comes from the myriads of meteors that hourly enter the atmosphere. The second is due to occasional volcanic explosions of great violence which, like those of Asama in 1783, Krakatoa in 1883, Katmai in 1912, and many others, hurl powdered rock far beyond the levels of the highest clouds.

The heavier dust particles of the turbulence layer quickly settle of their own weight. To a less extent that is true also of the dust in the convection layer. In the main, however, the finest particles in the troposphere are carried down by condensation, either of water vapour directly onto them or as a result of being picked up by falling drops or drifting snow flakes. In this way the whole lower atmosphere from the surface of the earth to the tops of the highest clouds—a layer 6 to 10 miles thick—is literally washed, or scrubbed, as such processes are called, by rains and snows. If the air were so dry that there could be no precipitation it quickly would become suffocatingly filled with fine dust. In fact, it is believed by some that our sister planet, Venus, has just such a waterless, dust-filled atmosphere.

The dust of the stratosphere is not so fortunate, if we may put it that way, as that of the lower levels. It must get down through this region by itself, for there is not enough vapour up there to lend it any aid on its earthward journey. Often it is years in getting out of this arid realm, but once it has covered that part of its course the rest of the trip is quick and easy by way of the snowflake and raindrop routes. Even the dust of the earth, therefore, reveals a considerable structure of the atmosphere—at least four distinct layers. And it falls into still other great divisions according to this or that basis of separation. Nearly all the foregoing concerns the troposphere. The little that follows relates to the stratosphere, about which our knowledge still is very slight.

**Upper Trades**

Since the stratosphere is much warmer, 30°F to 40°F, in the polar regions, than in the equatorial, it would seem that there must be an upper interzonal circulation of the atmosphere somewhat like a mirrored image of the lower—towards the poles in its under portion and from them in its upper levels. The rotation of the earth obviously would affect this upper interzonal circulation in the same manner that it does that of the troposphere, and therefore lead to east winds over the tropical and adjacent regions, and west winds over the higher latitudes. Furthermore, a little calculation based on the temperature of the atmosphere at various levels in high and low latitudes shows that this upper circulation must begin at the height of 10 to 12 miles. This calculation further shows that at that level the average pressure must be nearly the same everywhere, and therefore the average wind at this level very light. Both these conclusions, namely, that the winds 10 to 12 miles above the surface of the earth must be light, and above that level from the east in tropical and adjacent regions, are supported by all the observational data (a fair amount) we have on the subject.

**Twilight Top**

A small beam of sunshine in a darkened room because there are dust particles in the air that scatter the light. That is why a few whisks over the floor with a dry broom makes the beam brighter. This explains, too, why we can see the shaft from the searchlight, and the focusing "streams" when the sun is "drawing water."

In each case the contrast is between the myriads of illuminated motes and the shaded, hence darker, portions of the surrounding air. Not only the dust particles, but also, though to a far less degree, all the gas molecules of the atmosphere, are luminous in sunshine. This air luminosity is the chief factor in the blue of the sky, and an important factor in other sky colours.

It accounts also for the twilight arch the visible boundary between the shadow of the earth and the illuminated atmosphere that rises above the eastern horizon as the sun sinks beneath the western.

Since the observer is himself within the earth's shadow it is obvious that by noting the exact time this arch is directly over head, say, he may know, from certain astronomical tables, just how many degrees the sun is then below the horizon; and that from this value, in turn, and the radius of the earth he can compute the height of this arch, that is, the greatest height at which the density of the air still is sufficient to scatter a perceptible amount of incident sunshine.

Numerous measurements of this kind have been made, and that height thus found to be about 44 miles. In respect, then, to its efficiency as a light-scattering agent also the atmosphere has structure, an inner shell about 44 miles thick in which the scattering is appreciable, and an outer in which it is imperceptible.

**Auroral Base**

The polar lights, both northern (aurora borealis) and southern (aurora australis) divide the atmosphere into distinct parts, an inner, about 62 miles deep, into which auroras generally do not penetrate, and an outer of unknown thickness, but certainly hundreds of miles, in which they commonly do occur.

**Kennelly Heaviside Layer**

When radio-telegraphy over long distances was first attained we were much puzzled to know how it could be, for surely radio waves are just greatly magnified light waves, and light doesn't bend to the curvature of the earth in such manner that an object can be seen a thousand miles away. But if the air were highly transparent, and both the earth and the encircling sky excellent reflectors, then a powerful light at London say, might well be seen from New York, or any other place on the globe.

The light could not get through either reflector and therefore would keep on traveling between them until finally absorbed. The same is true also of electric waves, and for these the earth is a reflector. If therefore, the sky reflected them too we would expect long distance radio communication to be possible, but not otherwise. Hence, when such communication was accomplished, Heaviside and Kennelly told us that the sky must be a reflector of electric waves, that is, an electric conductor, and everybody answered. "Why, of course, it is a conductor." And then came the long search to find how it is made a conductor and at what level its conductivity is adequate to account for the phenomena observed.

We believe now that this conductivity is owing essentially to the presence in the upper atmosphere, 30 to 60 miles or more above the surface, of a million or so free electrons per cubic inch, due to solar radiation in the far ultra violet. The height of the under surface of this reflecting region, or Kennelly-Heaviside layer, as it commonly is called, is not sharply determined since it appears to vary with the wave-length of the incident wireless wave; nor is it constant for any given wave-length, but varies from day to night and from season to season.

But despite these variations the atmosphere, in respect to its electric state, and its relation therefore to wireless waves, always consists of two parts: a highly ionized, conducting and wave-reflecting outer shell, the Kennelly-Heaviside side; and a relatively non-conducting, but wave-transmitting inner shell. Electrically, also, the atmosphere has structure.

### Ozone Layer

The composition of the atmosphere, there is very little ozone (triatomic oxygen) in the lower air up at least to the level of the highest clouds, but certainly very much more somewhere beyond that height. From spectroscopic observations several persons have computed the height of the centre of gravity of the ozone to be 25 to 30 miles above the surface of the earth. But whatever the correct value of this Iteight, surely in respect to ozone also the atmosphere has its structure—an intermediate shell rich in ozone, and an inner one and an outer that contain practically none at all.

## DISTRIBUTION OF TEMPERATURE

The pull, as we call it, of gravity makes water run down hill. It also makes a heavy liquid underrun a lighter one in the same level; both are drawn in the direction of the bottom, but the pull on the heavier, or denser, is greater than on the lighter, and the stronger pull prevails.

Gravity also makes an isolated mass of liquid or gas in a heavier one go up, not down; it is pushed or buoyed up by a force equal to the difference between the weight of the lighter and that of an equal volume of the heavier. Clearly, then, whenever two masses of air of unequal density come into free contact with each other the lighter is pushed up and away, except in the case of properly adjusted winds. Now air rapidly increases in volume, and correspondingly decreases in density, with increase of temperature roughly 1 per cent per 5°F at ordinary temperatures. Hence the hot air in a chimney is lighter than an equal volume of the cold air on the outside, and therefore is pushed up by the latter which, in turn, is heated and itself pushed up, and so on as long as there is a fire in the grate to supply the heat. To be sure, the combustion alters the composition of the air (makes it richer in carbon dioxide if coal is used, and in both carbon dioxide and water vapour if wood or gas is the fuel, and poorer in oxygen) in such manner as to render that in the chimney heavier, at the same temperature, than that outside, but this increase in density through change in composition is small in comparison to its decrease in density by heating.

At most it could balance or offset a temperature increase of only about 40°F over coal, or 10°F over wood, while ordinarily the effect if much less, since commonly only part of the oxygen is consumed; hence the heating, being several times this maximum value, has, in any case, the best of the argument, as it were, and the chimneys keep on drawing.

Similarly, air in the open is underrun and pushed up by even slightly cooler adjacent air of the same composition, unless, as already explained, the two masses happen to be flowing past each other in the right positions and directions and with the proper velocities. Actually, the heated air expands as its temperature rises, and overflows above wherever its pressure is thus made greater than that of the adjacent atmosphere.

This overflow, or outflow, decreases the pressure at the bottom, and in the lower portions, of the heated air, and at the same time increases the pressure round about under the places of overflow—mass, hence weight, is removed from one place and added to others. This disturbs the balance. Gravity tends to restore it and thereby induces winds in the direction, initially at least, of higher to lower pressure.

If the heated region is very small, equilibrium is quickly established, unless the heating is maintained. But where the higher temperature covers a large the winds no longer flow directly towards the centre of lowest pressure but more or less round about it, owing to the rotation of the earth, in a manner seemingly most contrarious. This heating in innumerable cases is very local and of only a few hours' duration; in many others it is quite extensive and lasts days Weeks, and even all season long; while its greatest manifestation is the year after year and age after age continuously higher temperature in tropical realms and lower in the frigid zones.

This perpetual heating of the atmosphere over one great region, and its ceaseless cooling over another, or rather, two others, keeps it continuously out of balance and makes the winds, especially the trades and the westerlies, forever to blow—to blow dizzily over a rotating earth, and time and again violently and confusedly incident to the rapid, the all but explosive, delivery, by condensation, to a limited region of vast quantities of heat that had been slowly accumulated by evaporation from others afar off. The whole of the atmosphere to the tops of the highest clouds, that is, the whole of the troposphere, is a huge convection system, greatly complicated by the rotation of the earth and all but hopelessly confused by evaporation and condensation. The stratosphere, too, has its circulation, but as yet not much is known about it. Of course it is difference in pressure *at the same level* that pushes the air about, or makes the winds to blow, but, as explained, this difference in pressure depends, in turn, mainly on the distribution of temperature. That is one reason, but not the only one, why this distribution is so important. Perhaps some good physicist will insist that it really isn't difference of pressure at the same height above sea level that makes the winds blow, but difference of pressure over an "isentropic surface," or surface of "equal entropy."

Well, he would be right in respect to appreciable heights above the surface, because for the free air the isentropic surface is the "level" surface. But nothing short of a surgical operation can get the idea of entropy into the other fellow's head, and there is no rivet, weld, or hermetic seal that will keep it there. Besides, commonly (not always), there isn't much difference between the two after all—"same level" and "isentropic level"—and so we will stick to the one everybody knows and no one forgets, that is, "same level."

**Source of heat**

When we think of the source of heat, especially in the winter-time, we are likely to have in mind some sort of combustion, for that is the cause of the tropical climate we have indoors at that season.

But indoors is a mighty small place in comparison with all outdoors; and outdoors is heated, too, often very hot in summer, and always far above the 460° below zero Fahrenheit that would be its temperature if there were no heating at all. Almost every bit of this enormous amount of heating comes from just one source, the sun.

**Incoming Radiation**

The radiation from the sun is so great that if it all got through the atmosphere enough would fall on each square foot directly facing it to heat a gallon of water from the freezing point to the boiling point in just three and a half hours. But it does not all get through, and what does get through always comes in slopingly except wherever the sun happens for the moment to be directly overhead.

In fact, owing to the reflecting power of clouds, especially, and the surface of the earth, and to the scattering of light by the molecules of the air and by the myriads of dust motes, one-third, roughly, of the incoming radiation is thrown off to space without producing any effect whatever on the temperature of the atmosphere or of the earth beneath.

Another one-third, again roughly, of the incoming solar radiation is absorbed by the atmosphere, and the remaining portion by the earth. These statements apply to the earth as a whole. The ratios between loss by reflection and scattering, air absorption, and earth absorption, vary widely from place to place and season to season, owing mainly to differences in humidity, cloudiness, and elevation of the sun above the horizon, and differences in the character of the surface of the earth—whether land, water, snow or ice, bare soil or vegetation.

### Clear Sky Radiations

It is interesting to note that the amount of radiation reaching the earth from a clear sky is equal to a considerable fraction of that which reaches it from the sun directly. At sea level the amount of sky radiation onto a horizontal surface of any particular size, a square foot, say, is equal to about 7.8 per cent of the amount of unaffected, or direct, solar radiation onto an equal area squarely facing the sun at the same time and place.

When the sun is directly overhead its supply of heat to a horizontal surface at sea level is nearly 13 times as great as that from the sky. When it is one-third of the way down from the zenith to the horizon its contribution of heat to the earth is only a little more than 6 times that from the sky, and each is then decidedly less than it is when the sun is in the zenith. Finally, the two sources are equal, though both are still further enfeebled, when the sun is above the horizon about one-twelfth the distance to the zenith.

The brightness of the clear sky is greatest near the sun, as even casual observations readily show, and decreases gradually with increase of distance therefrom over a large part of the whole area. Hence the total of sky radiation received per minute on a horizontal surface is greatest at noon, as is also the direct solar radiation. The intensity of sky radiation decreases, in general, with increase of height above sea level, while that of the direct solar radiation increases.

## RADIATION FROM AN OCERCAST SKY

When the sky is overcast, neither its brightness nor the total amount of radiation received on a given horizontal area is at all constant, even for the same height of the nun, because the clouds in question may be of any kind from the thinnest cirrus that just dims the sun, to the darkest nimbus that reduces even noonday brilliance to twilight. If, however, the sky is completely overcast by an approximately uniform cloud layer dense enough, but not greatly more than enough, to prevent the position of the sun from showing, then the total radiation from this cloud layer onto a horizontal surface is, on the average, slightly greater than that from a clear sky. Evidently, too, the amount received of this cloud-transmitted radiation generally must increase with increase of height above sea level of the place of reception. The brightness of the cloud layer is surprisingly close to uniform. It is greatest nearly overhead (just a little way off in the direction of the sun), but still nine-tenths as bright half way to the horizon, and half as bright almost at the horizon.

## DISPOSAL OF RADIATION OF SURFACE

There are just three things that can happen to radiation incident onto any extended object. It must be reflected, transmitted or absorbed. When the object is extremely small it more or less scatters incident radiation, and radiation that just grazes the boundary of an object suffers still another effect which we call diffraction. However, neither scattering nor diffraction occurs when the object is large and its edges are not involved. They do not, therefore, occur in the case of radiation incident on the surface of the earth.

Here then, the incident radiation is all used up by two processes, reflection and absorption, since there is no transmission—no passage of radiation through the earth and out at the other side. The portion reflected is about 70 per cent for snow-covered regions, and 7 per cent for the rest of the world.

The remainder is absorbed, that is, 30 per cent wherever there is snow, and 93 per cent at all other places, both land and water. That which is reflected is lost except in so far as it is absorbed by the air above. The absorbed portion goes largely to heating the upper layers of the soil or water, but not all of it, since a considerable part is consumed in maintaining evaporation, and a much smaller part in effecting plant growth and development. Another relatively small part merely melts snow and ice without raising their temperature above the freezing point. The heated surface in turn heats the soil or rock by conduction, but appreciably to a depth of only a few feet.

The heating of water extends to a greater depth owing partly to the penetration of the rays to some distance below the surface, and partly to the mixing of the water by wave action. The heated surface also warms the air above it both by direct contact and by radiation. Furthermore, the heated air through convection shares its warmth with other and colder air above; and the heat consumed in evaporation at one place is liberated, that is made sensible or temperature-producing, some other place, usually in mid-air, where condensation occurs, and far away.

Practically every bit of this heating of earth, ocean and air, and supply of energy for evaporation, plant growth, ice melting, and what not else, comes from the sun—all directly except about one part in half a million that reaches us after reflection by the full moon and the planets. A negligibly small amount comes from the fixed stars, enough to keep the average temperature of the out-doors air about two millionths of a degree Fahrenheit higher than it otherwise would be. Finally, another very small amount comes from the heated interior of the earth.

## QUANTITY AND EFFECTS OF HEAT

If the earth had no atmosphere, and if there were no sun or stars to send us a flood of radiation, the supply of heat from the interior (of which four-fifths, roughly, is from radioactive material) alone would keep up the surface temperatures to about 60° absolute, on the Fahrenheit scale, that is, -400°F, approximately. Hence the flow of heat from the interior of the earth per square foot of surface is sufficient to raise the temperature of a gallon of water about 1°F in 16 days, and the total flow through the whole surface out to space sufficient to heat 92,000 tons of water from the freezing to the boiling point every second of time; or enough, starting at room temperatures, to melt 1,000,000 tons of lead per second.

These are big figures, and yet all this flow of heat keeps the actual temperature of the surface of the earth only about 1/25 of a degree F. higher than it otherwise would be. The figures also tell us the surprising story that if 10,000 times as much heat came from the interior of the earth as now actually does come, or, what amounts to the same thing, if everywhere there was a sea of molten cast iron covered over with a layer of rock and dirt only 10 to 12 feet thick, the oceans above could rest thereon serene with no close approach to the boiling point, so excellent an insulator, or poor a conductor, is this material; and that if the dirt and rock crust were 20 feet thick we could go about over it ourselves in perfect comfort.

**Outgoing Radiation**

On the average, the earth loses to space, or emits to space, by radiation very approximately the same amount of heat each year that it absorbs of incoming radiation during the same time, plus, of course, the supply of heat that reaches the surface from the interior. We know that the loss is substantially equal to the gain because otherwise the surface would be growing warmer from year to year, and we know that this loss is by radiation as there is no other way for the loss to occur—there being no such thing as conduction to empty space. The amount of this loss, or radiation of the entire earth to space, can be estimated from our knowledge of the incoming radiation and the fraction of it that is ineffective through scattering and reflection, especially by clouds.

It can not be measured directly because we have no means of getting out beyond the atmosphere and from that ideal place pointing our heat-gathering apparatus towards the earth. But, as implied above, we can make a pretty close estimate of the average rate at which radiation is going out from the earth as a whole, and the conclusion is that it is very nearly the same as that from a perfect radiator, or "black body," at the absolute temperature 454° on the Fahrenheit scale, or -6°F. This is sufficient, per square foot of surface, to heat a gallon of water from the freezing to the boiling point in about 20 1/2 hours.

The radiation from the surface of the earth often is very much greater than this value, even twice as great, or more, because the temperature of the surface frequently is far higher than the -6°F, here assumed. On the other hand, at times and places, owing to very low temperatures, it is much less. On the average, however, the radiation from the surface is much in excess of that which finally gets away to space-greater by the amount of return radiation it absorbs (nearly all of it) from clouds and the atmosphere.

The surface of the earth radiates at a relatively high temperature, hence in comparative abundance. Some of this radiation goes directly through the atmosphere, but ordinarily most of it is absorbed by the water vapour and clouds in the lower air, and a little by other things, especially ozone (when the sky is clear, for it is above all clouds) and carbon dioxide. That which is absorbed in the lowest layers is, in general, reradiated, but at a lower temperature than that of the surface. This reradiation is in every direction, half of it downward, some of which is absorbed on the way, and the rest by the surface whose initial temperature and radiation it thus helps to maintain; and half upwards to the next higher layers; and so on up and up from layer to layer, but always with decreasing absorption by the air still above and increasing absorption by that below until the entire atmosphere is left behind.

**Clouds do not Check Radiation**

It is a well-known fact that during still clear nights the surface of the earth, and, through it, the adjacent air, cool to a much lower temperature, especially over level land and in valleys and bowl-like depressions, than they do when either the sky is clouded, or the wind is strong. Furthermore, the lower the clouds, other things being equal, the less the cooling. That seems very simple, and would be but for one little fly in the ointment—there isn't a word of truth in it. It might do perhaps as a dose of mental paregoric for a kid with the quizz colic, but it is no good for anything else. Just one thing alone ever reduces radiation, and that is decrease of temperature. No Clouds do not check in the least radiation from the surface below. They are, however, themselves good radiators; and as their temperature, when they are low, is nearly that of the earth, they send down to it almost as much radiation as it itself emits, and as practically all this cloud radiation is absorbed by the earth, it follows that the surface temperature remains substantially constant.

It isn't that the radiation from the earth is checked in the least, but that it receives from the cloud canopy and absorbs wellnigh as much as it itself gives out. Neither is the approximately constant temperature maintained by an appreciable wind owing to any check whatever exerted by it on surface radiation, but to the fact that the net loss of heat thus sustained, and on clear nights it is considerable, is distributed by turbulence through such a deep layer and great quantity of air that the fall in temperature is small even when the total loss of heat is large.

**Temperature of Surface Air**

When we talk about the "surface" air it often is advisable to explain just what air we have in mind, for this term is quite flexible. We might mean only that air which is in actual molecular contact with the surface, or that which at most is within a few inches of it, or finally, all below the height of eight or ten feet, the air to which we chiefly are exposed while outdoors.

The temperature of the surface air, in any one of these senses, is determined mainly by that of the surface itself. Whatever the temperature of the surface, that also is the temperature of the contact air, and very nearly the temperature of all that air which by turbulence or otherwise is frequently brought into contact with the surface. On the stillest of nights this layer at places may be only a few thick.

During the daytime, however, especially when there is sunshine to induce thermal convection, and whenever, day or night, there is a measurable movement of the air, it is certain to be at least a good many feet thick. The essential point in this: The temperature of the air near the surface (how near varies with the circumstances) depends more on contacts with that surface than it does on the amount of solar radiation to which it may be exposed. The surface temperature of course does vary with the intensity and duration of the sunshine, and so therefore does also that of the surface air, but indirectly through contact with the surface and not directly by absorption of solar energy.

### Relation Between surface Temperature and Temperature of Surface Air

As above stated, the temperature of the actual contact air must be the same as that of the surface (of the substance, ground or what not, at its surface) against which it rests. If this surface air remained fixed, as we often are told that it does, then it would seem that the air next in contact with it also should become fixed in position, and so on indefinitely. But we know that fixity of position of the air molecules does not extend to a measurable distance from any solid, for we can blow smoke past it and see the motion of the air. We therefore are forced to the conclusion that fixity of position does not apply, at least not for any appreciable length of time, even to the contact molecules. The way out of the difficulty appears to be this: The actual contact molecules of the air are at rest, like a liquid film, but they do not stay at rest. They evaporate, and as they leave the surface others condense thereon—are adsorbed—a continuous process the details of which are not yet all known. In this way the contact molecules, during the extremely brief interval of their contact, are fixed in position, but they are continuously reverting to the gaseous state, and therefore the atmosphere at ordinary temperatures is always fluid however measurably near it may be to the surface in question.

Since the air is directly heated chiefly by contact with the surface of the earth, and indirectly by the sharing of this heat, through convection, with colder air above, it follows that in general wherever the temperature of the lower atmosphere is increasing, that is, over nearly all snow-free land, and particularly during the day time, there the average temperature of the surface is higher than that of the free surface air. This is in accordance with what physicists call the second law of thermodynamics, and what everybody else knows without calling it anything, namely, that the temperature of the heater is higher than the temperature of the thing heated. Similarly, where the lower air commonly is cooled, as it is over snow-covered regions, there the average temperature of the surface is lower than that of the surface air—the cooler is colder than the thing cooled.

**Maximum and Minimum Temperatures**

Obviously if the heater changes temperature, so also will the heated, and the heater will be the first to change and the first to reach its extreme values—maxima and minima. There is no surprise, therefore, in the fact that the daily maximum temperature of a snow-free land surface occurs earlier, about 1 o'clock P.M., than that of the air above it, which is delayed until around 3 o'clock.

The air and surface minima, occurring near daybreak, are much closer together, owing partly to the slow cooling of the soil through the night. Over the ocean the temperature of the air normally is a little higher, a degree or so, than that of the water, and the time of its maximum value, near 1 o'clock P.M., a little earlier than that of the water.

This is due to the fact that here the surface air is humid and also "dusty" with salt particles and therefore absorbs a large amount of radiation, so much indeed that its daily range of temperature is more dependent on this direct absorption than it is on conduction and convection from the surface. The minimum temperatures of air and water occur simultaneously, or nearly so.

**Periodic Temperature Changes**

Nearly 150 different periods of temperature and other weather changes have been published, ranging from 24 hours to 744 years. Nearly all of them, however, have a shorter period than 40 years, and half of them a period of 8 years or less. Of this great number of periods there are only two, the 24-hour or daily period, and the 12-month or annual period, that everybody accepts. There is one other, the so-called 11-year or sunspot period, that is widely, though not universally accepted; and still another the 35-year, or Brückner, period that many believe to be real. No credence was ever given to any of the others save perhaps by their discoverers, and in most cases even that must have been half-hearted.

The daily period is everywhere conspicuous (save for part of the time in polar regions, when the sun is continuously above or continuously below the horizon), and in respect to temperature, gives, on the average, a maximum in the early to mid afternoon and a minimum shortly before sunrise. Over the oceans this diurnal range is only 1°F to 3°F as a rule. It also is small in the humid and cloudy portions of the continental tropics, owing to the large amount of return radiation from the clouds and water vapour.

In desert regions, especially at high altitudes, where the sky is clear and the humidity very low the diurnal range of temperature is at its maximum. In extreme cases this range is of the order of 100°F, from distinctly below freezing to decidedly over 100°F—both in the shade. The annual range also is extremely conspicuous in most parts of the world. In this case the exception does not occur at and near the poles, but at and for some distance on either side of the equator. At the equator the "year," as it were, counted from the time the sun is overhead at noon until farthest away (23 1/2°), and then back again is 6 months.

Next beyond the equator on either side there obviously are two such "years" but of unequal length. At one distance they are 5 and 7 months, at another 4 and 8, and so on until at the Tropics, Capricorn and Cancer, only one is left, and that one 12 months in duration, the same as from there on to the pole. The times of occurrence of the annual maxima and minima vary widely from place to place, but always they are after maximum and minimum reception of heat from the sun. The delays are least over inland deserts and greatest over mid to high latitude portions of the oceans.

The sunspot period, approximately 11.1 years, is most pronounced at high levels within the tropics. Here the average temperature during the year or two around sunspot minima, or when the spots are fewest and smallest, is about 2°F higher than the average temperature during the time of spot maxima. The same relation appears to hold, in general, for the middle and higher latitudes but with decidedly less contrast.

The Brückner period is very irregular in length, varying from roughly 20 years to perhaps 50, and the amplitude of its temperature range uncertain but always small. Its irregularity in length deprives it of practically all forecasting value, and indeed makes its very existence as anything other than a fortuitous recurrence highly doubtful.

There are two other known and real periods in respect to average temperatures and other climatic elements, but they are far too long to consider in any business affairs. One concerns the slow change of the season of the year when the earth is nearest the sun, due to the combined effect of the motion of the perihelion and the precession of the equinoxes. Just at present the earth is nearest the sun the first week of January and farthest away the first week of July; and this difference in distance is sufficient, if long continued, to vary the average temperature of the earth by at least 7° or 8°F. That is, at present the winters of the northern hemisphere are shorter and milder, and the summers longer and less hot, than they would be if we were nearest the sun the first week of July and farthest from it the first week of January, as we were about 10,500 years ago, and, in the same length of time, will be again. This is one reason, and the unequal distribution of land and water another, why the average temperature for the year is about 2°F higher in the northern hemisphere than in the southern, and why the thermal equator is north of the geographic equator.

Beyond question this particular period is of great climatic importance, but we know all about its course and its cause and the changes it effects come about so slowly that practically they do not concern us at all. The other period referred to is that of the changes in the ellipticity of the earth's orbit or variations in the difference between the annual maximum and minimum distances of the earth from the sun. But the length of this period, roughly 100,000 years, keeps it out of every business equation however prudently constructed.

Just to make the list complete it may be worthwhile to mention a few utterly unimportant but entirely real temperature periods. Those are the periods of the changes in light and heat received from the moon-maximum at full moon, minimum at new moon; changes in the distance of the earth from the sun due to the pull of the moon in its orbit about the earth; and similar but far less changes of and by the planets. The sum total of the effects of all the planets is about equal to that of the moon alone, that is, a change in the average temperature of the earth of about.02°F, due almost wholly to variations in our distance from the sun or as we say, to perturbations in the earth's orbit. But, as already stated, this change in temperature is too small to bother about.

**Temperature Lag**

It was stated above that the hottest time of the day is not noon, when the sun is most effective, but two to four hours later; and similarly, that the coldest weather does not come with the shortest days, but generally a month or so later. In proverb form: "As the days grow longer the cold grows stronger." In the early morning of a clear day following a cloudless night, say, the earth and surface air are relatively cool. Then with sunrise they begin to warm up, but not rapidly, even when there is no wind, because it requires an appreciable amount of heat to warm even a pound of soil 1°F, and several times as much to equally warm a pound of water.

But as the sunshine continues, the soil at first gets hotter and hotter, and as its temperature rises the rate at which it loses heat by radiation rapidly increases. However, since the soil, including of course its covering, warms slowly, owing to its large capacity for heat, its loss by radiation falls more and more behind its gain by absorption as the sun rises higher in the heavens, and therefore catches up with the latter only in the afternoon when the insolation is distinctly less than it was at midday.

Hence the diurnal maximum temperature, whether of the lower air or of the surface of the earth (an earlier phenomenon) necessarily lags behind the maximum intensity of the sunshine. Similarly, the annual maximum temperature occurs several weeks after the days are longest and the heating strongest. Very similarly too, because the earth can give off stored up heat when the supply becomes deficient, the minimum temperature comes several weeks after the shortest days. During this period, as the days grow longer the cold grows stronger. The diurnal and annual heating and cooling, and lagging of temperature extremes, may be likened to the alternate rise and fall of the water level in a reservoir having a continuously open drain pipe at the bottom and a periodically variable inflow, now greater, now less, than the then rate of outflow, but so regulated that the reservoir may never become empty.

**Day Degrees**

Not only are we interested in the values and times of occurrence of maximum and minimum temperatures but also, and even more, concerned in the occurrence of certain critical temperatures. For instance, we are very much interested in the temperature at which frost can occur until it does occur, after which, if it has been a "killing" one, we are no longer much concerned as there is nothing left for the next one to injure. Another critical temperature is 42°F as that closely marks the boundary between growth and dormancy for most vegetation of the temperate zones. In fact it is customary to call the difference between the average temperature of a given day, if higher than this value, and 42°F, its day degrees. The sum of these daily values over a week, month or season, is the number of day degrees for that period, and is an important index to what might have been the vegetable growth during the time in question. Similarly, engineers and others interested in artificial heating of buildings, count day degrees relative to a temperature of 65°F.

**Occasional Extremes**

Once in a while an exceptional combination of conditions brings to a given place an abnormally high or low temperature, usually for only an hour or two, or a day at most, but sometimes for several days together, and even a month or longer. It is always easy to know from the current maps of weather distribution exactly what caused the extreme in question, but it never is possible to trace them farther back than two or three steps at most, nor very long to foresee their coming. Some of them one never forgets, and a few continue for a century or more to put disconcerting humps or depressions on our statistical curves.

**Wind Direction and Temperature**

The effect of wind direction on the temperature of a place depends on its location. Well within the Tropics, and also near the poles, the effect of wind direction obviously is small because the temperature is pretty nearly the same round about in every direction. In middle latitudes, however, the situation is quite different, partly because here the temperatures commonly are not the same in every direction, and partly also, in fact mainly, because here each section of the cyclone and of the anticyclone has its own wind direction, and some of them a wind system entirely distinct from that of the others.

In the forward or eastern portion of the anticyclone the winds are from the region of higher latitudes, and having come a long ways often are distinctly cool to cold for the place and time of year.

Similarly, the winds of the western segment, having come from much nearer the equator, usually are relatively warm. In the cyclone, or widespread disturbance, all that segment of 90°, more or less, lying between one line running east, to southeast, from the storm centre and another generally south to southwest (in the northern hemisphere; east to northeast, and north to northwest, in the southern hemisphere) is occupied by a great current of warm air from low latitudes. The rest of the storm area is covered with cold winds from the east, north, and northwest, in succession as one in the northern hemisphere passes from the front to the rear of the storm centre on the poleward side; from the east, south, and southwest, in the southern hemisphere.

In general, all these cold winds in each hemisphere are of polar, that is, high latitude origin. Clearly then, the temperature of the air in a cyclonic region is likely to change with the direction of the wind. In one portion of this disturbance, namely, along a narrow strip that meteorologists call the cold front, or wind shift line, and which commonly runs west of south (west of north in the southern hemisphere) from the storm centre, the wind direction rapidly changes from southwesterly to northwesterly, with, as a rule, a sharp drop in temperature as the tropical breezes give way to polar blasts. Hence in middle latitudes air temperature is closely dependent upon wind direction, both in cyclones and anticyclones; and that means the greater portion of the time, for usually we are in the midst of one or the other of these disturbances.

## LOCATION AND TEMPERATURE

Every one knows that the average temperature of the tropical regions is higher than that of the polar areas, but it is not a familiar fact that, nevertheless, int he course of a year the temperature reaches 90°F or over (in the shade) on more days in central Alaska than at Panama. It also is a surprising fact to most of us that the average temperature through January at St. Louis is the same as that in southern Iceland; and that the average temperature for the entire winter, December, January and February, at Sitka, Alaska, is about the same as that at Washington, D. C. And there are lots of other similar surprises, as, for instance, the fact that semi-tropical vegetation that would be killed by frost is now growing wild on the Scilly Islands in the latitude of northern Newfoundland. Clearly, then, difference in longitude may be accompanied by nearly as great a variation in temperature as is difference in latitude. During winter especially the lines of equal temperature run far poleward over the oceans, and equatorward over the continents.

Another matter of great influence on the local temperature is the nature of the surrounding area, both near and distant. A fair inland point, for instance, becomes much hotter in summer and greatly colder in winter than does a mid-sea island at the same latitude. Also a coast where the prevailing winds are on-shore has a more nearly even temperature than one of the same latitude with off-shore winds. The first is bathed in ocean breezes of relatively equable temperature; the second in winds that have traveled far over land and that therefore are characterized by its temperature irregularities and extremes.

It is interesting in this connection also to note that the temperature is a little higher, and in some cases quite noticeably higher, in a city than in the adjacent country; and further that in the country the forest is cooler in summer, and slightly warmer in winter, than the open fields.

In some cases the foot of a high mountain occasionally is very peculiar in respect to temperature. It may happen that the air is quite cold when all of a sudden there comes across and down the mountain a roaring hot wind that within a few hours clears away every trace of even a deep snow. This wind went up the other side of the mountain saturated and rainy, hence it cooled relatively little with ascent owing to the latent heat of vaporization there rendered sensible by condensation. As it came down, however, it was dry and therefore heated rapidly with descent and consequent increase of pressure. Such are the famous foehn winds on the northern side of the Alps, and the chinook winds of the Rocky Mountains.

**South Pole Colder than the North**

We often are asked which is the colder, the north pole or the south. One answer might be that, as we have no records longer than a few hours at either, we don't know. That is true enough so far as bare statistics are concerned, but in this case we can reason the matter out from sure and simple premises:

- Normally, the greater the height the lower the temperature, other things being equal;
- The south pole is 10,000 feet, roughly, above, and the north pole at, sea level. Hence we should expect the south pole to be the colder of the two.
- The greater the height the less, in general, the amount of cloud, water vapour and other gas to radiate back to the earth and thereby help to keep up its temperature;
- The south pole is much higher than the north. The south pole, therefore, because it gets less return radiation than does the north, should have the lower temperature.
- The faster heat is supplied by the surface to the lower air the higher the temperature of that air, other things being equal;
- By actual measurement the ice over the Arctic Ocean gives off enough

heat in 24 winter hours, coming from the relatively warm water below, to increase the temperature of a layer of air 450 feet thick by 20°F, while that given off at the south pole is only a small fraction of this amount, owing to the much poorer conductivity of the snow and the far greater depth to a temperature equal to that of the arctic water.

Then for this reason also the south pole must be colder than the north. It seems therefore that even without the actual observations we may feel reasonably certain that the temperatures of the two poles are not the same, especially their night temperatures since these occur when they can be but little affected by the variations in our distance from the sun, and that the south pole is distinctly the colder of the two.

**Relation of Surface Temperature to Height**

Those who live in mountains regions know by personal experience that, in general, the higher up the mount iin the lower the temperature. Instrumental records show that this relation is true not only for mountains but also for hills and even plateaus, and that the approximate numerical values are 1°F decrease per 330 feet ascent on a mountain, 365 among hills and 455 on plains.

The reason for this difference in favour of the plateau is the fact that the air is heated mainly by the surface of the earth which, in turn, is heated by the sunshine. That is, in the case of the plateau the surface which is the heater of the air is at the height in question all around as far as the level area extends, while the air on the mountain is affected in part by the temperature of the free air, especially when there is an appreciable wind, and this free-air temperature is lower, except on still clear nights, than surface air at the same altitude.

**Relation of Temperature to Height in the Free Air**

According to observations the temperature of the air normally decreases with increase of latitude from the surface of the earth up to the height of several mile, roughly 6 to 7 in middle latitudes. Beyond this level the temperature remains substantially constant up to the greatest heights yet attained, probably around 18 miles. The average rate of decrease of temperature of the free air with increase of height is about 1°F per 300 feet, from the surface up to the level at which appreciable decrease ceases—up to the "tropopause," or limiting reach of convection.

In general this temperature decrease is most rapid in the upper half to two-thirds of the depth under consideration and least rapid in the lower one-third. Immediately above a land surface the change of temperature with height is widely variable, from a *decrease*, perhaps ten fold the above average value, to an *increase* ten fold that rate, at least through the first 20 feet or so, above suitable regions, such as plains, valley bottoms and bowl-shaped depressions, and, of course, during still clear nights.

Few phenomena of the atmosphere are as often "explained" as is the fact that, in general, temperature decreases with increase of height, and hardly any other as inadequately, not to say erroneously, explained. The facts are:

- Half, roughly, of the sunshine that is not lost by reflection or by scattering, that is, half of it that does any heating at all, gets entirely through the air and is absorbed by the surface of the earth, which, on being thus heated, heats in turn the adjacent or lowest air.
- The other half of the effective radiation from the sun (portion used and not immediately lost) is absorbed mainly by the water vapour, and as the density of this vapour generally decreases very rapidly with increase of height it follows that by the direct absorption of sunshine the heating of the air likewise is greatest in its lowest levels.
- The surface of the earth loses heat (it doesn't keep on getting hotter and hotter indefinitely) not only by conduction to the adjacent air, but also by radiation, a kind of radiation greatly absorbed by water vapour. Therefore in this third way, too, as in each of the others, the atmosphere is more and more strongly heated with decrease of height.
- Although the surface air is most heated it is not equally heated everywhere. Hence the warmest and lightest portions are pushed up—forced to rise—by the cooler and denser air round about as a cork is bobbed up when let go of under water.
- The ascending (pushed-up) warm air comes under less and less pressure with increase of height by the weight of the air left below. It therefore continuously expands while rising, and all the time against pressure—the weight of whatever air is still above it. But this expansion against pressure is work, and work at the expense of the only supply of energy the rising air has—its heat. Hence as it rises it must and does become cooler.
- This cooling of the air by convection does not go on to absolute zero, nor, as explained, does the surface air keep on getting hotter and hotter. And these limitations are owing to the fact that the atmosphere loses heat by radiation. The free air therefore is all the time gaining heat by absorption of radiation and losing heat by emission of radiation, while the surface air is gaining heat also by contact with the warmed earth.
- But loss of heat by radiation decreases very rapidly with fall of temperature, while the power to absorb radiation does not change. Hence as the air ascends higher and higher, and thereby gets colder and colder, it presently comes to a temperature at which its loss by radiation is equal to its gain by absorption. Beyond this level it can not rise, because if it did so it instantly would become colder and denser than its environment and fall back again. From this level on up the temperature of the air must remain roughly constant except as modified,

perhaps, by change of composition. This is the isothermal region, or stratosphere, which every planet must have whatever the extent and composition of its atmosphere.

- Since the cold air of the stratosphere is losing heat by radiation at the same rate that it is gaining it by absorption it follows that the warmer air of lower levels is losing by radiation faster than it is gaining by absorption, the net difference at each *level* (not moment to moment for the moving air) being made up by heat brought there by convection from the surface, either directly as such, or indirectly through evaporation and subsequent condensation. Thus the whole of the troposphere, or, in other words, the whole of that portion of the atmosphere in which clouds can and do occur, is continuously being heated below and cooled above. In this way, that is, by heating below and cooling above, convection and, in general, a decrease of temperature from bottom to top of the convective layer, is maintained without the air as a whole getting either warmer or colder.

**Temperature Changes in the Stratosphere**

Although the stratosphere, that portion of the atmosphere beyond the highest clouds, has no immediate contact with the warming and cooling surface, nevertheless its temperature at any given locality often varies 10°F to 20°F, and even more, not vertically, as a rule, but horizontally, or from day to day. Thus the temperature of the stratosphere commonly is distinctly higher over the forward and central portions of a cyclonic area than it is over the corresponding portions of an anticyclone. The cause of this change is not definitely known.

**Temperature Inversions**

The structure of the atmosphere, during still clear nights the surface air becomes so cold over level land areas, in valleys and in basins, that often there is a rapid rise of temperature with increase of height through the first 10 to 100 feet, or more. This is the surface inversion, so favourable to production of frost, and without which orchard heating commonly would be unnecessary and, moreover, in general impracticable. Every wind of appreciable strength so thoroughly mixes up the lower air by turbulence that the temperature of the top portion of the agitated stratum is decidedly lower than that of the undisturbed air immediately above it. This is the turbulence inversion, which, because of its low temperature often is accompanied by a stratiform cloud.

It might seem that a similar inversion should occur at every interface between over- and under-running air currents, but such inversions, so far as they exist at all, are too slight to be of any particular importance. This is owing to the smallness of the friction between free air currents and the consequent all but complete absence of turbulence. The highest temperature inversion in the atmosphere, of which we have any actual and direct measurement, is that at the base of the stratosphere during the passage of an anticyclone. As already stated the cause of this particular, and often very pronounced, inversion is not yet definitely known.

## THERMAL BELT, OR GREEN BELT

There are various ways of protecting fruit from frost, but the best of them all is the proper selection of the orchard site. In a hilly or mountainous region that best location is neither on the floor of the valley nor, generally, on the top of a ridge, but, as a rule, some distance, not too far, up one or the other of the slopes. A strip along a hillside or mountainside at this level is known as the thermal belt, green belt, verdant belt, frostless belt, etc., because on still, cloudless nights this level is warmer than any other above or below it, hence least likely to have frost, and most likely to show green and uninjured vegetation.

After sundown, when the sky is clear and there is no wind, the surface of the earth everywhere cools much more rapidly than the free atmosphere, and in turn correspondingly chills the nearby air either directly through actual contact, or indi- rectly by turbulence mixing with that which had been so chilled. On the side of a hill, then, this air, because it is denser (being colder) than that at the same level over the adjacent valley, flows down slope much as would a sheet of water. However, as this drainage air reaches lower levels it evidently is subjected to increase of pressure to the extent of the weight of the air passed below. It therefore is compressed—work is done upon it—and its temperature made higher than it otherwise would have been. In the early evening, and well up on the hill, this heating of the down flowing air causes it to become warmer and warmer with descent, almost to the valley bottom. Here the gain of heat through increase of pressure not only makes up for all that was lost by contact with the cold surface, but actually warms the air to a higher temperature than it had before it was first chilled. But this heating is pronounced only where the descent is rapid. Air already at the bottom of the hill is not thus heated. It does not come under any greater pressure for there is no lower place to which it can rapidly drain. It therefore just gets colder and colder as the surface temperature continues to fall incident to the net loss of heat by radiation.

It also gets colder, but not so rapidly, near the bottom where the slope is gentle and the current sluggish; and less and less rapidly with increase of height and gain in the speed of flow. There must therefore be some level along either valley wall at which, for the time being, the heating of the descending air by compression is just equal to its cooling by contact with the chilled surface. Above this level the air evidently gets colder with ascent and below it colder with descent.

During most of the night the flood of cold air grows gradually deeper, carrying the level of maximum temperature—the level of its crest-higher and higher up the valley sides. Near morning, however, it becomes practically stationary, and where it then is frost obviously is least likely to occur. This is the place to plant your orchard—here along the thermal belt where the temperature is highest and the chance of frost the very least; where vegetation may pass through the night unharmed, while all above is frozen stiff and all below white with frost.

## COMPOSITION OF ATMOSPHERE

The most primitive man—the wildest savage—recognizes a wind when out in it, but it took a Greek philosopher to tell us what it is: air in motion. What motion is we know, at least well enough for practical purposes, but what is air? What is that invisible and odourless something we breathe and therefore call atmosphere, the thing that affords us all our weather perceptions and whose states and conditions we have learned to measure and even to foretell, and what was its origin?

### ORIGIN

To start as nearly as possible at the beginning, how and when did the earth ever get an atmosphere—its gaseous envelope? Well, there are two, and so far as we know, only two basic substances, the electron, a certain extremely minute quantity of negative electricity, plus something else, maybe; and the proton, an equally small quantity of positive electricity, plus something else, also maybe. The neutron or chargeless mass is omitted as so little is known about it.

Of the origin of these entities no one has the slightest idea. They can exist separately, in which case they are electrically very active; or variously grouped together in equal numbers, with increase of inertia but almost total loss of electric force, at least on things external.

**Fig.** Kilauea, an Air Factory

Every such group that is stable, or even measurably durable, is a chemical element. There are no other elements, and this number is limited. Furthermore, though occurring in unequal quantities, all these elements appear to be distributed throughout the universe.

Most of those known on the earth have been found in the sun, for instance, and we believe the others are there, too, even if in such relatively small amounts as to be difficult of detection. Hence, when that other star, according to one cosmic theory, some three billion four hundred million years ago, passed so near our sun as to drag off from it by tidal action the masses that coalesced into the planets, there were present, and came off together, all the possible elements.

That is, the primordial material of the earth, as it was pulled off from, or out of, the sun consisted of all the elements that now make it up, so that at the very beginning of the independent existence of the earth it had, if not an actual atmosphere, at least the makings of one.

It is believed, by those who hold to this theory that as the earth mass drew together in a molten sphere the heavier and more refractory substances formed mainly the inner core, while the lighter and more volatile elements and compounds formed the rocky shell and the gaseous envelope. In this way, they claim, the earth soon had oceans and an atmosphere of some kind. However, even if all the gases, hydrogen, oxygen, nitrogen, and others, that can combine with various elements and compounds and form solids had then so com- bined, leaving neither water nor air, it is certain that before long the earth would have begun accumulating both.

It is doing so now through every volcano, every fumarole, and every bubbling spring, and must always have done so since before even the first crust began to form. But it seems unlikely that our present atmosphere actually did come entirely from the molten interior, whether by way of volcanic activity or otherwise, because volcanoes do not, so far as we know, give off free oxygen—volcanic gases captured before there has been any chance for admixture with the air show no trace of it. Furthermore this element could not exist uncombined in the presence of hydrogen and sulphur at high temperatures, both of which are abundant in volcanic vapours.

But we have oxygen; where did it come from? It is known that highly developed green plants extract, under the stimulus of light, a great deal of oxygen from carbon dioxide, a gas abundantly emitted by volcanos. But most primitive plants do not, they consume it; hence this promising source of free oxygen appears, on close examination, to be quite uncertain. However, some forms of lower life thrive in the absence of free oxygen and yet, in the presence of light, evolve it from certain of its compounds, especially water.

Again, lightning, which must have occurred from the beginning, frees a little oxygen from water; and so also may ultra-violet light. There have been, then, continuously active means of obtaining free oxygen from its compounds since the beginning of the world. Nevertheless, it seems most likely that at the beginning there was more oxygen present than was necessary to use up the free hydrogen and to combine with the available surface materials—enough to do all this and to have a goodly amount left over as free oxygen of the atmosphere.

Presumably, too, there were present other primitive gases, especially nitrogen, argon, carbon dioxide and water vapour. But whatever the primitive state of the atmosphere when the earth first formed, it may be regarded as practically certain that during the whole of the three billion four hundred million years since that time it has been continuously depleted by combination with many things in and of the crust, and also as continuously repleted by their decomposition. It is always changing, but, except in respect to water vapour, the change is so small in comparison with the whole that we are not ordinarily aware of it.

It has been argued that the could not have retained an atmosphere when molten, or even when dull red. But this is true of only the lightest two gases, hydrogen and helium, and of them only if no other gases were present in large amounts. A deep atmosphere of water vapour, for instance, would catch any light gas that might leave the earth beneath with an escaping velocity. It could escape only if the outer portions of the atmosphere also were quite hot.

At any rate, as soon as a crust, however thin, formed over the earth the supply of heat from beneath was so reduced (the crust being a good insulator) that the upper air necessarily became cool enough to retain the lightest gases. Also water must soon have begun gathering on the surface. Even if, up to this stage, all helium and free hydrogen had been driven wholly away from the earth, a condition that seems unlikely, there has been since then, and still is, abundant opportunity to accumulate both of them—hydrogen from volcanos, and helium from radioactive materials everywhere.

Presumably, therefore, the atmosphere is primitive in part—pulled off from the sun with all the other elements—and in part, at least, certainly regenerated inasmuch as every volcano is an active air factory.

## Apparent Simplicity of Air

Presumably every one usually thinks of air as being just air, a homogeneous and single thing. Many of us always think of it that way, as the ancients did, when we think of it at all.

Indeed so far as its behaviour and most of its physical properties are concerned it shows no obvious complexity. It is just air in motion that is responsible for a thousand familiar things from the stir of a leaf to the wreck of a house; and just air that floats the balloon and sustains the aeroplane. In all these matters it usually is quite satisfactory to regard the air as the single substance it ordinarily seems to be.

## Evidence of Complexity

This apparent oneness of air does not, however, extend to all physical processes. When we try to liquefy it, for instance, evidences of its complexity soon become amazingly conspicuous. If untreated air is forced through the cooling coils they quickly become choked with ice; and they still clog up when even the driest air is used, if nothing but the water vapour has been removed—this time with solid carbon dioxide. Then, too, the liquid air itself shows abundant evidences that it is a mixture and not a simple substance like water. What then are the known constituents of the atmosphere, and how and when were they discovered?

## Water Vapour

The earliest considerations of the composition of the air that have come down to us are those of the Greeks. In their speculative philosophy on the composition of objects, they considered air to be one of the four elements (fire, earth and water being the other three) that, singly or variously combined, make up all substances.

From this it might seem that these Greek philosophers regarded the atmosphere as strictly a single thing—the "element" air. Yet it appears that by "air" they meant anything gaseous, and not necessarily the atmosphere. At any rate, Aristotle 250 years B.C., says very distinctly, in his work on meteorology, that cloud and rain are caused by condensation from the atmosphere of water vapour that had gotten there by the evaporation of water at the surface of the earth. He thus makes it very clear that the air consists of at least two things, and that water vapour is one of them.

Water vapour, then, was the first constituent of the atmosphere to be explicitly recognized. Aristotle mentions it, but it is not certain that this discovery was original with him. However, his is the earliest record we have of it, and for that reason, there being no evidence to the contrary, we regard him as one of the discoverers the earliest one of the constituents of the air.

## DISCOVERIES

For more than 22 centuries, therefore, and perhaps for much longer, it has been known that the air we breathe consists of at least two things, water vapour and whatever is left after the water is removed. And for more than two thousand years after the days of Aristotle this is all that was known about its composition. Indeed it was practically impossible to push our knowledge of the atmosphere any farther without something of the facilities and methods of the modern labouratory, nor before there had been acquired—very slowly and tediously it was—a fair concept of chemical elements and pure substances. Not until the beginning, then, of the 18th century was it reasonably possible for any constituent of the air to be discovered in addition to water vapour, nor indeed was any discovered until long after that. The chief obstacles that prevented such discovery for more than a hundred years after enough advance for that purpose had been made in labouratory technique, for that had been adequate from the beginning of the 17th century, were:

- The fixed idea that all gases are alike, all just air, and that any differences between various samples are due only to greater or less modifications of one and the same thing.
- The completely misleading and faulty concept that flame or combustion is the escape of something, phlogiston they called it, from within the burning object.
- The failure to recognize that change of weight incident to strong heating, during combustion, or under any other circumstances, was a matter of importance or had any scientific significance whatever.

A century before any constituent of the atmosphere, except of course water vapour, was recognized and collected in an approximately pure state, and while the faulty notions just listed were still prevalent, two people, working entirely independently, came near to finding one or more of its elements. The first of these was Robert Boyle a wealthy bachelor, chemist and theologian; discoverer of the fact, known as "Boyle's Law," that doubling the pressure on a gas reduces its volume by one-half, of course for the same temperature. It seems very probable that Boyle would have discovered some of the constituents of the air if he had carried to completion certain experiments that he definitely listed. But there is nothing in his voluminous writings to show that he ever got them beyond the paper stage, despite the fact that, so long as his health permitted, he was a persistent worker.

After many and well-devised experiments Mayow concluded that the air consists of at least two portions, one that supports combustion and sustains life, and another part that does neither. The former he called "fire-air," because it keeps a flame going. He also said that it consists of "nitro-aerial particles," that is, particles in a gaseous form of the kind that makes a mixture of niter and charcoal, or other combustible, burn, when lighted, in the absence of air—even under water.

All this is true enough, but his proof that the air consists in part of a special constituent, different from all the rest, was not complete. He did not collect "fire-air," the gas we now call oxygen, in a practically pure form and show that it is identical with the "fire-air" of the atmosphere. However, he recognized the incompleteness of some of his arguments, and it seems likely that if he had lived a few years longer his proofs would have been perfected, and our knowledge of the composition of the atmosphere set forward almost a hundred years. But this near attainment to the goal, and also even the direct route to it, appear to have been lost sight of for nearly a century. Indeed "fire-air," that came so near to being the first constituent of the atmosphere to be discovered, except, of course, water-vapour, and whose properties make it the most conspicuous, turned out to be the very last of all the major ones, save only argon.

**CARBON DIOXIDE**

The first of the permanent gases of the air of nearly constant quantity to be clearly discovered was not, as would seem most likely, either oxygen or nitrogen that together make up nearly the whole, but carbon dioxide, that is present as scarcely more than a trace—three parts in ten thousand. At that time alkalies were given to persons suffering with urinal calculi, and certain physicians recommended lime water for the same purpose. Which the view of finding something still better Black undertook investigations with *magnesia alba*, a form of carbonate, as we now know, or, more exactly, basic carbonate, of magnesium.

On strongly heating this substance a gas is given off, and Black turned his attention particularly to that gas, or air, as all gases were then called. He tried calcining, or burning to a powder, various substances, such as limestone, that we now know to be carbonates, and studying the gas thus obtained. In the end he found that this gas is much heavier than ordinary air, that it will not support life or combustion, that it will recombine with the calx, or powder, produced by the strong heating of the original substance, and that heat will again expel it as before. This gas seems to be fixed, or somehow fastened, in the objects from which it may be obtained.

**Nitrogen**

The composition of the air in a closed vessel after it no longer would support combustion or life, and to determine the cause of its unwholesomeness. After burning charcoal, phosphorus, or other combustible, in a closed volume of air, as long as possible, he removed the "fixed air" (carbon dioxide), if any had been formed, by means of lime, or an alkali, all in accordance with the previous investigations of Black, and then examined the remaining gas. He showed that this residue is not ordinary air because it supports neither life nor combustion, and that it is not fixed air for the alkalies do not absorb it. He called this residue "mephitic air," because it does not support life.

We now know and say that this residue had been obtained by burning out the oxygen of the confined air and then absorbing the carbon dioxide thus produced. And we know, too, that this residue, this "mephitic air," was nearly pure nitrogen. But at that time oxygen had not yet been discovered, and of course Rutherford could not talk in terms of things and chemical reactions then unknown. He did know, however, that it was obtained by combustion in an inclosed or limited volume of ordinary air, and therefore concluded, after the philosophy of his day, that it was atmospheric air combined with, or modified by the addition of, phlogiston—a mysterious fire substance whose escape from an object commonly is manifested by flame. Nevertheless, and no matter what his ideas as to its nature, Rutherford did obtain reasonably pure nitrogen and did record some of its properties, and therefore may be regarded as the discoverer of this constituent of the air, the most abundant of all.

**Oxygen**

Almost immediately after the discovery of nitrogen, the other major constituent of the atmosphere, oxygen, was independently found by Joseph Priestly. Priestley was a preacher by occupation and a chemist for recreation. At one time, while waiting for the building of the parsonage to be finished, he had the good fortune to live next door to a brewery—good fortune, because he was induced thereby to take up the chemistry of gases, his chief avocation for many years, on which he published several volumes, and which made his name famous. In the course of this work he obtained a red powder by heating mercury in the presence of air, and then on more strongly heating this powder with a burning lens he got a gas which supported combustion much better than ordinary air. He had gotten oxygen, as we now know, by first burning mercury to an oxide and then decomposing that oxide by raising it to a high temperature.

This element of the atmosphere he called dephlogisticated air. Combustible things burned in it readily; their phlogiston, or fire principle, rushed into it as air into a vacuum. It was air, that is, a gas, but air deprived of phlogiston. But regardless of what he called it Priestley had discovered a new constituent of the atmosphere, the one we now call oxygen, the one that sustains life and supports combustion. On the other hand, Scheele was a professional chemist, so completely absorbed in his subject that he had little time for anything else, whether occupation or diversion.

His incentive to study the air was his wish to solve the riddle of fire, a thing essential to so many chemical processes. He found substances, such as phosphorus, which on burning reduced the volume of the confined air (at the same temperature and pres sure) in which they were burned by about one-fifth. In all cases the remaining air would not support combustion.

He also found many ways of getting a gas that would support combustion far better than ordinary air, that was heavier than the air left after combustion, and that made burnt air indistinguishable from ordinary air when mixed with it in the proportion of one volume to four, or thereabouts. The residual or burnt air, the nitrogen, essentially, as we now name it, he called "vitiated air." The other portion, the part that supports combustion, he called "fire-air," just as Mayow had called it a century before.

Scheele, like Priestley and every other chemist of his day, except the immortal and tragic Lavoisier, interpreted fire phenomena in terms of phlogiston. He therefore considered "fire-air," which we call oxygen, to be a combination of phlogiston and a subtle acid substance. But whatever his opinions may have been as to its possible composition, he did find this constituent of the atmosphere, and that too even before it was found by Priestley.

Another student of the atmosphere that must be mentioned in connection with the discoveries of its composition was Henry Cavendish. He may not generally be credited with the discovery of any one of these constituents, and yet he appears to have found nitrogen about the same time that Rutherford did, if not earlier, by the simple process of passing the same confined air back and forth over red hot charcoal and then removing the fixed air (carbon dioxide) with an alkali. He did not publish this—he appeared always to be indifferent about publishing his investigations—but described it in a letter to Priestley in 1772, the year Rutherford published his discovery of mephitic air.

Lord Rayleigh had been making careful measurements of the densities of different gases, and thus came upon the fact that the nitrogen of the atmosphere, or rather that residual then regarded as pure nitrogen, was a little denser than the nitrogen obtained from any one of several chemical compounds. This was capable of three or four different interpretations, one of which, then regarded as the least likely, was the presence in the atmosphere of an unknown gas denser than nitrogen.

At that stage Sir William Ramsay joined Lord Rayleigh in a common attack on this problem. Each tried removing from the air the then known constituents, using entirely different methods for getting rid of the nitrogen, and each found conclusive evidence of the presence in the atmosphere of an unknown gas to the extent of about one per cent. It proved to be chemically inert, and for that reason was called argon.

### Other Inert Gases

During the years 1895-1898, that is, immediately after the discovery of argon, four additional inert gases, helium, neon, krypton and xenon, were found by Sir William Ramsay and his assistant. Travers, to be constituents of the atmosphere. Helium was found as that portion of the atmosphere that is still gaseous while the rest has been chilled to liquids or solids; and the others, neon, krypton and xenon by fractional distillations of large quantities of liquid air and liquid argon, impure, of course.

## HYDROGEN

Lord Rayleigh and others also found free hydrogen always present in the air, but in amounts that varied from, roughly, one part in 1,000,000 to one 5,000. Perhaps this gas can not be regarded as a permanent component of the atmosphere in the same sense that oxygen and nitrogen are, but rather as an accidental and variable impurity. At any rate it is irregularly added to the air, at least by volcanoes, and more or less irregularly removed from it.

**Traces and Impurities**

Varying traces of ammonia, nitric and nitrous acids and their compounds, sulphuric and sulphurous acids and their compounds, oxides of nitrogen, hydrogen dioxide, and ozone are among the innumerable things always in the air if not of it. Minute particles of sea salt from evaporating spray, fine earth dust caught up by winds, pollen of every description and spores of many kinds are among the coarser and ever present pollutions of the atmosphere from the earth beneath. Shooting stars furnish a continuous, though invisible, shower of dust the world over from outer space.

Also radioactive products of radium, thorium and other elements are pouring into the atmosphere continuously and contributing to the maintenance of its electrical state. This state means, in part, around 20,000 electrified particles, or ions, per cubic inch in the lower air. The number of ions 60 miles or so above the surface is far greater, at least 1,600,000 per cubic inch, as we infer from the phenomena of radio communication.

In short, while the things of the atmosphere are but few the things in it are innumerable. All these latter are relatively very small in amo t, but some of them are exceedingly important. The salt particles and some others are essential to condensation and rainfall; the ammonia and other nitrogen compounds, brought down by precipitation, add much to the fertility of the soil; the ions of the high atmosphere make distant radio communication possible; and ozone, also in the upper air, shields us from that portion of the ultra violet radiation that would destroy our eyesight, as we are now constituted, and otherwise do us irreparable harm.

**COMPOSITION OF AIR**

If the atmosphere is freed from all its numerous impurities and also freed from water vapour, there will remain pure dry air whose percentage composition at the surface of the earth is, very closely:

| | |
|---|---|
| Nitrogen | 78.03 |
| Oxygen | 20.99 |
| Argon | 0.9323 |
| Carbon dioxide | 0.03 |
| Hydrogen | 0.01 |
| Neon | 0.0018 |
| Helium | 0.0005 |
| Krypton | 0.0001 |
| Xenon | 0.000009 |

The amount of water vapour in the atmosphere, or better, perhaps, the amount of the water vapour constituent of the atmosphere, varies from scarcely more than a trace, at extremely low temperatures, to at least 5 per cent by volume on the hottest and most humid days. At and above the height of 5 miles, say, the amount of water vapour always is small, even when saturation obtains, owning to the very low temperatures at these levels. That is, water vapour is confined almost wholly to the lower atmosphere. Its average value, the world over, is such that if it were all condensed it would be the equivalent of a layer of water about one inch deep over the entire earth.

## FORMS OF OXYGEN

Oxygen also is peculiar in its distribution, and it occurs in three different forms. All but about one part in 400,000 is ordinary oxygen, or, in the language of the chemist, diatomic oxygen. Triatomic oxygen, or ozone, occurs chiefly beyond the highest clouds, its greatest density being, apparently, at the levels of 20 to 30 miles. It occurs in the lower air only as a trace. In the highest air, 60 miles and more above the surface, the oxygen appears to be at least partly monatomic, according to the spectrum of the aurora.

### Importance of Ozone

Every one knows that life would be impossible if there were no ordinary or diatomic oxygen in the atmosphere, and that without it nearly if not quite all vegetable life also would be impossible. But it is not so well known, though equally true, that animals, including the human species, could not exist as now constituted if the air did not contain a small amount of triatomic oxygen or ozone. And yet, paradoxical as it may if seem, ozone were just a little more effective in its goodness, again life, as now constituted, could not last. These surprising facts come about in this way: The radiation from the sun includes not only every colour, that is, the whole of the visible spectrum, but also extends indefinitely beyond the red into the long-wave-length invisible region, and likewise, in the opposite direction, well beyond the limit of the violet.

But in this ultra violet portion of the spectrum the radiation from the sun ceases far short of the limit to which that from an electric arc, for instance, can be followed; and ceases, not because no radiation beyond that limit is given out by the sun, but because it is absorbed by the ozone in the upper atmosphere. Now, much of that particular radiation which ozone absorbs is destructive to the eye, and when intense probably injurious to other tissues as well.

On the other hand, it also absorbs a large part of the radiation that is effective in preventing rickets, but, and this is of the utmost importance, it does not absorb quite all of this antirachitic portion. Enough is let through to keep us in proper health. This particular limiting, then, of the solar radiation that reaches the earth is amazingly located. If it was a little farther out in the ultra violet, eyes, as now constituted, could not have developed; and if a little nearer the visible, again, animals, as they now are, could not have come into being. Of course ozone, in moderate amount, presumably was in the air long before there were animals of any kind to be affected by its action on solar radiation. This radiation, therefore, was not adapted to them, but they developed in adaptation to it; nevertheless the fit is so close and of such a surprising nature as to give us decided pause for thought.

## Origin of Ozone

Ozone is produced by the action of extreme ultra violet light on oxygen, and therefore at great heights; a little where the oxygen is very rare, then more and more with decrease of height and increase of the oxygen supply until by absorption the ozonizing rays are considerably enfeebled. As the density of the oxygen increases the intensity of the effective radiation decreases, hence the rate of production of ozone by this process must be very slow in the outermost air, increase to a maximum with decrease of height and then rapidly fall off at still lower and lower levels.

The total range of ozone, top to bottom, probably is not less than 100 miles, with its maximum concentration around 25 miles, perhaps, above the surface, and yet all told, and despite its great importance, it is the equivalent of a layer of this gas only about one-tenth of an inch thick at atmospheric pressure and room temperature. Ozone is produced also by electric discharges through oxygen. Lightning certainly produces some ozone, and it may be that the auroras form it too, though the great heights at which they occur, 60 miles and more above the earth, seem to render this conclusion doubtful.

## Distribution of Ozone

As already stated, we know that at most there is only a trace of ozone in the atmosphere up to the level of the highest clouds, and that it exists to an appreciable extent at considerably greater heights. But this is not all. The total amount of ozone vertically over one to the limit of the air appears to increase, in general, with increase of latitude; to be greater during winter and spring than summer and fall; and to be greater in winds from high latitudes than in those of the opposite direction. This complicates the problem of the origin of ozone. If it is produced by ultra violet radiation alone why should it not be most abundant in tropical regions, and elsewhere in the summer time?

If we surmise that it is produced largely by auroral discharges, how, we ask, can these discharges, 60 miles up and more, reach a sufficient supply of oxygen? And if there is enough oxygen at these great heights how can the ozone subsequently become concentrated at the intermediate levels? These are some of the questions to be answered by future observations and studies.

# 8

# Environmental Pollution

## POLLUTION

Environmental pollution is any discharge of material or energy into water, land, or air that causes or may cause acute (short-term) or chronic (long-term) detriment to the Earth's ecological balance or that lowers the quality of life. Pollutants may cause primary damage, with direct identifiable impact on the environment, or secondary damage in the form of minor perturbations in the delicate balance of the biological food web that are detectable only over long time periods.

Until relatively recently in humanity's history, where pollution has existed, it has been primarily a local problem. The industrialization of society, the introduction of motorized vehicles, and the explosion of the human population, however, have caused an exponential growth in the production of goods and services. Coupled with this growth has been a tremendous increase in waste by-products.

The indiscriminate discharge of untreated industrial and domestic wastes into waterways, the spewing of thousands of tons of particulates and airborne gases into the atmosphere, the "throwaway" attitude towards solid wastes, and the use of newly developed chemicals without considering potential consequences have resulted in major environmental disasters, including the formation of smog in the Los Angeles area since the late 1940s and the pollution of large areas of the Mediterranean Sea. Technology has begun to solve some pollution problems, and public awareness of the extent of pollution will eventually force governments to undertake more effective environmental planning and adopt more effective antipollution measures.

## POLLUTION PREVENTION

Pollution prevention is reducing or eliminating waste at the source by modifying production processes, promoting the use of non-toxic or less-toxic substances, implementing conservation techniques, and re-using materials rather than putting them into the waste stream. Pollution prevention means "source reduction," as defined under the Pollution Prevention Act, and other practices that reduce or eliminate the creation of pollutants through:

- Increased efficiency in the use of raw materials, energy, water, or other resources, or
- Protection of natural resources by conservation. The Pollution Prevention Act defines "source reduction" to mean any practice which
- Reduces the amount of any hazardous substance, pollutant, or contaminant entering any waste stream or otherwise released into the environment prior to recycling, treatment, or disposal; and
- Reduces the hazards to public health and the environment associated with the release of such substances, pollutants, or contaminants. The term includes: equipment or technology modifications, process or procedure modifications, reformulation or redesign or products, substitution of raw materials, and improvements in housekeeping, maintenance, training, or inventory control.

**Specific Pollution Prevention Approaches**

Pollution prevention approaches can be applied to all pollution-generating activities, including those found in the energy, agriculture, Federal, consumer, as well as industrial sectors. The impairment of wetlands, ground water sources, and other critical resources constitutes pollution, and prevention practices may be essential for preserving these resources.

These practices may include conservation techniques and changes in management practices to prevent harm to sensitive ecosystems. Pollution prevention does not include practices that create new risks of concern. In the agricultural sector, pollution prevention approaches include:

- Reducing the use of water and chemical inputs;
- Adoption of less environmentally harmful pesticides or cultivation of crop strains with natural resistance to pests; and
- Protection of sensitive areas. In the energy sector, pollution prevention can reduce environmental damages from extraction, processing, transport, and combustion of fuels. Pollution prevention approaches include:
- Increasing efficiency in energy use;
- Substituting environmentally benign fuel sources; and
- Design changes that reduce the demand for energy. Some Types of Pollution are: Air, Water and thermal, Radioactive.

## ENVIRONMENTAL HAZARD

'Environmental hazard' is a generic term for any situation or state of events which poses a threat to the surrounding environment. This term incorporates topics like pollution and Natural Hazards such as storms and earthquakes.

I feel that this sight has no information on what I am looking for! There are five types of environmental hazards:

1. Chemical
2. Physical
3. Mechanical
4. Biological
5. Psychosocial

The term can also refer to biological hazards; a large algal bloom is an environmental hazard because it makes the lake uninhabitable for other organisms.

## Natural Hazards

A natural hazard is a threat of an event that will have a negative effect on people or the environment. Many natural hazards are related, *e.g.* earthquakes can result in tsunamis, drought can lead directly to famine and disease. A concrete example of the division between hazard and disaster is that the 1906 San Francisco earthquake was a disaster, whereas earthquakes are a hazard. Hazards are consequently relating to a future occurrence and disasters to past or current occurrences.

*Notable avalanches include:*

- The 1910 Wellington avalanche
- The 1954 Blons avalanches
- The 1970 Ancash earthquake
- The 1999 Galtur Avalanche
- The 2002 Kolka-Karmadon rock ice slide

## Earthquakes

An Earthquake is a sudden shaking or vibration of the Earth's crust. The vibrations may vary in magnitude. The earthquake has point of origin underground called the "focus". The point directly above the focus on the surface is called the "epicentre".

Earthquakes by themselves rarely kill people or wildlife. It is usually the secondary events that they trigger, such as building collapse, fires, tsunamis and volcanoes, that are actually the human disaster. As many of these could be avoided by better construction, safety systems, early warning and evacuation planning, the term unnatural disaster is not unwarranted. Earthquakes are caused by the discharge of stress accumulated along geologic faults.

*Some of the most significant earthquakes in recent times include:*

- The 2004 Indian Ocean earthquake, the second largest earthquake in recorded history, registering a moment magnitude of 9.3. The huge tsunamis triggered by this earthquake cost the lives of at least 229,000 people.

- The 7.6-7.7 2005 Kashmir earthquake, which cost 79,000 lives in Pakistan.
- The 7.7 magnitude July 2006 Java earthquake, which also triggered tsunamis.
- The 7.9 magnitude May 12, 2008 Sichuan earthquake in Sichuan Province, China. Death toll at over 61,150 as of May 27, 2008.

**Lahars**

A lahar is a volcanic mudflow or landslide. The 1953 Tangiwai disaster was caused by a lahar, as was the 1985 Armero tragedy in which the town of Armero was buried and an estimated 23,000 people were killed.

**Natural Disaster**

It is a form of tuberculosis where you die in a car and madeline kills you in the middle of the day. This understanding is concentrated in the formulation: "disasters occur when hazards meet vulnerability." A natural hazard will hence never result in a natural disaster in areas without vulnerability, *e.g.* strong earthquakes in uninhabited areas. The term *natural* has consequently been disputed because the events simply are not hazards or disasters without human involvement.

## POLLUTION AND POLLUTANTS

The change in any component of the environment which leads to its deterioration is termed as pollution of the environment. It is the undesirable change in the physical or biological components which adversely alters the environment.

*Pollutants*:

- The substances which are present in harmful concentration and is the agent who causes pollution is termed as the pollutant.

## CLASSIFICATION OF POLLUTANTS

*On the basis of existence in nature*:

- *Quantitative Pollutants*: The substances which are already present in the environment, but are termed as pollutants when their concentration increases in the environment, *e.g.* $CO_2$ is present in the environment in greater quantity than normal and is hence termed as a quantitative pollutant.
- *Qualitative pollutant*: The substances which are not normally present in the environment and are added by human beings and are pollutants by nature. *E.g.* insecticides, pesticides

*On the basis of the form in which they persist*:

- *Primary Pollutants*: The substances which are directly emitted from the source and remain in that form are termed as primary pollutants eg, smoke, fumes, ash, dust, nitric oxide and sulphur dioxide

- *Secondary pollutants*: The substances which are formed by chemical reaction between the primary pollutants and constituents of the environment.

*On the basis of disposal*:

- *Bio-degradable pollutants*: The pollutants which are decomposed by natural processes eg domestic sewage.
- *Non bio-degradable pollutants*: The pollutants which don't decompose naturally or decompose slowly *e.g.* DDT, aluminium cans.

## NATURE-MAN INTERACTION

Kerala is known as 'the God's own land'. Its environment, culture, and practices also endorse this truth. Kerala is blessed with plural culture, marvelous natural setups, and beautiful people. The life of the people is very calm and quite, and shows high adaptability, and international standards, at least in the sphere of health and education. All these highly appreciable achievements are due to its natural conditions, and, 'Nature-Man interaction'. We, anthropologists, are very keen about this kind of interactions, and strongly believe that culture, and all behaviour patterns are evolved out of this interaction.

And, it also helps us to understand who we are and who other people are or, who is who and what is what, the identity of people. So, I would like to examine how the identity is evolved from the 'Nature-Man' interaction with reference to the unique tradition, the teyyam performance of North Malabar. According to Kurup "The Teyyam or Teyyattam is a popular cult in Malabar which has become an inseparable part of the religion of the village folk".

*Damodaran says that*:

- "Through teyyam, the people of North Malabar worship Nature, spirits, ancestors, gods, and goddesses as their local deities...Teyyam performance is the glory of this mode of worship...The teyyam performance is a complex process, which includes the observance of several rituals, and the appearance of the beloved deities in front of believers".

The very word, teyyam, can bring forth in the mind of a listener a fascinating, as well as a colourful picture, particularly, its charming attires, superb dance and enchanting music, and equally important, the commitment of the people. It is also true that as far as the people of North Malabar are concerned, the word, teyyam, has more than one meaning. According to them, the teyyam is everything. Absolutely they believe that, it is their present, past and future, as far as their society is concerned. More than that, teyyam is culture, tradition, and environmental relationship.

The devotees worship and perform teyyam, as an indispensable part of their religion, magic, and even politics. For this reason, it is deemed as a unique religious ritual. It is believed that the teyyam possess great power, and is sacred and divine, which is also at once both non-human and supernatural. The followers strictly follow its code, in terms of its rules and behaviour patterns because, they fear that its anger will lead to their destruction, and its pleasing will lead to their well being. People of North Malabar think that all activities are directly, or indirectly linked with the teyyam tradition.

For that reason, they consider it both as their designer, and destroyer. The term, teyyam, is used as a synonym to daivam. It is always grappled with symbols, and rituals, on the basis of local myths. The basic number of teyyam is only onnu kuraya nalpathu, meaning one less than 40 hence, 39 but, for a few, the number is still larger, onnurunalpathu teyyams, 140 forms. According to the information I have gathered from my fieldwork, and also from published materials, the number goes farther, and as many as three hundred to five hundred different forms of teyyams are performed in this area. We can classify teyyam, in terms of gender in general, and myth of origin and tõttam pãttu in particular.

So there are two groups, male and female teyyams, on the basis of gender. The females are the dominant ones, at least as far as the number is concerned. In terms of myth of origin and tõttam pãttu, it is possible to separate teyyams into five categories namely, Gods and Goddesses, Ancestors, Heroes and Heroines, Spirits and Devils, and Nature and Animals. There are foreign scholars who consider the teyyam as, 'nothing but devil dance', but it is not a 'devil dance'. It is, in fact, a part of people's sacred tradition.

There are others, who consider the teyyam as an 'art form', and they also do not know its sociocultural significance and ramifications. In addition to these views, there is also a common feeling which is prevalent among the people that teyyam is a 'caste-based occupation' of a few groups. Such assumption may be partly true. Actually, in reality, it presupposes that all that are endowed with life in North Malabar form the 'part and parcel' of the teyyam. Another important truth is that this social etiquette has been unceasingly flowing all the way through centuries, through generations, and it not only perpetuates the culture, but also maintains its 'identity'.

There are many rituals and performances all over the world, which are related to religions, but surprisingly, we cannot acknowledge a similar creation such as, the teyyam, anywhere in the world. The society of North Malabar is mainly stratified into three social groups, in terms of religion: Hindu, Muslim, and Christian. Jainism, and Buddhism are not having an influential presence in this area, today. The three social groups have their own life styles, and behaviour patterns. The Hindus are the dominant ones, and traditionally, they are said to be the real inhabitants, and the Mapila and Christian are the later migrant groups. It is the dominant Hindus, who came in close touch with the teyyam through an attempt at assimilation. This is why the Hindu society overlaps with that of the teyyam, that also having the elements of Animism, Animatism, and Nature worship.

There are three types of teyyam celebrations namely, prarthanakaliyãttam, kalpanakaliyãttam, and perumk-aliyãttam. During performances, several rituals are observed according to the rites and rules of the respective teyyam, and in a specific point of time, the performer performs the teyyam, dressed in a peculiar manner by using specially designed colourful and magnificent attires. They wear respective attire and decorations and perform certain kind of rhythmic dance.

## METHODOLOGY

The method and techniques executed for data collection and interpretation here in this document largely drawn from anthropology. Extensive fieldwork was the soul of this study since, both participant and non-participant observation, and unstructured and informal interview supplies bulk of the informations. Secondary sources and reports were also make use for obtaining information. The gathered informations were cross-checked and recorded.

And finally analysed, and interpreted according to the objective of this document. It has long been established that ecology plays a vital role in conditioning the culture of a given area, and that the geographical situation of a locale goes a long way in shaping the needs, customs, behaviour, and thoughts of the people.

According to Redfield, "both man and Nature are the twin-agents of the perennial revolution that shapes and re-shapes the face of the earth". It is not only the anthropologists who speak about the importance of environment in the evolution of the human society but also, even Indian philosophers have spoken about it.

For example, the Tolkappiam, written by Tolkappiar speaks about the three factors in relation with the formation of a society. These are space and time, local resources or things that are available in the immediate neighbourhood, and the cultural elements that evolve due to man's utilization of them. So also in the case of the teyyam, ecology plays a vital role. The interaction network between human beings and environment produces a culture. The teyyam is a proof to this. A study of the teyyam, from an anthropological standpoint, reveals us the fact that interaction also gave shape to new modes of adaptations that were necessary for different situations for easy survival.

## DISCUSSION

Absolutely, the very uniqueness of teyyam stands on the 'interaction' between man and man, and between man and his surroundings, *i.e.* man's dependency on environment. The environment of which I have spoken of includes both the physical environment, and the socio-cultural environment. This pattern of attachment also makes the people to deem teyyam as their 'science', and 'culture', inherited from the past, and their limited technology and know-how insist to them that they trust heavily teyyam for the common good of the people. It also helps the people to discard the stress and strains of dayto- day struggle for survival.

Therefore, I have the firm belief that this ritual, the teyyam, as a demand of the whole society, is considered as a necessity by the people of North Malabar. There was a time when teyyam performers directly depend on Nature for obtaining the essential commodities to make the teyyam attires. All the costumes, and other items, including those for offerings, were obtained from the surroundings, which were copious and cheap. At present, they find it hard to obtain the natural materials from the surroundings.

In addition to this, the increasing influences of the market also act as a force to change the traditional mode of environment-exploitation. Therefore, the performers now show a greater amount of dependency on market, which was not the case in the past, for obtaining essential teyyam commodities. The area, North Malabar, experiences three main seasons: the cold, the rainy, and the hot season every year. The Monsoon renders good supply of water, and the ponds, gullies, channels, rivers and dales, which are in plenty, act as the best reservoirs of water.

The area also consists of hilly uplands, valleys, and forests. Most of the mountainous regions are covered with thick forests. The forests provide in abundance with food, fuel, and other materials for building houses, medicines, and for other needs. The environment acts as a multifaceted resource, and contributes for the subsistence, and survival of the people. When the necessary resources are available to all, what should the people to do? Just exploit the resources that are available in their surroundings! Thus, they start to subsist on them, and therefore, the economy of the people pivots around the forest, *i.e.* the Nature. Subsequently, when the people felt that these facilities are not enough to fulfil their growing needs and drive, they readjusted their life style that minimized the degree of Nature-dependency.

Even when the degree of dependency decreased, they did not allow themselves to destroy Nature, but were vigil to preserve it. This had led to Nature-worship, and this offered deities a chance to reside, and perform in every nook and corner. In this regard, they came forward to preserve trees in the form of kãvu as one of the sacred centres of teyyam performance. Likewise, the references that are made in tottampãttu about certain practices like hunting, gathering, pastoralism, and similar activities definitely reveal the degree of such subsistence, and survival.

For example, the tottam pãttu of kuttychãthan teyyam runs like this: "ezhala kali yundalo Kalakaderku, a kali maiponae kayariduvonae". This indicates the then prevalent pastoralism. Similarly, we can find many references about punam krishi, nayattu, and meen pidutham. In that way the teyyam of North Malabar clearly tells us the fact that on a number of occasions humans interacted with the Nature. A number of teyyams were observed as the transfigured Nature-objects such as monkey, leopard/tiger, pig, and snake. For example teyyams like Bali, Pulikandan, Madayil Chamundi, and Naga Rajan represents monkey, leopard, pig, and serpent respectively.

Most of the attire and wearings of teyyam also made up as in the shape of Nature-objects. In such way, masks, and pseudo bosoms and nails are used during performances. The pattern of facial writings also borrowed from Nature. Kozhipushpam, sangum valum, anachuvadu, and kurangirutham are few examples for such Nature-based pattern of facial writings.

These facts clearly confirm there is a Nature-Man interaction exists in the sphere of teyyam. The sacred performance needs ritual functionaries, who can carryout various performances hence, considered as, 'sacred specialists'. The ritual processes of contacting, and offering in a particular manner are already established things, in respect to each deity. Altogether, the worship, the performance, and the offering exercise are of complex nature that needs a 'middleman'.

The people successfully surmounted the operational difficulty of approachability by assigning certain persons the task of establishing liaison with the supernatural, through their divine teyyam performance. However, teyyam has existed in harmony with ecological settings, it also made possible some sort of social relationships. Thus, it often describes symbolically the actual social relations, status, and the role of the individuals in the society.

This relationships, and social positions have been working on a special kind of social network in which each caste is tied up with some obligations. In another word, every one has his own role to play in each performance. For example, the celebrant and the devotees celebrate and worship, and the performer performs teyyam.

The traditional economic structure of North Malabar has a special kind of system that characterizes a 'service-return' relationship, and under this system, each caste within a territorial division is expected to give certain standardized service to the families of other castes. Every one works for the respective family or, group of families with which he has 'hereditary ties'. His forefathers worked for the same families, and their descendants will continue to work for them because, the caste is the determinant of the occupation or, service.

The server in return will get certain economic benefits. For example, the Malayan, a traditional performer of magic, medicine, midwifery, and the teyyam, renders his services to families residing in a territory with which he has hereditary ties. In return, he gets cash or, kind. The system has also been shown to follow the pattern that each and every caste constitutes two-way interaction: one way, they render service to others, and on the other way, they receive certain benefits. In that sense, the performance of the teyyam provides a means to its performer's existence. Earlier, this kind of master-servant relationship was quite common in Kerala but, latter this pattern could not continue forever, due to political reasons. The introduction of a revolutionary act, 'Kerala Land Reforms Act', by the government of Kerala, under E.M.S Ministry, disturbed the feudal set up, as well as the occupational homogeneity of the state. But surprisingly, it is noticed that the age-old pattern of extending services between castes remains, not in severity, in teyyam performance. That is also noted as a pattern of behaviour against the modern innovations, and way of living. It is true that the society and its culture have been subjected to change, and yet, the people in this area still blindly carry out almost all the rituals and practices related to the teyyam performance without diluting it.

Due to this reason, the 'socalled' untouchables are the only people permitted to perform the teyyam in different sacred centres. Hence, it reflects the performers' identity too, because, they are simply identified as, the teyyam 'executors'.

In vise-versa, the performances of the teyyam also reflect the 'caste identity'. So, a few teyyams are customarily identified as, the teyyams performed by certain castes. Malayante teyyam, Vannante teyyam, and Pulayante teyyam can be considered as the examples to this. For that reason Vishnumurthi teyyam, Puthiya Bhagavathi, Kundora Chamundi, and Pulimarnja Thondachan teyyam are considered as the teyyams of Malayan, Vannan, Velan, and Pulayan respectively.

The villagers continue to perform the teyyam because, it still encloses a number of positive functions. Once, teyyam had socio-economic significance, and only later on, with the evolution of a multi-caste, pluralistic, and complex society, it came to perform more elaborate duties, political, religious, communicative, and ceremonial, besides its earlier functions. Apart from the said functions, the teyyam also serves the society, as a critique of the sociocultural discrimination, and which is an instrument to condemn social evils, and is a form of protest and criticism.

Some of the teyyams are directly attacking social oppressions and exploitations and there are also teyyams that indirectly oppose such discriminations. This liberty to criticize discriminations such as, 'untouchability' and 'pollution', gives the oppressed people some 'satisfaction', and 'relief'.

Even though this, significance of the teyyam is fading away to certain extent, but still has its holds in the sociocultural milieu of North Malabar. The myth, the rituals, and the whole performance of the teyyam itself glorify a few of the past stories of man, and his socio-cultural activities such as, hunting-gathering, nomadism, pastoralism, agriculture, as well as other elements such as, social structure, social status, social discriminations, and protests. The endurance of this old tradition namely, the teyyam, connects the past with the present, and keeps up a cultural continuity of this region, the rural and the urban areas. The sacred centres of the teyyam also act as linking agents to connect the performance and the people, and the society and its culture, even today. On the other hand, it is also alive as a more popular cultural phenomenon in North Malabar, which helps in maintaining the socio-cultural identity through observing certain widespread rituals and performances, that have been inherited from the distant past, and evolved from the Nature-Man interaction. Thus, for that reason, the teyyam is apt to survive as an important and living part of the socio-cultural sphere.

## SEDIMENT AS A CHEMICAL POLLUTANT

The role of sediment in chemical pollution is tied both to the particle size of sediment, and to the amount of particulate organic carbon associated with the sediment. The chemically active fraction of sediment is usually cited as that portion which is smaller than 63 μm (silt + clay) fraction. For phosphorus and metals, particle size is of primary importance due to the large surface area of very small particles. Phosphorus and metals tend to be highly attracted to ionic exchange sites that are associated with clay particles and with the iron and manganese coatings that commonly occur on these small particles.

Many of the persistent, bioaccumulating and toxic organic contaminants, especially chlorinated compounds including many pesticides, are strongly associated with sediment and especially with the organic carbon that is transported as part of the sediment load in rivers. Measurement of phosphorus transport in North America and Europe indicate that as much as 90 per cent of the total phosphorus flux in rivers can be in association with suspended sediment.

The affinity for particulate matter by an organic chemical is described by its octanol-water partitioning coefficient ($K_{OW}$). This partitioning coefficient is well known for most organic chemicals and is the basis for predicting the environmental fate of organic chemicals. Chemicals with low values of $K_{OW}$ are readily soluble, whereas those with high values of $K_{OW}$ are described as "hydrophobic" and tend to be associated with particulates. Chlorinated compounds such as DDT and other chlorinated pesticides are very hydrophobic and are not, therefore, easily analysed in water samples due to the very low solubility of the chemical. For organic chemicals, the most important component of the sediment load appears to be the particulate organic carbon fraction which is transported as part of the sediment. Scientists have further refined the partitioning coefficient to describe the association with the organic carbon fraction ($K_{OC}$).

Another important variable is the concentration of sediment, especially the <63 μm fraction, in the water column. Even those chemicals that are highly hydrophobic will be found in trace levels in soluble form.

Where the suspended load is very small (say, less than 25 mg/l), the amount of water is so large relative to the amount of sediment that the bulk of the load of the chemical may be in the soluble fraction.

Unlike phosphorus and metals, the transport and fate of sediment-associated organic chemicals is complicated by microbial degradation that occurs during sediment transport in rivers and in deposited sediment.

Nevertheless, the role of sediment in the transport and fate of agricultural chemicals, both for nutrients, metals, and pesticides is well known and must be taken into account when monitoring for these chemicals, and when applying models as a means of determining optimal management strategies at the field and watershed level.

For this reason, models using the "fugacity" concept (uses the partitioning characteristics of chemicals as a basis for determining the environmental compartment - air, sediment, water, biota - in which the chemical is primarily found) has proven effective in predicting the environmental pathways and fate of contaminants.

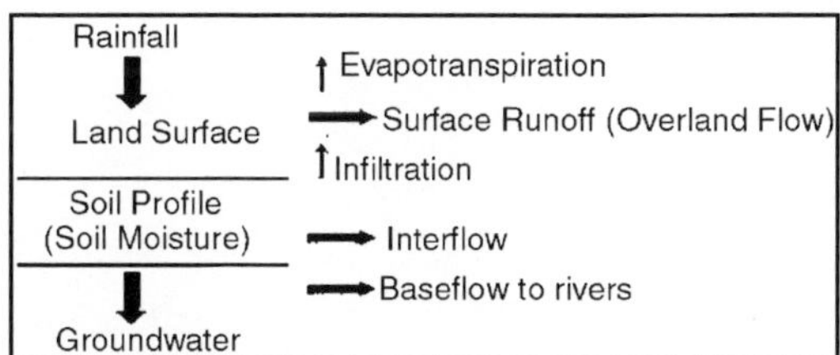


**Fig**. Schematic Diagram Showing the Major Processes that Link Rainfall and Run-off

- *Conclusion*: The role of sediment as a chemical pollutant is a function of the chemical load that is carried by sediments.

Organic chemicals associated with sediment enter into the food chain in a variety of ways. Sediment is directly ingested by fish however, more commonly, fine sediment (especially the carbon fraction) is the food supply for benthic (bottom dwelling) organisms which, in turn, are the food source for high organisms. Ultimately, toxic compounds bioaccumulate in fish and other top predators. In this way, pesticides that are transported off the land as part of the run-off and erosion process, accumulate in top predators including man.

## KEY PROCESSES: PRECIPITATION AND RUN-OFF

The major characteristic of non-point source pollution is that the primary transfer mechanisms from land to water are driven by those hydrological processes that lead to run-off of nutrients, sediment, and pesticides.

This is important, not only to understand the nature of agricultural pollution, but also because modelling of hydrological processes is the primary mechanism by which agriculturalists estimate and predict agricultural run-off and aquatic impacts. Except where agricultural chemicals are dumped directly into watercourses, almost all other non-point source control techniques in agriculture involve control or modification of run-off processes through various land and animal (manure) management techniques. In large parts of the world, precipitation is in the form of rain. However, in those areas where precipitation is in the form of snow, the science becomes more complex. Nevertheless, control measures, whether for areas subject to rain or snow can be easily summarized. Therefore, for the purpose of this publication, focus will be on the relationships between rainfall and run-off. While the practice of hydrology can be quite theoretical, the principal concepts are easily understood.

**Rainfall**

The primary controlling factor is the rate (intensity) of rainfall. This controls the amount of water available at the ground surface, and is closely related to measures of energy that are used in many mathematical formulations to calculate soil detachment by rain drops. Soil detachment makes soil particles available for sediment run-off.

**Soil Permeability**

Permeability is a physical characteristic of a soil and is a measure of the ability of the soil to pass water, under saturated conditions, through the natural voids that exist in the soil. Permeability is a function of soil texture, mineral and organic composition, etc..

In contrast, "porosity" is the measure of the amount of void space in a soil; however, permeability refers to the extent to which the porosity is made up of interconnecting voids that allow water to pass through the soil. As an example, styrofoam is highly porous but impermeable, whereas a sponge is both porous and permeable.

**Infiltration**

Infiltration rate, the rate at which surface water passes into the soil (cm/hr), is one of the most common terms in hydrologic equations for calculating surface run-off. Infiltration is not identical to permeability; it is mainly controlled by capillary forces in the soil which, in turn, reflect the prevailing conditions of soil moisture, soil texture, degree of surface compaction, etc. Infiltration will vary between and within rainfall events, depending upon factors such as antecedent soil moisture, nature of vegetation, etc. In general, infiltration rate begins at a high value during a precipitation event, and decreases to a small value when the soil has become saturated.

**Surface Run-off**

This is the amount of water available at the surface after all losses have been accounted for. Losses include evapotranspiration by plants, water that is stored in surface depressions caused by irregularity in the soil surface, and water that infiltrates into the soil. The interaction between infiltration rate and precipitation rate mainly governs the amount of surface run-off. Intense rainstorms tend to produce much surface run-off because the rate of precipitation greatly exceeds the infiltration rate. Similarly, in areas of monsoonal rain and tropical storms, the length and intensity of precipitation frequently exceeds infiltration capacity. Destruction of protective surface vegetation and compaction of the soil, especially in tropical environments, leads to major erosional phenomena due to the amount of surface run-off. Except for nitrogen which is usually found in groundwater in agricultural areas, surface run-off is the primary contributor of agricultural chemicals, animal wastes, and sediment to river channels.

**Interflow**

Sometimes called "throughflow") Because soil horizons have different levels of permeability not all water in the soil will move downward into the groundwater. The residual water in the soil will move along the soil horizons, parallel to the ground surface. Interflow usually emerges near the bottom of slopes and in valley bottoms.

Therefore, identification of these hydrologically active zones is an important part of agricultural non-point source control measures. Interflow is the mechanism which has also been linked to soil piping, a potentially destructive characteristic in some soils by which shallow "pipes" form naturally in the soil and are enlarged by interflow to the point where they collapse causing gullies in the agricultural surface.

**Groundwater**

Groundwater is supplied by water which passes through the soil horizons into the parent material and/or bedrock underlying the soil. Groundwater tends to flow towards rivers channels where it emerges and supports stream flow during periods of little or no rain. This component of stream flow is called "base flow". The chemistry of base flow reflects the soil and bedrock geochemistry, plus any agrochemicals that have been leached into the groundwater.

**Snowmelt**

The phenomenon of snowmelt greatly complicates prediction of agricultural pollution using conventional hydrologic models. Snowmelt, by itself, is not normally a major producer of surface run-off.

However, the combination of spring rain and snowmelt on frozen or thawing soils can produce serious erosional problems. Snowmelt tends to contribute greatly to agricultural non-point source pollution by carrying to adjacent streams the animal wastes, sludges, and other wastes that were spread on frozen agricultural soils during the winter period. Correct management of animal wastes in regions of frozen ground has major beneficial effects on water quality.

## THE LIMITS OF GLOBAL ENVIRONMENTAL MANAGEMENT

Although steps have been taken between Rio 1992 and New York 1997 in the direction of the globalization of environmental protection, the world seems to be further away from sustainability today than it was then. With environmental globalization and the consensual concept of sustainable development, the perception of an ecological crisis, or at least of ecological limits to development, appears to have vanished: acting for environmental protection increasingly tends to be seen as a technical problem, the task of increasing efficiency and of better using resources such as science, technology, information, capital, and institutions. The cause of environmental problems is no longer perceived as linked to industrial development and ever-increasing material accumulation, but it has become the very existence of human beings. Environmental problems are understood as unavoidable, as side effects of human activities, and efforts are then directed at solving these problems.

Indeed, global environmental management and sustainable development can be seen as "problem solving" concepts, in Robert Cox's terminology, as they only represent a strategy to allow the pursuit of present lifestyles and standards. Following Strange's call for a critical International Political Economy and the need to address the question of "who gets what, how and why," the analysis of the evolution of the system of global environmental management has revealed that it tends to strengthen the mechanisms of exclusion and inequality. Global environmental management and sustainable development tend to be uneven concepts, as they do not aim at promoting the correction of global disparities. They attempt to offer a universal framework in which the global society is the unit of analysis and a large share of the blame for environmental degradation rests on the Third World.

Instead of stressing affluence, over-production and over-consumption in advanced countries as the main causes of environmental degradation, it tends to suggest that problems arise from poverty. Environmental degradation is transformed from a problem of affluence into a problem of poverty. The responsibility is shifted from major polluters and industrialized countries' abuses to all inhabitants of the planet. For Chatterjee and Finger (1994), the only different element in this approach is that development is now looked at from a global perspective, making the development discourse universal.

And the New York Summit represented a step further in that direction, asserting the desirability of the globalization process and underlining its beneficial aspects. It also consolidated the role of business and industry as privileged partners of the United Nations, establishing permanent contact and consultation on environmental issues. The regulatory situation relating to TNCs and business in general has worsened greatly in the past five years. Already in 1992, the US government successfully pressured for downsizing the UN Centre on Transnational Corporations (UNCTC), which had been set up to monitor the social, economic and environmental impacts of corporate investment in developing countries. Today, the UN is considering cutting a sub-group of the UN Human Rights Commission which addresses the impacts of corporations on a broad spectrum on rights issues. The main international initiatives and institution for establishing guidelines for the behaviour of corporations, which could together lay down a code of obligations and rights of TNCs and states, have disappeared. In their place has come a strong and growing opposite trend to reduce and remove regulations that governments have over corporations, to grant them increased rights and powers, and to reduce the authority of states to control their behaviour and operations.

The Uruguay Round for example has already granted far higher standards of intellectual property rights protection to corporations, thus facilitating further their global monopolization of technology and ability to make profit through higher prices. There are also strong pressures from Northern governments at the World Trade Organization to grant foreign companies the right of entry, establishment and national treatment in all WTO member states.

In addition, the partnership between the UN and global corporations seems to have been further strengthened in the past two years. In part due to the difficult financial and political situation in which the UN finds itself as a result of the US government's refusal to pay the US $1.6 billion it owes, the UN is now openly seeking political and economic support from corporations. At the last Davos Economic Forum in Switzerland, UN Secretary General Kofi Annan called for a human face to the global market and challenged business leaders to adhere to universal values defined by the UN and contribute to global environmental protection, indicating a broader trend of growing UN collaboration with transnational corporations.

Recently, the United Nations Development Programme (UNDP) has solicited funds from global corporations with poor records on human rights, labour and the environment, such as Dow Chemical, Citibank and Rio Tinto Plc, in exchange of special UNDP sanctioned logos for use by corporate sponsors. Called the "Global Sustainable Development Facility (GSDF)," the plan calls for corporate sponsors to funnel donations to a separate entity which they will manage. In the words of the UNDP, the GSDF "brings together leading global corporations and the UNDP, to jointly define and implement a new facility to eradicate poverty, create sustainable economic growth and allow the private sector to prosper through the inclusion of two billion new people in the global market economy."

According to the internal memo, sponsors will benefit from the advice and support of UNDP through a special relationship, allowing corporations unprecedented access to UNDP's network of offices, high level governmental contacts and the knock-on effects of its reputation. The plan, revealed through a leaked internal UNDP memo, has been heavily criticized by observers and

NGOs, who warn that the interests of global corporations are often at odds with the basic economic and social needs of the world's poor and the values of human rights and environmental protection the UN is meant to protect.

According to Ward Morehouse, President of the US Council on International and Public Affairs, the UN should be monitoring the human rights and environmental impacts of corporations in developing and industrialized nations, not granting special favours... Increasing collaboration will lead to a reluctance to criticize corporations which are central players in the human rights, environmental and developmental dramas unfolding every day across the globe.

To conclude, the UN and international organizations in general seem to be moving towards the adoption of a market-oriented global model of environmental governance, which sees economic globalization as a positive and integrative process. The key actors of economic globalization, transnational corporations, are taking the leading role and consolidating their influence on the system of international environmental governance. With the adoption of this project of global environmental management, one particular understanding of the world, the one promoted by business and large corporations in Western affluent societies, becomes hegemonic and appears to be universal.

Environmental concerns have been incorporated as a mere dimension of the "globalization project," understood in McMichael's definition as "an emerging vision of the world and its resources as a globally organized and managed free trade/free enterprise economy pursued by a largely unaccountable political and economic elite". This project advocates a universalized model of production, of consumption, and thus of dealing with problems of environmental protection resulting from these activities. By assuming its universality, it tends to marginalize other knowledge and other solutions to problems of environmental protection. Interestingly enough, it was unanimously recognized that the most positive result and follow-up of UNCED was without doubt realized at the micro-level, within the Local Agenda 21 framework. In an effort to implement Agenda 21 locally, social groups have worked together with local authorities to make sustainable development a reality at the local level, often on a truly participatory basis and reflecting grassroots concern and involvement.

From a critical point of view, global management's failure, exemplified at the New York Summit, was not entirely unexpected. Indeed, the global management approach inspired by business perspectives and propagated by the international development establishment tends to strengthen the globalization process at work today, failing to counter its effects in terms of social exclusion and environmental destruction. It tends to weaken social protection and environmental protection in the name of economic efficiency.

It stands at odds with the commitment to social change and to equity that lies at the root of a critical, political economy view of global environmental politics as inserted within the dynamics of economic accumulation and social structures. The question of the ownership of natural resources, for example, is not addressed. However, environmental problems in the South are often linked to problems of resource ownership and equity.

Sustainable Development as defined in Rio and reasserted in New York has been practically translated into technocratic responses to what are in reality political problems. The purpose of this article was to analyze and evaluate the process of "mainstreaming" of environmental concerns. Recalling the radical and transformative origins of the ecological project, it has provided evidence that environmental concerns have been remodeled by the joint action of technocratic environmentalists, the international UN-related development establishment and business and industry sectors.

Examining the results of international cooperation, the article has questioned the nature of the sustainable development consensus, a consensus deeply marked by the growing access and influence of global corporations on UN activities. Today, market-oriented perspectives to environmental problems seem to be prevailing over more transformative views, especially at the international level, within the framework of international organizations and institutional agreements.

The article has suggested that the mainstream approach to sustainable development tends to reduce ecology to a set of managerial practices aiming at resource efficiency and risk management. In doing so, it tends to address a civilizational impasse as a mere technical problem. The mainstream approach proposes that environmentalists should operate using the language and the worldview of Western economics in approaching ecological concerns.

Instead of designing cultural and political limits to development, the project of "global environmental management" tends to become part of a technocratic effort to sustain industrial development in the age of economic globalization. Environmental protection, together with democracy, human rights and free market economics, becomes a universal consensus, a universal consensus which, as Baudrillard remarks, arouses suspicion, since it is about values that have become devalued, values becoming emptied at the very moment of their hegemony. Environmental concerns become just another element in a process leading to global uniformity, a uniformity of cultures, lifestyles, mentalities, but also of relationships with nature. This market-oriented agenda may provide a starting point for dealing with global environmental problems.

The documents, which emerged from international environmental negotiations from Rio to New York, replete with inconsistencies, represent a complex mix of disagreements, hopes and compromises. One may concede that conventions and obligations reflect the need for government negotiators to find the minimum agreeable grounds to initiate a large open-ended process on major environmental issues. In this sense, they only produced a general framework for negotiations, steps on the way to building international regimes. They do not form a series of real commitments representing an effective consensus on how to deal with global environmental issues.

Yet considering the amount of time, energy and resources invested in this process of international environmental regime building, one might have hoped for more concrete, positive results. The failure of the present framework to effectively promote sustainability, which became evident in the 1997 New York summit, is recognized even by one of the major promoters of this path, Maurice Strong, ex-secretary general of UNCED, today President of the Earth Council. For Strong, unfortunately, the economic, social and demographical forces that lead to unsustainable development still prevail. Strong sees the lack of political will from governments as the main cause of this failure.

The present framework appears to contain many contradictions that limit the ability of the international and national communities to solve satisfactorily environmental problems. Adopting an international political economy perspective, this article has argued that the main problem that international efforts to protect the environment have to address is the issue of the impact of economic globalization. Economic globalization has in a sense helped to create conditions for the development of policy mechanisms and institutions that will universalize and promote the concept of "sustainable development." Global change is exerting a structuring influence on the redefinition of environmental politics. However, the kind of sustainable development being promoted seems to represent more the consolidation of a global project of "environmental management" than a real shift away from destructive practices. Globalization is consolidating a market-friendly view of sustainable development, a view that gives priority to the sustainability of "global growth" and to the correction of environmental damage. This tends to be carried out at the expense of the competing alternatives and participative view of sustainable development as stressing not only development but also social equity and decentralized participation. The "globalization project" has shaped and redefined both the content of environmentalism and environmental policies and structured the international political economy in a way that makes sustainability more difficult to achieve.

## ENVIRONMENT POLICY AND ENVIRONMENT MANAGEMENT SYSTEM

Driven by its commitment for sustainable growth of power, NTPC has evolved a well defined environment management policy and sound environment practices for minimising environmental impact arising out of setting up of power plants and preserving the natural ecology.

### NATIONAL ENVIRONMENT POLICY

At the national level, the Ministry of Environment and Forests had prepared a draft Environment Policy (NEP) and the Ministry of Power along with NTPC actively participated in the deliberations of the draft NEP. The NEP 2006 has since been approved by the Union Cabinet in May 2006.

## NTPC ENVIRONMENT POLICY

As early as in November 1995, NTPC brought out a comprehensive document entitled "NTPC Environment Policy and Environment Management System". Amongst the guiding principles adopted in the document are company's proactive approach to environment, optimum utilisation of equipment, adoption of latest technologies and continual environment improvement. The policy also envisages efficient utilisation of resources, thereby minimising waste, maximising ash utilisation and providing green belt all around the plant for maintaining ecological balance.

## ENVIRONMENT MANAGEMENT, OCCUPATIONAL HEALTH AND SAFETY SYSTEMS

NTPC has actively gone for adoption of best international practices on environment, occupational health and safety areas. The organization has pursued the Environmental Management System (EMS) ISO 14001 and the Occupational Health and Safety Assessment System OHSAS 18001 at its different establishments. As a result of pursuing these practices, all NTPC power stations have been certified for ISO 14001 and OHSAS 18001 by reputed national and international Certifying Agencies.

## POLLUTION CONTROL SYSTEMS

While deciding the appropriate technology for its projects, NTPC integrates many environmental provisions into the plant design. In order to ensure that NTPC comply with all the stipulated environment norms, various state-of-the-art pollution control systems / devices as discussed below have been installed to control air and water pollution.

### Electrostatic Precipitators

The ash left behind after combustion of coal is arrested in high efficiency Electrostatic Precipitators (ESPs) and particulate emission is controlled well within the stipulated norms.

### Flue Gas Stacks

Tall Flue Gas Stacks have been provided for wide dispersion of the gaseous emissions (SOX, NOX etc) into the atmosphere.

### Low-NOX Burners

In gas based NTPC power stations, NOx emissions are controlled by provision of Low-NOx Burners (dry or wet type) and in coal fired stations, by adopting best combustion practices.

**Neutralisation Pits**

Neutralisation pits have been provided in the Water Treatment Plant (WTP) for pH correction of the effluents before discharge into Effluent Treatment Plant (ETP) for further treatment and use.

**Coal Settling Pits / Oil Settling Pits**

In these Pits, coal dust and oil are removed from the effluents emanating from the Coal Handling Plant (CHP), coal yard and Fuel Oil Handling areas before discharge into ETP.

**DE and DS Systems**

Dust Extraction (DE) and Dust Suppression (DS) systems have been installed in all coal fired power stations in NTPC to contain and extract the fugitive dust released in the Coal Handling Plant (CHP).

**Cooling Towers**

Cooling Towers have been provided for cooling the hot Condenser cooling water in closed cycle Condenser Cooling Water (CCW) Systems. This helps in reduction in thermal pollution and conservation of fresh water.

**ASH DYKES AND ASH DISPOSAL SYSTEMS**

Ash ponds have been provided at all coal based stations except Dadri where Dry Ash Disposal System has been provided. Ash Ponds have been divided into lagoons and provided with garlanding arrangements for change over of the ash slurry feed points for even filling of the pond and for effective settlement of the ash particles. Ash in slurry form is discharged into the lagoons where ash particles get settled from the slurry and clear effluent water is discharged from the ash pond. The discharged effluents conform to standards specified by CPCB and the same is regularly monitored. At its Dadri Power Station, NTPC has set up a unique system for dry ash collection and disposal facility with Ash Mound formation. This has been envisaged for the first time in Asia which has resulted in progressive development of green belt besides far less requirement of land and less water requirement as compared to the wet ash disposal system.

**ASH WATER RECYCLING SYSTEM**

Further, in a number of NTPC stations, as a proactive measure, Ash Water Recycling System (AWRS) has been provided. In the AWRS, the effluent from ash pond is circulated back to the station for further ash sluicing to the ash pond. This helps in savings of fresh water requirements for transportation of ash from the plant.

The ash water recycling system has already been installed and is in operation at Ramagundam, Simhadri, Rihand, Talcher Kaniha, Talcher Thermal,

Kahalgaon, Korba and Vindhyachal. The scheme has helped stations to save huge quantity of fresh water required as make-up water for disposal of ash.

**Dry Ash Extraction System (DAES)**

Dry ash has much higher utilization potential in ash based products (such as bricks, aerated autoclaved concrete blocks, concrete, Portland pozzolana cement, etc.). DAES has been installed at Unchahar, Dadri, Simhadri, Ramagundam, Singrauli, Kahalgaon, Farakka, Talcher Thermal, Korba, Vindhyachal, Talcher Kaniha and BTPS.

**Liquid Waste Treatment Plants and Management System**

The objective of industrial liquid effluent treatment plant (ETP) is to discharge lesser and cleaner effluent from the power plants to meet environmental regulations. After primary treatment at the source of their generation, the effluents are sent to the ETP for further treatment. The composite liquid effluent treatment plant has been designed to treat all liquid effluents which originate within the power station *e.g.* Water Treatment Plant (WTP), Condensate Polishing Unit (CPU) effluent, Coal Handling Plant (CHP) effluent, floor washings, service water drains etc.

The scheme involves collection of various effluents and their appropriate treatment centrally and re-circulation of the treated effluent for various plant uses. NTPC has implemented such systems in a number of its power stations such as Ramagundam, Simhadri, Kayamkulam, Singrauli, Rihand, Vindhyachal, Korba, Jhanor Gandhar, Faridabad, Farakka, Kahalgaon and Talcher Kaniha. These plants have helped to control quality and quantity of the effluents discharged from the stations.

**Sewage Treatment Plants and Facilities**

Sewage Treatment Plants (STPs) sewage treatment facilities have been provided at all NTPC stations to take care of Sewage Effluent from Plant and township areas. In a number of NTPC projects modern type STPs with Clarifloculators, Mechanical Agitators, sludge drying beds, Gas Collection Chambers etc have been provided to improve the effluent quality. The effluent quality is monitored regularly and treated effluent conforming to the prescribed limit is discharged from the station. At several stations, treated effluents of STPs are being used for horticulture purpose.

**ENVIRONMENTAL INSTITUTIONAL SET-UP**

Realizing the importance of protection of the environment with speedy development of the power sector, the company has constituted different groups at project, regional and Corporate Centre level to carry out specific environment related functions. The Environment Management Group, Ash Utilisation Group and Centre for Power Efficiency and Environment Protection (CENPEEP) function from the Corporate Centre and initiate measures to mitigate the impact of power project implementation on the environment and preserve ecology in the vicinity of the projects. Environment Management and Ash Utilisation Groups established at each station, look after various environmental issues of the individual station.

### Environment Management During Operation Phase

NTPC's environment friendly approach to power has already begun to show results in conservation of natural resources such as water and fuel (coal, oil and gas) as well as control of environmental pollution. As already mentioned earlier, NTPC has chalked out a set of well defined activities that are envisaged right from the project conceptualisation stage so that during the entire life cycle of the power plant, NTPC is fully compliant with various environment regulations and a pristine environment and ecological balance is maintained in and around its power station and townships. Following is brief description of some of the measures taken during the operation phase of the stations. Performance enhancement and up-gradation measures are undertaken by the organisation during the post operational stage of the stations. These activities have greatly helped to minimise the impact on environment and preserve the ecology in and around its power projects. These measures have been enumerated as follows.

### Monitoring of Environmental Parameters

A broad based Environment Monitoring Programme has been formulated and implemented in NTPC. All pollutants discharged from the power plant such as stack emission, ash pond effluent, main plant effluent, domestic effluent and Condenser Cooling Water (CCW) effluent are monitored at the stipulated frequency at the source itself and at the points of discharge. In addition to the above, ambient air, surface water and ground water quality in and around NTPC plants are regularly monitored to assess any adverse impacts as a result of operation of the power plant.

### On-Line Data Base Management

In order to have better control on pollution and to achieve effective environment management in and around NTPC stations, it is imperative to have an on-line, reliable and efficient environment information system on the plant operational and environmental performance parameters at all three levels i.e generating Stations, Regional Headquarters and Corporate Centre. In consideration of the above, a computerized programme, namely "Paryavaran Monitoring System" - PMS, which could provide reliable storage, prompt and accurate flow of information on environmental performance of Stations was developed and installed in NTPC. This software facilitates direct transfer of environment reports and other environment related information from stations to the Regional Headquarters and Corporate Centre. The PMS has already been implemented at Corporate Centre, the Regional Headquarters and most of the Stations. This system has helped in achieving continuous improvement in NTPCs environment performance through improved monitoring and reporting system by using the trend analysis and advanced data management techniques.

**ENVIRONMENT REVIEWS**

To maintain constant vigil on environmental compliance, Environmental Reviews are carried out at all operating stations and remedial measures have been taken wherever necessary. As a feedback and follow-up of these Environmental Reviews, a number of retrofit and up-gradation measures have been undertaken at different stations. Such periodic Environmental Reviews and extensive monitoring of the facilities carried out at all stations have helped in compliance with the environmental norms and timely renewal of the Air and Water Consents.

**Upgradation and Retrofitting of Pollution Control Systems**

In order to keep pace with the changing norms and ensure compliance with statutory requirements in the field of pollution control, NTPC keeps an open mind for Renovation and Modernisation (R and M) and Retrofitting and Upgradation of pollution monitoring and control facilities in its existing stations. It is important to mention that such modifications/retrofit programmes not only helped in betterment of environment but also in resource conservation. High efficiency Electro-Static Precipitators (ESPs) of the order of 99.5 per cent and above have been provided at NTPC stations for control of stack particulate emissions. However, the ESPs of a number of stations were built prior to the promulgation of the Environment (Protection) Act, 1986 and notification of emission control standards under this Act. Remedial measures have already been taken up and implemented to improve the efficiency of the existing ESPs at various NTPC stations. ESP performance enhancement programme by adopting advanced microprocessor based Electrostatic Precipitator Management System (EPMS) was installed at its power stations at Singrauli, Ramagundam, Korba, Farakka, Rihand, Vindhyachal and Unchahar.

Additional ESPs were retrofitted in the older power stations, namely at Badarpur and Talcher Thermal.

As a result of the above retrofits, the emission of Suspended Particulate Matter (SPM) has been brought down appreciably at the above stations and is maintained within the present statutory limit of 150 mg/Nm$^3$. In new projects, the ESPs have been designed for a maximum permissible outlet dust emission of 50 mg/Nm$^3$ to meet the likely stringent emission norms in the near future.

## RESOURCES CONSERVATION

With better awareness and appreciation towards ecology and environment, the organization is continually looking for innovative and cost effective solutions to conserve natural resources and reduce wastes.

*Some of the measures include*:

- Reduction in land requirements for main plant and ash disposal areas in newer units.
- Capacity addition in old plants, within existing land.
- Reduction in water requirement for main plant and ash disposal areas through recycle and reuse of water.
- Efficient use of Fuel (Coal, Natural gas and Fuel oil)
- Reduction in fuel requirement through more efficient combustion and adoption of state-of-the-art technologies such as super critical boilers

## WASTE MANAGEMENT

Various types of wastes such as Municilal or domestic wastes, hazardous wastes, Bio-Medical wastes get generated in power plant areas, plant hospital and the townships of projects. The wastes generated are a number of solid and hazardous wastes like used oils and waste oils, grease, lead acid batteries, other lead bearing wastes (such as garkets etc.), oil and clarifier sludge, used resin, used photochemicals, asbestos packing, e-waste, metal scrap, C and I wastes, electricial scrap, empty cylinders (refillable), paper, rubber products, canteen (bio-degradable) wastes, buidling material wastes, silica gel, glass wool, fused lamps and tubes, fire resistant fluids etc. These wastes fall either under hazardous wastes category or non-hazardous wastes category as per classification given in Government of Indias notification on Hazardous Wastes (Management and Handling) Rules 1989 (as amended on 06.01.2000 and 20.05.2003). Handling and manegement of these wastes in NTPC stations have been discussed below.

### Municipal Waste Management

Domestic or municipal waste is generated in households at townships. This waste is seggregated into bio-degradable and non-biodegradable wastes at source itself in different coloured containers and thereafter the two types are disposed separately. Bio-degradable waste is spread uniformly in identified low lying areas and thereafter it is covered with soil for use later as manure after composting. The seggregated non bio-degradable waste is disposed off separately in other identified low lying areas and is spread out uniformly.

### Hazardous Waste Management

NTPC being a proactive organization, the handling and disposal of hazardous wastes are done as per the Hazardous Wastes (Management and Handling)

Rules 1989 (as amended in 2003) guidelines issued by Government of India for the treatment, storage and disposal of hazardous wastes. Scientific study on management and handling of hazardous wastes was carried out at a few NTPC stations to adopt the best practices so that there is a complete compliance with statutory requirements. In NTPC sataione, the Hazardous Wastes (Recyclable) are sold / auctioned to registered recyclers / refiners. The other hazardous wastes such as the activated carbon resins, used drums (hazardous) chromium (Cr-III electrolytes, used petro-chemicals, asbestos packings, used torch batteries, ribbon, toners / cartridges, mixed wastes (waste oil, water and cotton) filters, earth contaminated with synthetic oil (FQF) glass used and sodium silicate, lamps and tubes etc. fall under the category of Hazardous Wastes (Non-Recyalable). These wastes are small in quantity and are stored in properly identified locations. As per the notification, hazardous wastes (non-recyalable) are to be sent to State Pollution Control Board (SPCB) approved common treatment storage and disposal facility (TSDF).

**Bio-Medical Waste Management**

Hospital (or Bio-medical) wastes get generated from hospitals and they include urine bags, human anatomical wastes, plaster of aris waste, empty plastic bottles of water and glucose, blood and chemical mixed cotton, blood and urines tubes etc. these wastes are segregated and are placed in buckets of different colours as per the notification for Bio-Medical Waste (Management and Handling) Rules. The seggregated bio-medical wastes are either disposed through the SPCB approved agency or they are treated in autoclaves before disposal into bio-medical waste disposal pits. The treated bio-medical waste is spread uniformly and covered with 10 cm thick soil in bio-medical waste disposal pits.

**Reclamation of Abandoned Ash ponds**

The reclamation of abandoned ash pond sites is a challenging task. NTPC has reclaimed temporary ash disposal areas at some of its projects namely Ramagundam, Talcher Thermal, Rihand, Singrauli and Unchahar through plantation and converted these sites into lush green environments. Extensive plantations have also been undertaken on dry ash mound at NTPC-Dadri. It is planned to reclaim all the abandoned ash disposal areas by plantation.

**Green Belts, Afforestation and Energy Plantations**

Whats more, in a concerted bid to counter the growing ecological threat, NTPC is undertaking afforestation programmes covering vast areas of land in and around its projects. Appropriate afforestation programmes for plant, township and green belt areas of the project have been implemented at all projects. In order to enhance green cover in the areas around our projects, as a responsible corporate citizen, NTPC till date has planted more than 18 million trees at its projects throughout the country. The afforestation has not only contributed to the aesthetics but also has been serving as a 'sink' for the pollutants released from the station and thereby protecting the quality of ecology and environment in and around the projects. Thrust has also been given to bio-diesel plantation and around 4.8 lakh energy plants including Pongamia and Jatropha have already been planted. A pilot project for extraction of seeds from these bio-diesel plants has also been set up.

## ECOLOGICAL MONITORING AND SCIENTIFIC STUDIES

NTPC has been a leader in the industrial sector of India in undertaking scientific studies related to thermal power generation. NTPC has pioneered several scientific studies in collaboration with national/ international institutions to develop an environmental databank *e.g.* Detailed Geohydrological Studies to understand the impact of ash pond leachate on ground water and Ecological Impacts Monitoring through Remote Sensing Data have been carried out at its operating stations as discussed below.

### Environment Impact Asssement Studies

Environmental Impact Assessment (EIA) Studies are inevitably undertaken to evaluate potential negative impacts as well as to formulate Environmental Management Plans to overcome the identified impacts. Based on the recommendations of Environmental Impact Assessment Study and Environmental Management Plan (EMP) and the conditions stipulated in the clearances from Ministry of Environment and Forests and State Pollution Control Boards, These studies consists of impact assessment in the area of the land use, water use, socio-economic aspects, soil, hydrology, water quality, meteorology, air quality, terrestrial and aquatic ecology and noise.

These studies are conducted before starting the construction as well as after operation of the plant and gives comprehensive status of the environment as existed before construction as well as in the post operational stages of the project. The EIA involves stage-by-stage evaluation of various parameters which affect the environment. Based on EIA study, wherever required, specific scientific studies are also conducted to scientifically assess the likely impact of the pollutants on the sensitive flora and fauna in the surroundings, as also, to take preventive and mitigatory measures, wherever required. Apart from project specific EIA studies, Regional Environmental Assessment studies have been conducted for Integrated Development of Singrauli, Korba and Ramagundam areas. Such studies are of first of their kind in India and probably very few such studies have been undertaken in other countries.

### Socio-economic Studies

Detailed socio-economic studies are undertaken to establish the socio-economic status of project affected persons and rehabilitation and resettlement plans are drawn in consultation with the state government. Rehabilitation and resettlement options include land for land (subject to availability), limited jobs with NTPC and contractors and self employment schemes. In addition, NTPC also undertakes community development activities in the surrounding villages.

**Ecological Monitoring Programme**

NTPC has undertaken a comprehensive Ecological Monitoring Programme through Satellite Imagery Studies covering an area of about 25 Kms radius around some of its major plants. The studies have been conducted through National Remote Sensing Agency (NRSA), Hyderabad at its power stations at Ramagundam, Farakka, Korba, Vindhyachal, Rihand and Singrauli. These studies have revealed significant environmental gains in the vicinity areas of the project as a result of pursuing sound environment management practices.

Some of these important gains which have been noticed are increase in dense forest area, increase in agriculture area, increase in average rainfall, decrease in waste land etc. In general, the studies, as such, have revealed that there is no significant adverse impact on the ecology due to the project activities in any of these stations. Such studies conducted from time to time around a power project have established comprehensive environment status at various post operational stages of the project.

**Geo-hydrological Studies**

NTPC has conducted several geohydrological studies of the ash disposal areas at its projects (Singrauli, Rihand, Vindhyachal, Korba, Farakka and Talcher) through reputed institutions like Indian Institutes of Technology, Roorkee; Indian Institutes of Technology, Mumbai, Centre for Studies on Man and Environment, Calcutta.

All these studies conclude that the leaching of heavy metals from ash occurs only under pH 4 or below. In practice, the pH of the ash water is either neutral or alkaline (7 or above) and hence the leaching of heavy metals is highly unlikely.

**USE OF WASTE PRODUCTS AND SERVICES -ASH UTILIZATION**

Ash is the main solid waste which is put into use for various products and services. NTPC has adopted user friendly policy guidelines on ash utilisation. In order to motivate entrepreneurs to come forward with ash utilisation schemes, NTPC offers several facilities and incentives. These include free issue of all types of ash *viz.* Dry Fly Ash / Pond Ash / Bottom Ash and infrastructure facilities, wherever feasible. Necessary help and assistance is also offered to facilitate procurement of land, supply of electricity etc from Government Authorities. Necessary techno-managerial assistance is given wherever considered necessary. Besides, NTPC uses only ash based bricks and Fly Ash portland pozzolana cement (FAPPC) in most of its construction activities. Demonstration projects are taken up in areas of Agriculture, Building materials,

Mine filling etc. The utilisation of ash and ash based products is progressively increasing as a result of the concrete efforts of these groups.

**Advanced / Eco-friendly Technologies**

NTPC has gained expertise in operation and management of 200 MW and 500 MW Units installed at different Stations all over the country and is looking ahead for higher capacity Unit sizes with super critical steam parameters for higher efficiencies and for associated environmental gains.

At Sipat, higher capacity Units of size of 660 MW and advanced Steam Generators employing super critical steam parameters have already been implemented as a green field project. Higher efficiency Combined Cycle Gas Power Plants are already under operation at all gas-based power projects in NTPC. Advanced clean coal technologies such as Integrated Gasification Combined Cycle (IGCC) have higher efficiencies of the order of 45 per cent as compared to about 38 per cent for conventional plants. NTPC has initiated a techno-economic study under USDOE / USAID for setting up a commercial scale demonstration power plant by using IGCC technology.

These plants can use low grade coals and have higher efficiency as compared to conventional plants. With the massive expansion of power generation, there is also growing awareness among all concerned to keep the pollution under control and preserve the health and quality of the natural environment in the vicinity of the power stations. NTPC is committed to provide affordable and sustainable power in increasingly larger quantity. NTPC is conscious of its role in the national endeavour of mitigating energy poverty, heralding economic prosperity and thereby contributing towards India s emergence as a major global economy.

# Bibliography


Siddhartha Gautam: *Advance in Air Pollution: Its Causes and Contro*, Vista International , 2010.

Kanika Jain: *Air Pollution*, Mohit Pub, 2007.

Aradhana Salpekar: *Air Pollution*, Jnanada Prakashan, 2008.

S.K. Agarwal: *Air Pollution*, APH, 2009.

M.G. Chitkara: *Air Pollution*, APH Pub, 2012.

A.K. Tripathi: *Air Pollution*, APH, 2012.

A K Srivastava: *Air Pollution : Biopollutants in Air*, APH, 2008.

Yashpal Wadhwa: *Air Pollution : Causes and Contro*, Cyber Tech Pub, 2009.

M.P. Singh, Meenakshi Singh and B.S. Singh: *Air Pollution and Environment*, Enkay Pub, 2011.

Sumit Malhotra: *Air Pollution and Its Control*, Pointer, 2004.

Arvind Kumar: *Air Pollution and Its Control*, Shree Pub, 2011.

G. Gaur: *Air Pollution and Its Management*, Sarup & Sons, 1997.

Shilpa Shyam, H.N. Verma and S.K. Bhargava: *Air Pollution and It's Impact on Plant Growth*, New India Pub, 2006.

P Bhaskar: *Air Pollution and Plant Biotechnology*, Manglam, 2009.

Omasa: *Air Pollution and Plant Biotechnology: Prospects for Photomonitoring and Phytoremediation*, Springer 2010.

S. P. Mahajan: *Air Pollution Control*, TERI, 2009.

T K Ray: *Air Pollution Control in Industries (2 Vols-Set)*, Tech Books International, 2006.

Quamrul Islam Chowdhury: *Air Pollution in Asian Cities*, Forum of Environmental Journalists of Bangladesh, 2000.

Avinash Chiranjeev and Anil K. Jamwal: *Air Pollution, Acid Rain, Ozone Depletion and Sea Level Rise*, Jnanada Prakashan, 2010.

Yogesh T. Jasrai and Arun Arya: *Air Pollution: Development at What Cost*, Daya, 2003.

R.K. Trivedy and P.K. Goel: *An Introduction to Air Pollution*, BS Pub, 2005.

R.K. Trivedy and P.K. Goel: *An Introduction to Air Pollution*, ABD Publishers, 2003.

T I Khan: *Atmosphere and Air Pollution Control Technologies*, Aavishkar, 2004.

Arun Arya, S.J. Bedi and V.S. Patel: *Eco-Degradation Due to Air Pollution*, Scientific Pub, 2009.

G.R. Chhatwal: *Encyclopaedia of Environmental Air Pollution (3 Vols-Set)*, Anmol, 2003.

V.K. Prabhakar: *Environmental Air Pollution*, Anmol, 2001.

Klein Gomes: *Environmental Air Pollution*, Oxford Book Company, 2009.

P.N. Prasad and T.R. Amarnath: *Environmental Air Pollution : Causes, Effects and Control*, Crescent Publishing Corporation, 2010.

S. Gandhimathi: *Impact of Air Pollution on Health in India*, Serials Publications, 2015.

Sanjay Kumar Agrawal: *Politico-Administrative Dimension of Air Pollution*, Vista International, 2006.

J.R. Mudakavi: *Principles and Practices of Air Pollution Control and Analysis* , I.K. International, 2010.

S.C. Bhatia: *Textbook of Air Pollution and Its Control*, Atlantic, 2007.

A.P. Rastogi.: *Air Travel Ticketing and Fare Construction*, Aman Publication, Delhi, 2007.

Jagmohan Negi.: *Air Travel*: *Ticketing and Fare Construction (With Examination Questions)*, Kanishka Publications, Delhi, 2004.

Jitendra K. Sharma.: *Flight Reservation and Airline Ticketing*, Kanishka Publications, Delhi, 2009.

L K Singh.: *Foreign Exchange Management and Air Ticketing*, Isha Books, 2008.

---

# Index